Edward Jarvis

Physiology and Laws of Health

For the Use of Schools, Acandemiess and Colleges

Edward Jarvis

Physiology and Laws of Health
For the Use of Schools, Acandemiess and Colleges

ISBN/EAN: 9783337232122

Printed in Europe, USA, Canada, Australia, Japan

Cover: Foto ©berggeist007 / pixelio.de

More available books at **www.hansebooks.com**

PHYSIOLOGY

AND

LAWS OF HEALTH.

FOR THE USE OF

SCHOOLS, ACADEMIES, AND COLLEGES.

BY EDWARD JARVIS, M.D.

NEW YORK:
A. S. BARNES & CO., 51, 53 & 55 JOHN STREET.
SOLD BY BOOKSELLERS, GENERALLY, THROUGHOUT THE UNITED STATES.
1866.

Entered, according to Act of Congress, in the year 1865, by

EDWARD JARVIS,

In the Clerk's Office of the District Court of the District of Massachusetts.

PREFACE.

Every human being is appointed to take the charge of his own body. He must supply its wants, direct its powers, regulate its actions, and thus sustain his life. This responsibility for the care of health takes precedence of all others, and requires the earliest attention to prepare to meet it. Before any one can have any use for other knowledge, he must know how to live. He should, therefore, learn the nature and the wants of his frame, and of its various organs, even earlier than he studies the features of the earth, the science of numbers, or the structure of language; for, before he can put these to any practical use, he must eat, and breathe, and move, and think. Physiology should, then, be included among the subjects of all school education, and thus preparation be first made for the fulfilment of the first responsibility of life.

The great and sole object of this work is to teach the laws of health, the powers of the several organs, the limit of their strength, the way in which they are to be developed and sustained, their proper uses, and the certain and evil consequences that follow their misapplication.

For these purposes, it is necessary to learn, first, the general structure of the organs and parts which are submitted to our management; second, the law of their being and action; and third, their application to, and connection with, the common affairs of life.

The first is limited, and is only preparatory to the second. The second also is limited, and subordinate to the third. The last is almost infinite in extent and variety, and requires much study.

In sustaining the body with food, and drink, and air, — in defending it with clothing and shelter, — in the use of our muscles and brain, — in applying the body and the mind to whatever purpose, we use some or all of our organs; and the health and strength, or pain and weakness, the good or the evil consequences, must be in accordance with the wisdom and faithfulness with which we govern ourselves in these matters. Therefore the knowledge of the Physiological Laws, and of their requirements, becomes of practical importance in every moment, and in all the circumstances of our being.

This book, in a somewhat different form, was formerly published under the name of *Practical Physiology*. But as the main purpose of the work was to teach the Laws of Health, and as the whole of it has been revised, and much of it re-written, the title is changed to PHYSIOLOGY AND LAWS OF HEALTH, which better describes the work.

EDWARD JARVIS.

DORCHESTER, MASS., September, 1865.

CONTENTS.

PART I.

DIGESTION AND FOOD.

Chap.		Page.
1.	Food supplies the Growth and the Changes of Particles of the Body.—Digestive Apparatus.—Mouth......................	9
2.	Saliva.—Mastication.—Pharynx.—Œsophagus.—Swallowing.	13
3.	Stomach.—Gastric Juice.—Digestion.—Beaumont's Observations.	18
4.	Gastric Juice flows during Eating.—Measure of Food to be eaten.—Hunger...	24
5.	Motions of Stomach.—Heat during Digestion.—Pyloric Valve.	29
6.	Time required for Digestion of various Kinds of Food.........	33
7.	Drink with Food impedes Digestion.—Effect of Bulk on Digestion.—Light and heavy Bread.............................	36
8.	Chyme.—Alimentary Canal.—Lacteals.—Chyle differs with Food.	39
9.	Man selects his Food.—Healthy Digestion comfortable.—Hunger needs the Attention of the Brain........................	45
10.	Appetite affected by State of Mind—Not identical with digestive Power or Taste....................................	47
11.	Great Eaters.—Hunger recurs when Blood wants more Chyle.—Intervals between Meals.—Regular Hours of Eating best.....	50
12.	Breakfast should be before Labor or Exposure.—Dinner.—Lunch.—Supper, injurious if late.....................	55
13.	Quantity of Food should be larger for the Active and Laborious..	60
14.	Quantity of Food determined by Wants of System and digestive Power.—Measured thus when we eat slowly.—Excessive Food oppressive.—Rapid Eating	64
15.	Appetite allowable, but not a Guide; injurious if it governs Eating.	68
16.	Exercise before or after Eating impedes Digestion.............	74
17.	Cheerfulness at Meals.—Consequences of Abuse of digestive Organs...	78
18.	Animal and vegetable Food.—Climate and Season should affect Diet	80
19.	Temperaments—should affect Diet	83
20.	Food according to Difference of Constitution, Age, and Habit...	87
21.	Digestibility and Nutritiousness of Food not identical.—Condiments and Alcohol exhaust Sensibility of Stomach.—No Rule of Diet for all	89

PART II.

CIRCULATION OF THE BLOOD, AND NUTRITION.

1.	Apparatus of Circulation.—Heart.—Arteries..................	93
2.	Veins.—Capillaries.—Pulmonary Arteries and Veins.—Double Circulation	100
3.	Power of Heart.—Valves.—Arteries.—Quantity and Flow of Blood varies with Circumstances, Exercise, States of Mind, and Feelings	105

1*

4. Object of Eating.—All Parts of Body formed out of Blood—Composition of Blood and Flesh.—Atoms of Body enjoy short Life, and are removed.—Absorbents.................................. 110
5. Nutrients and Absorbents, active in Laborer, who needs more Food.—Wens and Swellings.. 113
6. Young and Active have new Atoms of Flesh.—Old and Inactive have old Atoms.—Unerring Precision of Nutrition.............. 117

PART III.

RESPIRATION.

1. Wasted Particles carried out of the Body.—Composition of Blood in right Side of the Heart.—Chest.—Ribs.—Spine.......... 120
2. Movement of Ribs.—Diaphragm.—Inspiration.—Expiration.—Size of expanded and contracted Chest............................ 123
3. Lungs.—Organ of Voice.—Air-Vessels.—Coughing.—Respiration.. 127
4. Waste Particles.—Carbon.—Air.—Oxygen.—Nitrogen.—Carbonic Acid.—Blood absorbs Oxygen from Air and gives Carbonic Acid.. 132
5. Venous changed to arterial Blood.—Oxygen consumed, and Carbonic Acid and other Matter given out, in Respiration.—Offensive Breath.. 135
6. Respired Air unfit to be breathed again.—Air should have full Proportion of Oxygen.—Air loaded with Carbonic Acid or Water can take no more from Lungs............................. 138
7. More Oxygen consumed, and Carbonic Acid given out in cold, than warm Air; by strong and cheerful than by consumptive and melancholy Persons... 141
8. Lungs must have Room to breathe................................ 144
9. Natural Shape of Chest most beautiful.—Corresponds to Size of Body... 147
10. Frequency of Respiration.—Size of Chest and Quantity of Air inhaled, correspond to the Waste to be carried away........... 150
11. Quantity of Air spoiled by Loss of Oxygen, by Carbonic Acid Gas, and Vapor of Lungs and of Skin................................ 152
12. Want of fresh Air in Houses, Parlors, Sleeping-Rooms, Cabins.. 154
13. Foul Air in Workshops, Churches, Public Halls and Schoolhouses.—Ventilation... 157
14. Connection between Fulness of Life and Respiration............ 162
15. Breathing foul Air creates Weariness, depreciates Life, impairs Constitution.—Drowning.—Consumption......................... 165
16. Privation of Air.—Supply of Air for all Animals.—Plants give Oxygen, and consume Carbonic Acid Gas........................ 168

PART IV.

ANIMAL HEAT.

1. Animal Heat not changed by Temperature of surrounding Matter.—Blagden's Experiments... 171
2. Animal Heat generated within.—Warm and cold blooded Animals.—Animal Heat connected with Respiration.................. 174
3. Latent and sensible Heat.—Internal Combustion................. 177
4. Animal Heat increased by Exercise.—Dependent on good Air and Food.—Well-fed warmer than ill-fed.—Effects of Diet and Alcohol on internal Heat...................................... 180

CONTENTS.

5. Effects of Disease, Fatigue, Age, Sleep, on Heat.—Amount of Heat prepared in a Day.—More in warm than cold Seasons. Winter and Summer Constitution.................................. 185

PART V.

THE SKIN.

1. Skin, Cuticle, thickened by Friction.—Blisters.—Corns....... 191
2. Cuticle.—Nails.—Hoof.—Horns.—Seat of Color.—True Skin contains perspiratory Apparatus.................................. 194
3. Perspiration, sensible and insensible; Quantity................. 198
4. Perspiration affected by State of Air.—Oily Excretion of Skin.—Effect of tight Clothing.. 201
5. Connection between Skin and Stomach, Lungs and Muscles.—Effect of Cold... 204
6. Skin absorbs Food, Liquids, Medicine, Contagion, Poison....... 207
7. Sense of Touch; impaired if Skin is foul; acute in the Blind.... 210
8. Skin regulates Heat by Perspiration.—Sensation of Cold and Heat... 213
9. Quantity of Clothing affected by Habit......................... 216
10. Clothing, Need of, affected by Food, Digestion, Health, Air, Habits, Age.. 220
11. Clothing should be loose.—Linen, Cotton, Silk, Wool.......... 224
12. Flannel.—Cutaneous Excretions.—Foul Clothing.—Airing Clothing and Beds.. 226
13. Bathing makes Skin soft and healthy........................... 230
14. Cold Bathing.—Laborers need Bath............................. 233
15. Effects of Cold Bathing on Health.—Time and Conditions of Bathing.. 235
16. Sense of Touch made more acute by Bathing and Friction...... 238

PART VI.

BONES, MUSCLES, EXERCISE, AND REST.

1. Bones, Composition of, in Childhood; in Old Age; have Blood-Vessels and Nerves; subject to Growth and Decay; grow strong by Use... 241
2. Skeleton.—Bones.—Head.—Chest.—Spine.—Pelvis............ 245
3. Arms.—Wrist.—Hand.—Leg.—Foot.—Arch.—Shape......... 250
4. Joints.—Elbow.—Knee.—Shoulder.—Hip.—Structure.—Ligaments.—Sprains.—Dislocations............................. 255
5. Muscles, Number, Action, Description, Use..................... 260
6. Muscles, Shape, Attachments, Arrangement, Action............. 268
7. Muscles, Situation, sacrifice Power, Coöperation................ 274
8. Muscles, Strength increased by Exercise........................ 278
9. Muscular Strength unequal; Action increases Power in other Organs... 282
10. Muscular Action strengthens Body and Mind.—Kinds of Exercise... 286
11. Exercise adapted to Strength, Effect of, on Weak; on Dyspeptics.—Gymnastic Exercises................................. 289
12. Kinds of Exercise.—Walking.—Exercise of Boys; of Girls.—Effects of.—Habits of English and American Women.—Time for Exercise... 292
13. Exercise, Place for.—Should be regular.—Needed by All....... 295
14. Labor, Limit of Power.—Proper Expenditure of Strength...... 298

15. Effect of great Efforts. — Proper Exercise of Children and Youth. 302
16. Laborer needs healthy Organs of Digestion, Nutrition, Respiration, and Skin. — Connection of Exercise with Brain...... 304
17. Effect of Hope and Confidence, Doubt and Fear, Cheerfulness and Melancholy, of Passion and Alcohol, on Strength........ 307
18. Attitudes. — Spine, supported by Muscles, very strong. — Porters. — Pedlers. — Burdens on Head. — Spine erect. — Centre of Gravity over Line of Support................................. 309
19. Erect Attitude best for Walking; for Labor; for Mechanics; Farmers. — Spine curved from side to side; Bent by stooping. — Students, Writers have curved Spine; and Girls more than Boys.. 315
20. Time for Labor. — Experiments. — Sleep, Time for; Effect of Loss. — Disturbed by difficult Digestion..................... 321

PART VII.

BRAIN AND NERVOUS SYSTEM.

1. Brain. — Nervous System. — Spinal Cord. — Nerves............. 325
2. Brain connected with all Parts by Nerves; receives Impressions through them. — Sensations in Brain. — If Nerve is cut, Sensation and Power cease.. 330
3. If Nerve is touched or diseased, Sensation is excited in the Brain. 333
4. Pain in amputated Limbs. — Voluntary and involuntary Motion. 336
5. All Organs and Functions impaired when Brain is impaired. — Brain not sensitive; subject to Growth and Decay; has large Supply of Blood; fatigued with Labor........................... 340
6. Day the Time for mental Labor. — Brain gains Power by Exercise; weakened by Over-Action; connected with other Organs — Effect of Alcohol... 344
7. Brain Seat of Mind and Affections. — Power of Mind limited by Power of Brain. — Mind impaired by Indigestion, Hunger, Excess of Food. — Cheerfulness favors Respiration, Digestion, and muscular Action... 349
8. Brain sustains physical and mental Actions. — Mind works best when Body is easy. — Uncomfortable Sensations interrupt Study... 353
9. Mental Action interrupted by moral Feelings, Anxiety, Fear. — Proper and improper Motives for Study......................... 357
10. Various mental Powers affected by Education. — Precocious Children. — Effect of Study on Health. — Ill Health of literary Men. 361
11. Inequality of Powers; affected by Education..................... 364
12. Habitual Actions easy. — Perfect Education. — Concentration of Mind. — Vacations... 368
13. If Inequality of Powers be disregarded, Mind may be deranged. — Insanity from Misuse of Mind, Dyspepsia, Cold, Over-Action.. 373
14. Insanity from misdirected Education, religious Anxiety, perverse Habits.. 375
15. Day-Dreaming, Passion, Intoxication, Fright may cause Insanity. — Various Grades of mental Health. — Sound Mind in sound Body... 377
16. Eye. — Composition. — Tumors. — Pupil. — Effect of Light. — Tears... 380
17. Near-sightedness. — Spectacles. — Far-sightedness. — Diseases of the Eye. — Use and Care of the Eye............................ 385
18. Ear. — Structure. — Hearing. — Deafness 387
Conclusion.. 391

PHYSIOLOGY AND HEALTH.

PART I.

DIGESTION AND FOOD.

CHAPTER I.

Growth of the animal Body, and the Changes of its Particles supplied by Food. — Food converted into Flesh by the Digestive Apparatus and the Blood-Vessels. — Digestive Apparatus. — Mouth. — Teeth.

1. THE animal body increases in size and weight from birth to manhood. The chief material which supplies this growth is the food we eat. Beside this, there is another demand for food. During the whole of life, there is an incessant change going on in the particles that compose the body. It is a law established by the Creator from the beginning, that life cannot continue long in animal matter. The atoms, which compose the living body, receive the principle of life when they enter their appointed places, and become a part of the animal frame. They retain this principle but a short period, and, while they retain it, they perform their part of the work of life. But soon their work is finished, and then they yield their vitality, and give up their places to other atoms, that come to enjoy life, and work a while, and yield and depart, as those that went before them had done.

2. This law of unceasing change is impressed upon all animal beings. In all, the particles are constantly going out through the skin and the lungs, and an equal quantity of matter must, therefore, be coming in through the mouth. If

this be not done, and the supply be unequal to the waste, the animal loses flesh, and the body diminishes in weight. This change of particles and waste of matter differ in different persons, and in different circumstances, as will be shown in the course of this book.

3. Here is a double necessity for the addition of matter to the body from without — the growth or increase of the animal body in its earlier years, and the waste consequent upon the changes of the particles through the whole period of existence. The food supplies both these demands, and is therefore necessary from the beginning to the end of life.

4. Food is not living flesh: much that we eat — bread, vegetables, fruit — has not even any resemblance, in its appearance or character, to flesh. Yet these matters — the vegetables and the lifeless meats — are converted into living flesh; and not only so, but into many and various kinds of flesh, as many and as various as enter into the composition of the human body. All this is done in part by the digestive apparatus, and in part by the blood-vessels.

5. The digestive apparatus effects the first change in the food; it grinds it in the mouth, dissolves it and converts it in the stomach into a pulpy material fit to supply the wants of the blood, and sends the nutritious portions to the heart. The blood-vessels carry this blood to all the portions of the body, and with this they supply the growth and the waste of all the organs and textures.

6. This apparatus consists of the mouth and pharynx, the œsophagus or gullet, the stomach, and the intestinal canal. In the mouth are the lips, the teeth, the tongue, and the salivary glands. The pharynx lies back of the palate, between the mouth and the gullet. The œsophagus is a tube, that connects the mouth with the stomach. The stomach is a large sack, in which the digestive process is mostly performed. From the alimentary canal, the lacteals or absorbent vessels open. These take up the nutritious portion of the digested food, and carry it to the veins. Beside these organs, there are some others, such as the liver and the pancreas, which

render some assistance in the work of digestion. All these organs together form the digestive apparatus, which is complete and perfect in itself; nothing is wanting for the work, and there is nothing unnecessary. Each one of these organs, or parts of organs, has a separate and distinct part to perform in the work of converting food into the nutriment of the blood.

7. The *mouth* is the first and only visible organ of digestion, and first receives the food from our hands and our tables. It is composed of several parts, all of which are employed in the preparatory work of digestion. The lips and the cheeks form the outward walls of the mouth, and retain the food after it is received. The teeth serve to divide and break down the morsel to a fineness suitable to the stomach. The tongue rolls the morsel about, and keeps it in its place between the teeth, while it is undergoing the process of mastication or chewing, and afterwards helps to propel it backward in the act of swallowing.

8. The *teeth* differ in various animals, according to the food which they eat. The carnivorous or flesh-eating animals have teeth fitted for seizing upon their prey, and for cutting up flesh. Hence they have sharp cutting or front teeth; and long, sharp and pointed canine or stomach and eye teeth; and grinders, with high and sharpened points, by which they chew or masticate their fleshy food with facility.

9. The vegetable-eating animals have short, blunt, and strong front or incisor teeth, by which they break off their food, either grass or foliage. As they seize no prey, they have no use for the sharp canine or stomach and eye teeth; therefore these teeth are very small, and in some scarcely seen; but their molar or grinding teeth are very large, broad, and flat. Their surface is covered with slightly-raised lines, to enable them to grind down their food, which requires more crushing than cutting.

10. Man is neither herbivorous nor carnivorous exclusively, but is either or both, as occasion requires. Commonly he is both. His food, for the most part, is a mixture of vegetables and flesh. He is therefore fitted with a set of

teeth partaking of the nature of each of these classes. He has sharp incisor teeth, but they are not so long and pointed as those of the dog. His grinders are not covered with points as prominent and sharp as those of the lion, nor are they so flat on the surface as those of the ox; yet they partake somewhat of the character of both. He can chew either meat or grains, as he may desire.

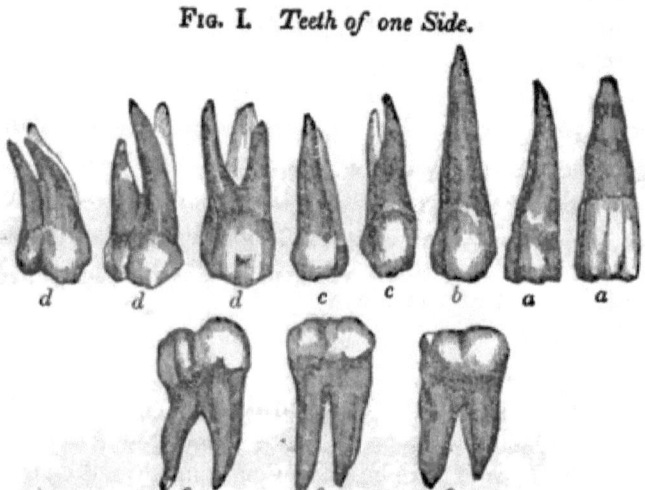

Fig. I. *Teeth of one Side.*

11. *Man has sixteen teeth in each jaw.* The four *incisor* teeth (Fig. I. *a, a*) stand in front; these are broad, flat, and somewhat sharp: with these he can bite or cut off his morsel of bread, meat, &c. Next to these are the *canine* teeth, (Fig. I. *b,*) one on each side; these are commonly called the *stomach teeth* in the lower jaw, and *eye teeth* in the upper jaw. Next to these are the *bicuspid teeth*, (Fig. I. *c, c,*) two on each side, with two fangs; and behind these are the *molar teeth*, (Fig. I. *d, d, d, e, e, e,*) or *grinders*, three on each side: these have three fangs, as shown in Fig. I., in the upper jaw, *d, d, d,* and only two in the lower jaw, *e, e, e.* These have pointed elevations sufficiently sharp to cut off meat, and sufficiently flat to grind vegetables and grains.

12. The teeth are firmly set in each jaw, with long fangs,

so that they are not easily started from their places. They are composed of soft bone within, but are covered on the outside with an enamel of very hard texture, which admits of an exquisitely fine polish. This enamel comes in contact with the food, the drinks, and the air. It will bear great variety of exposure, and resist the wear of great friction. Yet, when the mouth is neglected, the enamel is apt to decay. If any of the food, which has been masticated and mixed with the saliva, be suffered to remain about the teeth, this mixture undergoes a corrupting fermentation, and acts upon the hard enamel. After a while, a hole is eaten through this covering; and then, when this decay reaches the softer substance of the body of the teeth, it acts much more rapidly.

13. This decay is assisted also by the gathering of the secretions of the mouth. The salts and acids of the saliva combine with the food, and form tartar, which covers parts of the teeth with a hard crust. This can be easily prevented by washing the teeth frequently, and completely removing all the food and other gatherings of the mouth, after each meal, and also in the morning, after an interval of sleep.

14. The teeth have blood-vessels and nerves; they are endowed with life; and, as most people have occasion to know, they also have an exquisite sensibility. When they are sound, they seem to have little or no sensation; but when they are decayed, and exposed to extremes of temperature, to very hot or very cold matters, or even to the air, they suffer acute pain.

CHAPTER II.

Salivary Glands.— Flow of Saliva.— Mastication necessary.— Effect of imperfect Mastication.— Pharynx.— Œsophagus.— Epiglottis.— Swallowing.

15. THE teeth can only grind the food to powder. In this condition, it can neither be swallowed, nor is it fit for the

next stage of the process of digestion in the stomach. It must be not only crushed and divided, but it must be moistened and reduced to a pulpy consistence. For the purpose of supplying the necessary moisture, there is provided in the mouth a set of *glands*, which prepare and throw out the *saliva* or spittle, sufficient, in time of health, to moisten and soften all the food. These little glands are placed in the

Fig. II. *Salivary Glands.*

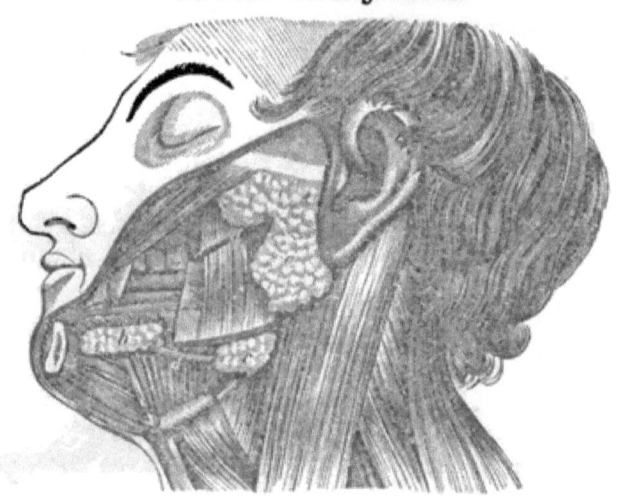

cheeks, (Fig. II. *a*,) and under the tongue, (Fig. II. *b*,) and under the jaws, (Fig. II. *c*,) and open, through very minute tubes and apertures, into the mouth, (Fig. II. *d,d*.) When the cheeks and tongue are still, these glands are inactive, and throw out no more liquid than enough to keep the mouth moist. The presence of any matter in the mouth, the chewing of our food, tobacco, &c., and any motion of the jaws, cheeks, or tongue, excite these glands, and induce the flow of their fluids. All these motions are entirely needless, except when we are eating or talking. They are under our control, and consequently the flow of saliva is under our command. If, then, we use the mouth only for its intended purposes of eating and conversation, it will be always moist, but never full of saliva; and then we should have no occasion to indulge in the unnatural and offensive habit of spitting.

16. The saliva flows, during the process of mastication, in some measure proportionate to the dryness of the food. The *salivary glands* have a very active sympathy with the appetite, and will sometimes send forth a flow of saliva at the mere presence of savory food, so that the common saying, that " one's mouth waters " at the sight of agreeable dishes, is physiologically true. These glands are important assistants in the masticating process, when they and the body are in health; but in fevers, and in some other diseases, these organs refuse to act, and the mouth then is dry and parched.

17. By the united operation of the teeth and the salivary glands, the food is first ground into small particles, and then made into a soft pulp, in the mouth. Both these operations are necessary, for we cannot swallow the morsels unless they are divided and moistened, either naturally or artificially. Let any one attempt to swallow a mouthful of dry bread without chewing, or powdered cracker, without saliva or other fluid to soften it, and he will find, that this process, which, when proper preparation is made for it, is one of the easiest and most agreeable, is now one of the most difficult and unpleasant. If the food be not divided while in the mouth, it cannot be done afterwards; there is no machinery nor power in the stomach to effect this division.

18. In the back part of the mouth is a second chamber, called *the pharynx*, separated from the anterior or front chamber by a movable curtain, called the *palate*. This curtain, hanging between these two chambers, is easily seen when the mouth is partly opened, and seems then to be the farthest boundary of the cavity. But when the mouth is stretched widely open, this curtain is raised, and the other chamber is disclosed behind it. A little knob or tongue of flesh hangs down from the middle of the palate, nearly, and sometimes quite, to the tongue, and partially or entirely divides the passage-way from the front to the back chamber. This gives the appearance of two passages, one on the right and the other on the left side.

19. Four passages open out from the back chamber; one

leads forward to the mouth, the second upward and forward to the nose, the third downward to the windpipe and lungs, and the fourth downward and backward to the stomach. The *œsophagus*, or *gullet*, that carries the food and drink to the stomach, opens from the farthest part of the mouth, and lies behind the windpipe, next to the back-bone.

20. The windpipe opens between the tongue and the gullet, in the front part of the throat. It is seen and felt in all persons, and in some it is very prominent. As the windpipe lies between the tongue and the gullet, the food, going from the mouth to the stomach, must pass over it, and would be liable to fall in if there were not an effectual protection provided against this accident. The windpipe is made of a number of stiff cartilages, and its mouth is always open; but there is placed over this mouth a little clapper, or valve, called the *epiglottis*, which is fixed by a hinge to the front edge toward the tongue, and opens toward the œsophagus behind.

21. This valve usually stands open to allow the passage of air into and out of the lungs. But it is exquisitely sensitive; when it is touched with any other matter, it falls down at once, and covers the aperture of the windpipe, and protects it from any intrusion. Whenever we swallow food or drink, the instant the morsel or the fluid reaches this valve, it falls, and allows it to glide over it, into the gullet behind, and then it rises again to give passage to the air.

22. *While we are swallowing we cannot breathe;* if we attempt to do this, or to speak, or do any thing which will cause this valve to open, some minute particle of food or drop of fluid may get into the windpipe, and cause painful irritation and coughing. This is a common accident, and may be easily prevented by not speaking while attempting to swallow food or drink. At the same time that the epiglottis falls to allow the food to pass safely over the windpipe, the soft palate, the curtain that hangs between the front and the back chamber of the mouth, is turned backward and upward, and covers the passage-way that leads to the nostrils, and defends them from the ingress of the food.

DIGESTION AND FOOD.

23. *The pharynx* connects the mouth with the œsophagus. It spreads out like a tunnel behind the palate, and is open to receive the food. The *œsophagus* extends from the pharynx to the stomach. It is a soft tube, about nine inches long, and rather less than an inch in diameter. It is covered with two layers of muscular fibres, one of which runs lengthwise, (Fig. III. *a, a,*) the other winds around it successively from top to bottom. (Fig. III. *b.*) These muscles have a power of contraction, or of drawing themselves up, like the earth-worm, and again of relaxing themselves, and being stretched out loosely. They draw around the gullet like the string of a work-bag, and thus, narrowing the passage, force onward whatever food or other matter there is within it.

Fig. III. *Section of the Œsophagus.*

a, a, Muscular fibres, running lengthwise.
b, Circular fibres.

24. When the food is thrust backward by the tongue, it passes into the pharynx, which closes upon it, and propels it downward into the œsophagus. Then the uppermost band of muscular fibres contracts, and closes its upper end, and prevents a return of the food backward. Then the next band contracts, and forces the food onward. Then the third band does the same. Thus, while each one is successively pressing upon the contents of the tube, these are forced onward and downward to the stomach. While one band is contracting, that which is next below it relaxes, to admit the entrance of the food. This is the process of swallowing, and is performed by the successive action of these circular muscles. These bands are so well adapted to each other, and work in such harmony, that we are not aware of the steps of this operation. Vomiting is performed upon the same principle, except that the order of contraction is reversed; the lower fibres first contract, and then those next above, and thus their action forces solid and fluid matters upwards, from the stomach to the mouth.

CHAPTER III.

Stomach.— Gastric Juice.— Processes of Digestion seen by Dr. Beaumont.

Fig. IV. *Stomach.*

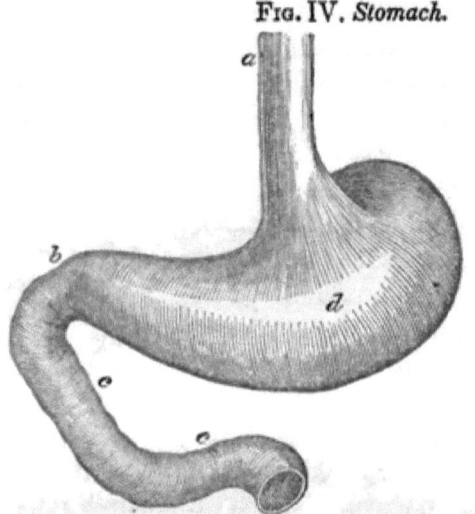

a. Œsophagus, opening into the stomach.

b. Pyloric orifice, opening into the alimentary canal.

c, c. Duodenum, or upper part of the alimentary canal.

d. Left end or great pouch.

25. THE *stomach* (Fig. IV.) is a long, round, and somewhat irregularly-shaped sack. It is placed on the left side of the abdomen, just below and within the lower ribs, and runs crosswise towards the right side, (Fig. V. *f.*) It has two apertures, one towards the left extremity, (Fig. IV. *a*,) where the gullet terminates, and the other on the right extremity, (Fig. IV. *b*,) where the stomach opens into the alimentary canal, (Fig. IV. *c*, *c*.) This organ is very expansive, and varies greatly in size, according to the quantity of matter contained in it. It is sometimes so much distended with a large meal, or with liquid or gas, as to hold two quarts or more; at other times it is so contracted as to contain less than a pint. It usually contracts down to its contents, however small; and is therefore always full, either of solid or fluid matters or of gas.

Fig. V. *Relative Position of the Organs of Digestion and Respiration.*

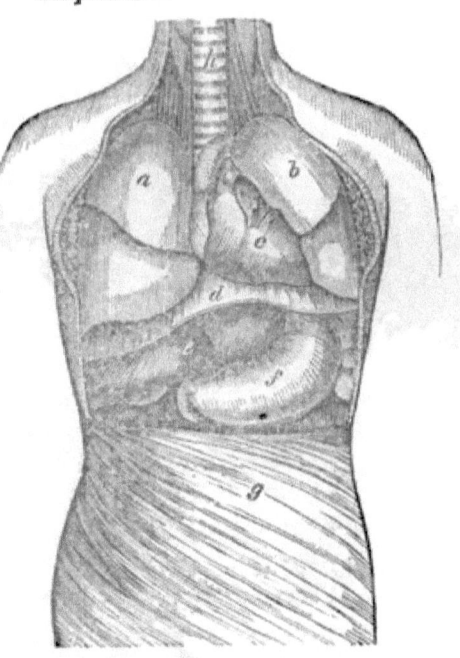

a. Right lung.

b. Left lung.

c. Heart.

d. Diaphragm.

e. Liver.

f. Stomach.

g. Front wall of the abdomen.

h. Windpipe.

26. *The average size of the stomach differs with the habits of men and the kind of food which they consume.* It is larger in those who live on vegetable food, which contains less nutriment in the same bulk, than in those who live on animal food, which is richer and more nutritious, and therefore occupies less space. This difference is more marked between the size of the stomach of the carnivorous, or flesh-eating animals, and that of the herbivorous, or vegetable-eating animals. The hare has a much larger stomach than the greyhound. The stomach of the cow is large, and her alimentary canal is twenty-four times the length of her body; whereas, in some of the carnivorous animals, this canal is not much longer than their bodies are. Men who are in the habit of gormandizing have very great stomachs,

and these are daily stretched to such an extent as would be very painful, and even injurious, to more temperate eaters.

Fig. VI.   *Inside and Coats of the Stomach.*

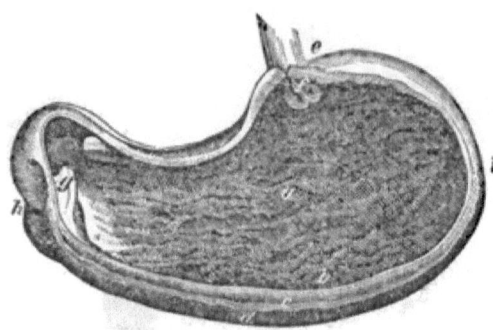

a, Mucous coat.
b, Its edge.
c, Edge of the muscular coat.
d, Peritoneal coat.
e, Œsophagus.
f, Its opening.
g, Pyloric orifice.
h, Right end.
i, Left end.

27. *The texture of the stomach is fleshy,* and very soft and flexible. Its thickness varies according to the quantity of its contents. It is thinner when it is expanded and its sides are stretched than when it is contracted and its sides are shrunken. It is composed of three *coats,* or layers. The outer or *peritoneal coat* (Fig. VI. *d*) is a part of that covering, which wraps about all the contents of the abdomen, and forms the outer coat of the whole alimentary canal, and lines the walls of this cavity. It is very tough and strong, and being attached to the back-bone and the sides of the abdomen, it holds with sufficient firmness all the inner organs, which it covers, and sustains them in their places; and yet it is attached in such a manner as to allow the expansions and motions of the stomach and of the canal.

28. The middle is the *muscular coat,* (Fig. VI. *c.*) It has two layers of fleshy fibres, or strings, which run crosswise of each other at right angles. One of these layers runs lengthwise from one end of the stomach to the other. The other layer surrounds the sack; and winding in successive circles from end to end of the organ, it covers the whole as a similar layer of fibres covers the gullet. This muscular coat has a cor-

tractile power, and when it draws itself down, it diminishes the capacity of the stomach so as to press upon its contents, however small they may be; and again, it expands so as to allow the sack to be enlarged for the reception of whatever food or liquid is then thrown into it.

29. The inner or *mucous coat* (Fig. **VI.** *a*) covers the inner surface of the organ. It is loose, soft, spongy, and porous. It is not elastic, and does not stretch and contract, when the stomach is enlarged or diminished. But when the organ is distended, this membrane is smooth; and when contracted, it is drawn into folds, like the skin of the palm of the hand, or of the inner side of the joints of the fingers when closed, or of the outside when open.

30. The anatomy of these three coats is somewhat familiar to most people who have eaten tripe, which is a preparation of the stomach of the ox or cow. The outer layer of this is a dense, tough covering of fatty matter, and comparatively strong. The next or middle layer is composed of reddish fibres. These are stringy, and are what is commonly called the *lean meat*. This layer differs in thickness in different parts of the stomach. The inner layer is a thick, soft, fatty matter, and filled with superficial cells. The human stomach has similar coats, which are arranged in the same manner, and perform the same duties.

31. These three coats, or coverings, constitute the stomach. They have each a distinct part to perform in the digestive process, and each one is fitted for its peculiar work, and for no other. The outer coat gives strength and support to the whole. The middle coat expands and contracts, to give due size to the sack; it produces the motions in the stomach, which agitate the food and promote the work of digestion; and finally this coat presses the sack down upon the contents, to expel them when they are digested. The inner or lining coat exudes upon its own surface a slimy, mucous substance, which protects it from the irritation of the matters that are put into the organ, and also prepares the gastric juice — a powerful fluid which dissolves the food.

32. The *gastric juice* is prepared within the walls of the stomach, and thrown out from its mucous or lining membrane in a manner similar to that in which the saliva is thrown out from the glands of the mouth, or the sweat is poured out from the skin. It exerts a powerful action on all proper and natural sorts of food — meat, bread, vegetables, fruits, mixed in every variety of combination, and cooked in all kinds of ways. It dissolves certain important elements which form a constituent part of nearly all our food. By this process it helps to reduce the finely-divided mass to a condition which is fitting for the absorption of its nutritious portions into the blood.

33. These elements, which are thus acted upon by the gastric juice, are called the *albuminoid* portions of the food, because they resemble the albumen or white of an egg in their properties and relations. These elements are combined, in different forms and proportions, with other constituents of the food, such as the starch, sugar, oily matters, salts, &c., and are among the most constant and nutritious of its elements. Thus, in form of *albumen* and *fibrine*, they constitute the basis of all our meats; as *cascine* they enter into milk; as *gluten* they exist in large proportions in the grains — wheat, corn, &c.; while as *legumine* they are found in peas and beans.

34. It has been ever easy to learn the structure of the stomach, and the arrangement of its coats, and its relation to other organs. But the method by which these operate, and the processes of digestion, have been left to inference or conjecture, until within a few years. Men could watch their own sensations of comfort or pain, and notice the results of strengthening and weakness from eating; but they could not see the steps, and had no means of knowing, for a certainty, what was going on in the stomach, until an opportunity was offered to Dr. Beaumont, of the United States army, in the year 1822, and afterwards. A young soldier, Alexis St. Martin, received a gun-shot wound in his left side, which laid his stomach open. The aperture did not close up, but left an opening, through which food could be put in and taken out, and the whole process of digestion observed. A flap of skin

and flesh hung over this unnatural passage-way, and closed it, and retained the contents of the stomach. But it could be raised at any time, and the cavity of the stomach exposed.

35. Dr. Beaumont took St. Martin into his family, and tried a great variety of experiments upon him, and made repeated observations in regard to his digestion, during the year 1825, and from 1829 to 1833. Dr. B. had thus an opportunity, which has been afforded to no other man on record, of watching the processes of digestion, and observing every step as it occurred. St. Martin masticated his food in his mouth, and swallowed it in the usual way. But it could be examined or taken out through the aperture, at any stage of digestion in the stomach. Food or fluids could be put into the stomach, gastric juice could be taken out, a thermometer passed in, and the temperature ascertained.

36. St. Martin took the various kinds of food usually eaten. These were prepared in all the common methods of cookery. Dr. B. closely watched them after they were eaten, and noticed the changes as they successively took place in each of the different articles which had been taken. He watched the effect of every kind of food upon the stomach, the flow of the gastric juice, the action of this fluid upon the matters presented to it, and the movements of the sack. He observed, also, precisely the time required to digest or to change each article into chyme. Subsequently Professor Lehmann, of Leipsic, and other chemists and physiologists of Europe, and Dr. Dalton, of New York, and others in this country, have tried many experiments, and made many observations on men and animals, in order to determine the law of digestion. These philosophers have generally confirmed the opinions of Dr. Beaumont; but they have made the farther discovery, that only parts of the elements of the food are digested in the stomach, while the others are digested in the intestinal canal below. All of these observers kept minute records of their observations, and have since published the results.

CHAPTER IV.

Gastric Juice secreted in the Stomach when Food is swallowed — This Quantity of Juice is the Measure of the Food to be eaten. — Hunger. — Gastric Juice combines with Food when swallowed, if eaten slowly.

37. But even these careful observers do not explain the mysteries of digestive power and action. They could only discover the steps by which nature accomplishes this wonderful transformation of dead food into the nutriment of the living blood. The food is masticated and swallowed. It then is carried to the left or larger end of the stomach, and there lodged in a great pouch. As soon as the food is swallowed, some gastric juice, sufficient to moisten it, is poured out from the lining membrane, and mingles with and softens it. The more completely the food has been masticated, and the more minutely it has been divided in the mouth, the more readily does this gastric juice enter into and combine with it. First it mixes only with the surface of each broken portion, and, as fast as the minute particles are softened by this fluid, they are separated by the continual motion of the stomach, and then the fluid has opportunity to mingle with other particles; and, these being removed, still other particles are exposed and moistened; and thus the work goes on, until all the food is wet and softened by this dissolving fluid.

38. The stomach is not always full of gastric juice. Usually there is none of this fluid in it, except when some food or other matter is there to excite the lining membrane to secrete it and pour it out. When we put a morsel into the mouth and begin to move the jaws, the salivary glands are stimulated to pour out saliva sufficient to moisten it. In a somewhat similar manner, the same morsel, when it reaches

the stomach, stimulates the mucous coat to throw out sufficient gastric juice to dissolve it. In neither case is fluid enough given out at once to mix with an entire meal. But as we see it in the mouth, so in the stomach it is given out, part by part, as often as a portion of food arrives and demands it. Even this small quantity is not poured out with a gush, but it oozes out slowly, as the perspiration oozes from the forehead, until there be enough to mix with the new morsel that is swallowed. This is not a rapid process; it takes a few moments to wet each morsel. If, then, we swallow more rapidly than the morsels can be wet with the juice, they must accumulate, and wait for the fluid to come.

39. This secretion and flow of gastric juice commences as soon as any food reaches the stomach, and then continues to flow, if stimulation by new morsels is successively repeated. But this secretion is not without end. This liquid cannot, like the saliva in the mouth, be made to flow as long as we wish. There is a limit to the gastric juice which will be secreted at any one time. Dr. Beaumont says, "When the alimentary matter is received into the stomach, this fluid then begins to exude from its proper vessels, and increases in proportion to the quantity of aliment naturally required and received." * It flows, then, not in proportion to the food which we may happen to eat, but in proportion to the quantity of nutriment which the body needs. When, therefore, so much of this juice is poured out as will dissolve what food we need at any one time, it will stop, and the mucous membrane will give no more. And, as only a definite proportion of aliment can be digested in a given quantity of this fluid, if more be eaten than this quantity can dissolve,—that is, if more food be swallowed than the body needs to supply its waste,—the excess either remains in the stomach undigested, and is there a cause of intense pain and oppression, or it is thrown out and onward, in a crude state, to disturb the organs beyond.

* Observations, p. 85.

40. Knowing, then, that the stomach can give out only a limited quantity of gastric juice at any one time, and that this fluid can dissolve only a limited quantity of food, and also that this corresponds with the wants of nutriment in the body, we have *a measure of the amount of food which should be taken at each meal;* that is, as much as the gastric juice can combine with in the stomach and digest. This is seemingly unknown, and therefore no guide for us. But it can be ascertained by watching the effects of eating. There is an apparent understanding between the general system and the healthy digestive organs; at least, there is such a sympathy between them that, when the one wants nourishment, the other is ready to digest it. This gives the sensation of hunger, which appears to be in the stomach. When this is felt, the lining membrane is ready to give forth the gastric juice to digest the food; and, as long as the hungry feeling continues, this fluid may flow. But when this ceases, there is no more need of food, no more sensation of want, and no more digestive power than sufficient to dissolve what has been already eaten. Then there is a feeling of satisfaction and ease in the stomach, for the appetite and craving are gone.

41. We can make use of this guide to the proper quantity of food, and measure it by the quantity of gastric juice which can be supplied at one time, only when we eat slowly, when the morsels which we swallow, and the fluid which is to dissolve them, keep pace with each other, and meet together in the stomach. By this cautious proceeding, we adapt the supply of food exactly to the wants of the body, and stop as soon as this want ceases. But if we eat more rapidly than this, and continue to eat at this rate as long as the gastric juice continues to be secreted, the food gathers in the stomach faster than the digesting fluid. There is an accumulation of food waiting to be moistened. And when, finally, as much of this liquid is given out as can be afforded, there is still this excess of food already in the stomach, so much more than is needed or can be digested.

42. *Hunger is the sensation of want of more nutriment in the body.* But this is felt, not in the body where nutriment is needed, but in the stomach, and there only when that organ is ready to give out sufficient and proper gastric juice to digest the food which is required. There must be a correspondence between the general frame and the stomach, to produce this feeling of hunger. The one must be in need of more nutriment, and the other ready and able to digest it. It is not enough that the body is in want. If the stomach cannot render aid in the supply, there is no hunger. In fever, the body wastes away and wants nourishment; but the stomach cannot digest, and consequently asks for no food. When nutriment is wanted, the body speaks to the stomach, and the stomach, if it have power of digestion, speaks to the nervous system. This is hunger. This sensation is felt in the stomach, and there it continues until all the gastric juice which will dissolve the needed quantity of food is poured out and combined with it, and then it ceases for the time.

43. An industrious man, who imagines the time spent at table to be lost, complains that he has a very great appetite, but a very weak and painful digestion. He says that he sits down to his dinner voraciously hungry, and eats very rapidly, without giving his mouth time to masticate his food. He eats much more than his companions, and yet he rises from the table hungry, and goes immediately to his work, from which he is absent a shorter time than his fellows, who eat less than he does. His hunger continues about twenty minutes or half an hour after he leaves the table, and then he is in pain. His food oppresses him; it lies like a weight in the stomach for several hours; and he is scarcely relieved of the distress before the time for another meal comes round.

44. The explanation is this: Mr. D. is a man of active habits, and his frame therefore wants nutriment. His stomach can digest and is ready to give out gastric juice sufficient to dissolve as much food as is needed, and he is consequently hungry. To gratify his keen appetite, and to save time at his meals, he has acquired the unnatural habit of eating

rapidly. He swallows his food faster than the gastric juice is prepared to mix with it. Before enough of this fluid is poured out for the first morsel, the second is swallowed, and before this is moistened, a third arrives; and thus every moment increases this mass of food waiting in the stomach for its dissolving fluid. As long as this juice continues to flow, there is an appetite. But before it ceases, he has eaten so much that his tardy reflections tell him that he has eaten enough, and more than enough. Obedient to his reflections rather than to his sensations, he rises from his table before he has satisfied his appetite, which continues for several minutes longer, until the gastric juice ceases flowing and mixing with the food. When this happens, there is yet in the stomach some food unmixed with the fluid, and this is so much more than can be dissolved. The stomach struggles painfully for several hours to digest or get rid of this excess, and hence comes the oppression that hangs heavily upon him through the interval between his meals.

45. *The work of digestion commences immediately after we begin to eat;* as soon as we swallow one morsel, some gastric juice is given out and combines with it. The second morsel excites the flow of more of this fluid, and enough for its own solution. In the same manner, each successive morsel provokes the secretion of as much of this fluid as it needs, until all shall be given out that can be given at that time. If, therefore, we eat slowly and naturally, by the time that we shall have finished eating, all the food will be moistened. In a few minutes more, all the gastric juice that can be prepared on that occasion is thrown into the stomach; and this is sufficient, in health, to dissolve as much food as the nutrition of the system then needs, and no more.

CHAPTER V.

Motions of the Stomach. — Digested Food homogeneous. — Heat of Stomach during Digestion. — Cold interrupts Digestion. — Pyloric Valve.

46. This mixture of the fluid with the food in the stomach is aided by the action of its muscular coat, and of the muscles of respiration. The muscular coat habitually reduces the size of the sack down to its contents. When more food or drink is received, it relaxes, and allows the sack to expand and give the new matter room. But this coat is uneasy. It not only relaxes, but it again contracts, and presses upon the food; and then again it loosens. This is repeated gently and continually; and by this means the food is kept in slight motion, as long as the process of digestion is going on. These movements are assisted by the muscles of respiration. The stomach lies in the upper part of the abdomen, (Fig. V. *f*,) and just below the chest, where the lungs are situated.

47. Immediately above, and in contact with the stomach, the *diaphragm* (Fig. V. *d*) stretches across the body from side to side, and from the back-bone to the breast-bone. This great muscle is the partition wall between the chest and the abdomen, and moves with every breath. Every time we draw our breath, this presses downwards upon the stomach, enlarges the cavity of the chest, and gives room for the air to flow into the lungs. Then, when we exhale, the abdominal muscles — those which principally constitute the front wall of the abdomen — contract, press upon the stomach, and throw it upon the diaphragm, and force the air outward. Thus the stomach is kept in incessant motion. Every inspiration of air presses it downward and outward, and every expiration presses it upward and inward. These combined movements keep the food in such agitation, that it becomes thoroughly mixed with the gastric juice, and the

digestive process is by this means very materially hastened.

48. The food is first mixed with and then dissolved in the gastric juice. Then all the peculiarities of the articles which we have eaten are lost. No trace of the original form of the digestible and digested matter can be found. Meat, bread, fruits — all are reduced to one homogeneous pulp. Whatever we may have eaten, the pulp, so far as we can discover, is the same. When we examine the digested food in the stomach of the man whose diet is mostly meat, in that of another whose diet is exclusively bread, and in that of the child, whose whole nutriment is milk, the eye and the fingers perceive no difference. This process is termed *chymification;* and the new pulp is called *chyme,* which is now ready to pass from the stomach to its second stage of digestion in the duodenum.

49. *This process of digestion requires the natural heat of the animal body.* Dr. Beaumont passed the thermometer frequently into St. Martin's stomach, and found that its temperature was 100°, which is two degrees higher than the natural temperature of the body. He also found that cold retarded the process of digestion. He threw into the stomach a single gill of water at the temperature of 50°, which reduced the temperature of the whole contents of the stomach 30°, that is, down to 70°. Digestion was then suspended, until the temperature was elevated to its natural standard.

50. Dr. Beaumont took a quantity of the gastric juice from St. Martin, and divided it into two portions. He put some pieces of meat into one portion of this, cooled down to 34°, and some into another portion, warmed to 100°, the natural temperature of the stomach during digestion. These were kept at these temperatures for many hours. At the same time St. Martin swallowed meat of the same kind. In one hour the meat in the stomach was partially chymified; that in the warm gastric juice out of the stomach was nearly in the same condition, while that in the cold gastric juice

was much less advanced. When the meat was entirely digested in the warm fluid, it was very little changed in the cold fluid. At the end of twenty-four hours the last was no more digested. But then he warmed this fluid to 100°, and kept it at that temperature, and digestion commenced and advanced regularly, as in the other parcels.

51. In his other experiments, Dr. Beaumont found that cold gastric juice acted not at all, or very imperfectly. Hence he concluded that heat to a certain degree is necessary, to give it the power of dissolving food. Dr. Carpenter thence infers that the practice of eating ice after dinner, or even drinking cold fluids or ice-water during dinner, or at any time of eating, is very prejudicial to digestion. If this be true, then all drinks which are colder than the stomach must interfere with the action of the gastric juice to an extent in proportion to their coldness. For, if a temperature of 100° is necessary for digestion, then any thing, whether food or drink, that cools the gastric juice below this degree, must suspend the digestive process until the heat of the body, or living power of the stomach, shall warm the fluid up to the necessary temperature.

52. The stomach (Fig. IV.) is large, and has its principal cavity at the left end, where the food is received through the œsophagus from the mouth. It grows smaller towards the opposite end, and finally opens, by a small aperture at its right extremity, into the *intestinal canal*. At this point of junction, it is surrounded by a strong circular band of muscular fibres, which, by contraction, can completely close the opening, as a string closes the mouth of a bag, and prevent the passage of any thing through. This is the *pyloric valve*, (Fig. IV. *b*,) which acts as a faithful sentinel, and retains the contents of the stomach during the process of digestion. While the food is undergoing the operation of churning and mingling with the gastric juice, it binds itself closely around the passage-way. The motions of the stomach, and the pressure of the respiratory muscles, would very naturally force the food out of this sack, if it were not thus effectually secured.

53. This ever-watchful door-keeper has a special duty to perform. Its business is to retain the food in the stomach until it be digested, and then to let it pass onward. For this purpose, it seems to be endowed with a kind of intelligence, by which it discriminates between the crude and the dissolved matters that present themselves to it, and with a sort of discretionary power, by which it opens and lets the finished chyme pass out, but closes and prevents the passage of that which is not so reduced to a pulp. While the stomach is empty, it may be relaxed, and the doorway left open; but as soon as any food is swallowed, it shuts the door, and holds it tight until this food be digested. As fast as any portion of the food is turned into chyme, it is carried by the motions of the stomach to the right end; then the valve relaxes, the door opens, and the digested food passes out. But, with a quick perception of the differences of condition of the food, it closes the moment the digested portion has gone out, and the undigested portions offer themselves.

54. *We sometimes eat food of a kind which the stomach cannot digest.* We sometimes eat more at a meal than the gastric juice can dissolve. In these cases, the stomach digests what it can, and makes great effort to digest the rest. When it becomes wearied with its unsuccessful efforts, it endeavors to relieve itself of the indigestible portion by thrusting it through the aperture at the right end into the intestinal canal. But as this crude matter is neither digested nor prepared for the action of the next organ, the valve refuses to open and let it go through. Again this is sent back, and again the stomach makes its fruitless attempts to digest it; and thus failing a third and a fourth time, presents it to this doorway for passage. This is refused over and over; the valve closes with a greater and even more painful force; until, at last, fatigued with the resistance, it yields to the importunity of the stomach, and permits the undigested and indigestible matter to go through. While this struggle is going on, we feel a distressing oppression about the right side, just below the short ribs. This usually happens within

two or three hours after eating, and is caused by the ineffectual effort of the stomach to convert the indigestible food into pulpy chyme, and the resistance of the valve to its passage outward.

55. But after a painful struggle of the stomach to get rid of that which it cannot master, and of the valve to hold back that which ought not to pass, the food is at length forced into the intestinal canal. There it is a strange matter; it is no more suited for this organ than it was for the stomach. These organs were made to receive, the one masticated and digestible food, and the other the chyme, or the food digested and reduced to a fine pulp. All other matters are foreign to them, and create disturbance and cause pain. The natural and healthy work of the stomach is preparatory to the work of the intestinal canal; and, unless the first organ has done its proper work upon the food before it enters the second, the last can do little, and generally it can do nothing with it. It has no more power over food that is not digested, than the other had over food that could not be digested. As long as it remains, then, in the body, it irritates the sack which contains it, and gives distress to the whole system.

CHAPTER VI.

Time required for Digestion in the Stomach.

56. DIGESTION commences in the stomach as soon as the food is swallowed, and continues from one to five hours — varying according to the kind of food, and the health of the person. Dr. Beaumont found that the various articles of diet differed very much as to the time required for their solution in the stomach. Pigs' feet and tripe soused were changed to chyme in one hour, while roasted fresh pork was not dissolved in less than five hours and a quarter. Other articles required various periods, ranging between these extremes. The average and usual time required for the complete diges-

tion and transmission of ordinary meals from the stomach is three hours and a half. Persons differ very much, according to their health and their habits. Those who have not abused their digestive organs — who have not overloaded them, nor tasked them with indigestible matter — have much more active stomachs than those who have misused these organs. The active and the energetic digest more vigorously than the sluggish and the inert. The following table, showing the time required for the stomach digestion of various articles of food, is taken from Dr. Beaumont's work, p. 269.

57. Mean Time required for the Digestion of various Articles of Food in the Stomach.

Articles.	Preparation.	Time. Hrs.	Time. Min.	Articles.	Preparation.	Time. Hrs.	Time. Min.
Apples, sour, hard,	Raw,	2	50	Corn, green, and beans,	Boiled,	3	45
———— mellow,	Raw,	2		———— bread,	Baked,	3	15
————, sweet, do.,	Raw,	1	30	———— cake,	Baked,	3	
Aponeurosis,*	Boiled,	3		Custard,	Baked,	2	45
Bass, striped, fresh,	Broiled,	3		Dumpling, apple,	Boiled,	3	
Barley,	Boiled,	2		Ducks, domesticated,	Roasted,	4	
Beans, pod,	Boiled,	2	30	————, wild,	Roasted,	4	30
Beef, fresh, lean, rare,	Roasted,	3		Eggs, fresh,	Boiled hard,	3	30
————, dry,	Roasted,	3	30				
———— steak,	Broiled,	3		————,	Boiled soft,	3	
————, with salt only,	Boiled,	3	36				
————, with mustard,	Boiled,	3	10	————,	Fried,	3	30
————, fresh, lean,	Fried,	4		————,	Roasted,	2	15
————, old, hard, salted,	Boiled,	4	15	————,	Raw,	2	
Beets,	Boiled,	3	45	————, whip'd,	Raw,	1	30
Brains,	Boiled,	1	45	Flounder, fresh,	Fried,	3	30
Bread, wheat, fresh,	Baked,	3	30	Fowl, domestic,	Boiled,	4	
————, corn,	Baked,	3	15	————,	Roasted,	4	
Butter,	Melted,	3	30	Gelatine,	Boiled,	2	30
Cabbage head,	Raw,	2	50	Goose,	Roasted,	2	30
————, with vinegar,	Raw,	2		Heart,	Fried,	4	
				Lamb, fresh,	Broiled,	2	30
————,	Boiled,	4	30	Liver, beef's, fresh,	Broiled,	2	
Cake, sponge,	Baked,	2	30	Meat hashed with vegetables,	Warm'd,	2	30
Carrot, orange,	Boiled,	3	15				
Cartilage,*	Boiled,	4	15	Milk,	Boiled,	2	
Catfish,	Fried,	3	30	————,	Raw,	2	15
Cheese, old, strong,	Raw,	3	30	Mutton, fresh,	Roasted,	3	15
Chicken, full-grown,	Fricas'd,	2	45	————,	Broiled,	3	
Codfish, cured, dry,	Boiled,	2		————,	Boiled,	3	

DIGESTION AND FOOD.

Articles.	Preparation.	Hrs.	Min.	Articles.	Preparation.	Hrs.	Min.
Oysters, fresh, ...	Raw,	2	55	Soup, beef, vegetables, and bread,	Boiled,	4	
———, ...	Roasted,	3	15				
———, ...	Stewed,	3	30	———, chicken, ...	Boiled,	3	
Parsnips, ...	Boiled,	2	30	———, marrow-bones,	Boiled,	4	15
Pig, sucking, ...	Roasted,	2	30	———, mutton, ...	Boiled,	3	30
Pigs' feet, soused, .	Boiled,	1		———, oyster, ...	Boiled,	3	30
Pork, fat and lean, .	Roasted,	5	15	Spinal marrow, animal, ...	Boiled,	2	40
———, recently salt'd,	Boiled,	4	30				
———,	Fried,	4	15	Suet, beef, fresh, ..	Boiled,	5	30
———,	Broiled,	3	15	———, mutton, ...	Boiled,	4	30
———,	Raw,	3		Tapioca, ...	Boiled,	2	
———,	Stewed,	3		Tendon,*	Boiled,	5	30
——— steak	Broiled,	3	15	Tripe, soused, ...	Boiled,	1	
Potatoes, Irish, ...	Boiled,	3	30	Trout, salmon, fresh,	Boiled,	1	30
———, ...	Roasted,	2	30	———,	Fried,	1	30
———, ...	Baked,	2	30	Turkey, domesticated, ...	Roasted,	2	30
Rice, ...	Boiled,	1					
Sago, ...	Boiled,	1	45	———,	Boiled,	2	25
Salmon, salted, ...	Boiled,	4		———, wild, ? ..	Roasted,	2	18
Sausage, fresh, ...	Broiled,	3	20	Turnips, flat, ...	Boiled,	3	30
Soup, barley, ...	Boiled,	1	30	Veal, fresh, ...	Broiled,	4	
———, bean, ...	Boiled,	3		———, ...	Fried,	4	30
				Venison steak, ...	Broiled,	1	35

* *Cartilage* is usually called *gristle*. *Aponeurosis* and *tendon* are very strong and tough parts of the flesh, somewhat similar to cartilage.

58. These are the results of many observations upon St. Martin. The time stated in regard to each article is the average time required to digest it. From these experiments we see that there is a very great difference in the time required for the stomach digestion of the various articles of food. Rice and souse are digested in one hour; salmon trout and sweet apples, in one hour and a half; beef's liver, codfish, and tapioca, in two hours; fresh beef and mutton, in three hours; veal broiled, and ducks, in four hours. Fresh pork, roasted, required five hours and a quarter for digestion. St. Martin's power of digestion may not exactly correspond with that of all other men. Some may require a longer, some a shorter time to digest these articles of food. Still he may be fairly considered as a representative of the average of mankind; and probably we, if in good health, shall digest these several articles in about the same time that he did.

CHAPTER VII.

Fluids drunk with Food impede Digestion. — Stomach acts more easily on a large than on a minute Quantity of Food. — Meat better digested if mixed with Vegetables. — Gastric Juice mixes easily with light, but not with heavy Bread. — Light Bread soaks readily in Water, but heavy Bread does not.

59. THE first work of the stomach in digestion is to get rid of all the fluid which has been swallowed with the food. In fifty minutes after Martin had dined on vegetables, soup, beef, and bread, Dr. Beaumont found that the fluid portion had been absorbed and carried away from the stomach, and the remainder was of a thicker consistence than usual, after a more solid food had been taken. This is necessary, in order that the gastric juice shall not be diluted and weakened, and its power of dissolving the food diminished.

60. Drink taken with food, then, must either reduce the power of the gastric juice, or postpone the work of digestion until the stomach shall have relieved itself of this needless matter, and in either case, suspend the digestive process. It is a common notion that those who have weak stomachs should take weak broths, soups, teas, &c., which seem to require less effort of digestion. But this is not always good advice. These liquids may require more effort, and are, therefore, inappropriate food for many of the feeble.

61. The natural secretions of the mouth afford sufficient fluid to aid in the grinding and softening the food, and to prepare it for swallowing; and the gastric juice in the stomach is sufficient for its solution there. Upon this principle, tea, coffee, or water with our meals cannot be of advantage. Dr. Warren says, "The quantity of drink required for health and comfort is very small. In cold weather, a pint of liquid in twenty-four hours is sufficient; in the hot seasons, this quantity may be increased; but this increase is rarely necessary when a reasonable amount of fruit can be ob-

ained."* Dr. Dalton thinks " a man in full health, taking free exercise in the open air, requires rather more than three pints of liquid daily." †

62. The stomach acts more easily upon a large than on a very small quantity of food. As the hands find it easier to grasp and hold a cane than a quill or a wire, and the arms can more easily clasp an armful of wood than a single stick, so the muscular coat of the stomach finds less difficulty in grasping and pressing upon a full meal than a little morsel. The quantity of nutriment in food is not always in proportion to its bulk; some kinds contain very much, and others very little, in the same space or weight. A pound of beef is more nutritious than a pound of bread, and a pound of bread contains more nutriment than a pound of roots.

63. Meats are very concentrated; that is, they contain great quantities of nutriment in small bulk; and if we were to live upon these alone, we should eat a small quantity — smaller than the stomach could manage with the greatest ease to itself. This difficulty is obviated by mixing meat with bread and vegetables. Some of the rude tribes in the extreme northern regions live upon the coarsest and most concentrated meats. But they find it better to mix this with bread, potatoes, or other roots, with bran, or even sawdust, for the purpose of facilitating the action of the stomach.

64. In order that the food should be mixed the most freely with the gastric juice in the stomach, it should be not only well divided in the mouth, but it should be of such a nature that the gastric juice can get access to all the minute particles, and separate them from each other. With light bread, that is thoroughly baked, and somewhat dried, this is easily accomplished. But heavy bread is cohesive, and the particles cling together and form a solid mass, so compact that it would be very difficult for any fluid to penetrate it.

65. The difference is easily shown, and is probably familiar to all; if not, the experiment can be tried in one moment, by throwing a piece of light, porous bread, that has been

---

* On Preservation of Health, p. 62. † Physiology, p. 113.

baked twenty-four hours, and is somewhat dried, into water. Immediately the water penetrates into all the cells, and fills all the pores; the bread absorbs more and more, and swells; and soon the mass is much enlarged, and is completely filled with water. In fact, it is itself mere pulp. If we divide it, we shall find that water has come in contact with every particle; every one, however minute, is wet.

66. Again, throw a piece of heavy, compact bread, that has no cells in it, into water, and let it remain as long as the other, and then examine it, and it will be found that it is as heavy as it was before; the water has not penetrated it — it has absorbed none; it is not enlarged, and the inner particles are not reached by the fluid; they are as dry as they were before the piece was thrown into water. The same effect takes place in the action of the gastric juice in the stomach. It finds it easy to penetrate among and wet the particles of the light, and hard to enter the heavy bread. This last then remains for a long time a solid, compact mass, or a mass of compact portions, which cannot be dissolved; or, if it be dissolved, it is not without much difficulty and pain, and after a long perseverance of the organ in its almost fruitless work.

67. We can determine this quality of bread even without the trouble of throwing it into water. New bread is almost always cohesive, and its particles disposed to cling together. If we take a piece of this, or of heavy bread, and roll it between the fingers, it forms into a compact ball or roll, so close that it is plain that it will not readily admit water to soak it. But if we try the same with old and light bread, it separates and falls into crumbs. It is impossible to make it into a ball. This would be easily soaked in water, and easily digested in the stomach, which is not the case with the new or the heavy bread.

CHAPTER VIII.

Chyme. — Intestinal Canal. — Mucous Membrane. — Lacteals. — Chyme in the Duodenum. — Chyle: differs with difference of Food. — Carried to Blood-Vessels. — Three Stages of Digestion.

63. When the stomach has finished its work, the food is converted into *chyme*. To the naked eye, it is the same in appearance throughout. All distinctions of the various kinds of aliment seem to be lost. No traces of the meats, bread, or vegetables are visible; all are reduced to an apparently homogeneous pulp. These were the former notions of science. But modern microscopic investigations and chemical analyses show that this is not homogeneous; of this a part only is digested, or converted to chyme; the rest, though finely divided and mixed with the fluids of the stomach, is yet to be submitted to a farther process of digestion in the small intestines. When the stomach has finished its work on the food, the pyloric valve opens, the muscular coat contracts, and, pressing upon the contents, forces this pulp through this passage into the *duodenum*, (Fig. IV. *c, c,*) which is the next link in the alimentary canal.

69. The *intestinal canal* is composed of three coats, which are similarly arranged and serve similar purposes to those of the stomach. The outer of these coats is strong and thick, and gives support to the whole canal; by this coat the organ is attached to the back-bone and held in its place. The middle coat is like the lean meat of tripe; it is composed of two sets or layers of fibres, one of which winds around the tube — the other runs lengthwise from end to end. The circular band regulates the size of the tube, by keeping it always pressed down upon its contents. When food is within this organ, these fibres contract, one after another, successively, and, pressing upon the matters contained within, force them onward. The longitudinal fibres shorten the

canal or its parts, and by this means they aid in carrying the food forward.

70. The inner lining, called the *mucous membrane*, secretes the slimy mucus with which the whole inner surface of the canal is moistened and protected from any irritating quality of the contents. This membrane is loose and flabby, and, when the canal is empty and contracted, it is drawn into wrinkles or folds, and seems to be too large for the sack. But when the canal is distended with food, or any other matter, it is drawn out, and lies more smoothly over the inner face of the sack. Besides these folds, which are made in the mucous membrane by the contraction of the canal, there are other folds, which run around the inner surface of the tube crosswise its length. These are permanent, whether the tube be contracted or distended. This membrane or lining of the intestine is furnished with a set of glands, which prepare and throw into the canal a peculiar fluid called the *intestinal juice*. This is another coöperator in the work of digesting the food and fitting it for the blood. It meets and combines with the chyme in this part of the alimentary canal, or the duodenum, and there it digests or changes some of the elements of the food that had not been so changed by the fluids in the stomach.

71. The *pancreas* is a large gland lying behind the stomach, and performs an important part in the work of digestion. It prepares another fluid, called the *pancreatic juice*, and sends it through a tube into the upper part of the duodenum. This juice enters into combination with still other elements of the food, especially the oily matters or fats, which had not been affected by the gastric or intestinal juices, and prepares these for the use of the blood.

72. Thus each of these three different digestive fluids or juices — prepared in the stomach, the duodenum, and the pancreas — performs its own and peculiar part in the work of digestion. They convert the nutritive elements of the food into a condition fit to enter and become a part of the blood. These elements constitute a milky fluid called *chyle*, which is yet in the alimentary canal; but it is destined to

nourish the whole body, and must first pass into and through the blood-vessels.

73. The mucous membrane, or inner lining of the intestinal canal, is soft, like velvet, and filled with myriads of pores, which perform an important part of the work of digestion. These numberless little pores, or tubes, have their open mouths upon the inner surface of the intestinal canal. They run from the inner channel outward through the walls of this canal. Their mouths are so small as to be invisible, except by aid of a powerful microscope; and yet they are so numerous as to cover over all the inner surface of this organ. Their duty is to absorb, or suck up, some of the nutritious portion of the digested food, which, when it is in these little tubes, has the appearance of milk. These are, therefore, called the *lacteal absorbents*. These tubes, when they first start from the inside of the canal, are almost inconceivably small, but they unite together, two and two, and more, and thus become fewer and larger. The larger tubes again unite, and form other and still larger ones, until they all are joined in one large tube, called the *thoracic lacteal duct*. This goes along the inner side of the back-bone, from the abdomen to the upper part of the chest, and opens into the great vein, at the right side of the heart. These mouths and tubes, small and large, and this duct, constitute what is called the *lacteal system*. Its object is to carry the nutritious portion of the chyme from the digestive organs to the blood-vessels.

74. Beside these lacteal absorbents there is another set of vessels that assist in doing the same work in the lining membrane of the intestinal canal. These are myriads of veins, as minute as the lacteals. They absorb other portions of the digested food, and carry it through their minute tubes into larger channels, and these pour their contents finally into the great vein near the heart.

75. In the *duodenum*, or the upper portion of the intestinal canal, the digested food is divided into two kinds — that which is to enter the blood-vessels, and the waste.

The former alone is needed for, or can give nourishment and strength to, the body. The latter is not only useless, but a burden on the whole system, and must be regularly excluded. Otherwise all the powers are oppressed, and health suffers.

76. The proportion of this chyle, which the absorbents are able to extract from the chyme, varies with the food; for one kind contains a much greater ratio of nutriment than another, as will be hereafter shown. It depends also upon the completeness of digestion in the stomach; and this, in great measure, upon the perfectness of the mastication and mixture with saliva in the mouth. Of course, then, the remote result of imperfect mastication and hasty eating must be, first, imperfect digestion; second, less chyle; and, consequently, less nutriment for the body.

77. Thus the work of digestion is shown to be performed by means of the mouth, the stomach, and the alimentary canal. The whole process is divided into three stages — mastication and insalivation in the mouth; the digestion, or conversion into chyme in the stomach; the separation of the nutritious and innutritious parts in the duodenum. It is necessary that each part should be well done, and in due order; else all that follow will be badly done.*

---

* Dr. Dalton says, "We find, then, that the digestion of the food is not a simple operation, but is made up of several different processes, which commence successively in different portions of the alimentary canal. In the first place, the food is subjected in the mouth to the physical operations of mastication and insalivation. Reduced to a soft pulp, and mixed abundantly with the saliva, it passes, secondly, into the stomach. Here it excites the secretion of the gastric juice, by the influence of which its chemical transformation and solution are commenced. If the meal consists wholly or partially of muscular flesh, the first effect of the gastric juice is to dissolve the intervening cellular substance by which the tissue is disintegrated and liquefied. In the small intestine the pancreatic and intestinal juices convert the starchy ingredients of the food into sugar, and break up the fatty matter into a fine emulsion, by which they are converted into chyle.

"Although the separate actions of these digestive fluids, however,

CHAPTER IX.

Digestive Process wonderful. — We are responsible for the Selection and the Preparation of our Food. — Healthy Digestion comfortable. — Hunger not owing to Emptiness of Stomach. — Brain and Nerves must be sound, to perceive Hunger.

78. This digestive process, which effects so great a change, is wonderful, as well as interesting. The food, which was of every sort, — meat, fish, bread, vegetables, and fruit, — mere lifeless matter upon our tables, — is now changed into chyle, that is homogeneous, and almost endowed with life. It was at first the food for the stomach; it is now nutriment for the blood. This change is a vital one; at least it is effected by the fluids which are within the living body, and which are the product of vital or living organs. By what unseen agency these fluids obtain this power, is known only to the all-wise Creator. It is not revealed to us, nor need it be. Enough is revealed for our government, to show us our duty in regard to food and digestion.

79. In this work, man has much to do. He is to provide food of suitable kinds, and must prepare it in a suitable manner. He is to determine the quality, and measure the quantity, which he shall eat. The times of his eating, and the intervals between his meals, are left to his discretion. The work of the mouth is under his control. But all the

commence at different points of the alimentary canal, they afterwards go on simultaneously in the small intestine; and the changes which take place here, and which constitute the process of intestinal digestion, form at the same time one of the most complicated, and one of the most important, parts of the whole digestive function." — *Human Physiology*, p. 156.

operations of the digestive organs, beyond the mouth, are not submitted to his direction, nor even to his observation.

80. When we are in good health, and the food is properly selected, prepared, and eaten, we are not conscious of the process of digestion in the stomach; but if the food be not properly selected or cooked, if we have not faithfully prepared the food, by complete mastication and mixture with the saliva in the mouth, for the next stage of digestion in the stomach, we are painfully conscious of the effort of the stomach to digest that which is unsuitable for its wants and its powers.

81. Although we are not conscious of the process of digestion from any feeling that we have in the stomach especially, yet there is always a pleasurable sensation throughout the whole frame, which accompanies the proper and healthy performance of this function; there is a feeling of comfort in the body, and satisfaction in the mind, and usually a glow of cheerfulness attending it. One feels better, and more disposed to be contented, after his meals.

82. Our part of this work is to select the food, and prepare it, by suitable combinations and cookery, for the mouth. Next, we are to masticate and moisten it in the mouth for the stomach. After this, nature takes care of it. We want, then, some guide to direct us in regard to the quality and quantity of this food, and the time and seasons for eating

83. Hunger and appetite are the first apparent guides, and with many, perhaps a majority of mankind, the only guides in the matter of eating. Hunger has been explained (§ 42) to be the sensation of want of nutrition in the general system, connected with the power of digestion in the stomach, and with the readiness of the stomach to supply the gastric juice for this purpose.

84. It is commonly supposed that hunger is a mere indication of emptiness of the stomach; that as soon as the last meal shall have passed out of this organ, more is wanted; and that as long as any food remains in it, there is no appetite. Neither of these suppositions is correct. The usual

meals are digested and carried into the intestinal canal, and the stomach is left without food, in about three hours and a half, (§ 56, p. 33;) and if they were composed of the most digestible articles, this time would be much less. Yet hunger does not usually return in less than five or six hours. This leaves the stomach empty nearly half the time, without any craving desire of food.

85. So certainly is the desire of food the result of wants of the system, in connection with the power of, and readiness for digestion, that in some diseased states of the system, where both these conditions exist, but no communication between the stomach and the blood-vessels, the appetite is ravenous.

86. The wants of the whole body for more nutriment are communicated first to the stomach, and thence to the brain. Here is the real sensation of appetite. It therefore is necessary, not only that the stomach should be in sound condition, but also that the brain be in a condition to recognize this feeling of want. This feeling is conveyed through the nerves from the stomach to the brain, and there perceived and recognized. If, then, the nerve of the stomach be diseased or divided, there can be no communication from this organ to the brain, and hence no sensation conveyed, and no hunger felt. Some physiologists have tried the experiment of cutting the nerves which connect the stomach with the brain in dogs. The consequence was, that the animals seemed to have lost all sensation of appetite, and although they had been long deprived of food, and were really in need of it, they did not appear to feel or to understand the want of it.

87. The appetite, then, is felt in the brain; but it is not perceived, nor are we conscious of it, unless we can give attention to it. It not unfrequently happens that one is so intently engaged in any pursuit, that he forgets his hours of eating and his own necessities. Students are sometimes so devoted to their books, that their meals do not occur to them. Men who are absorbed in any care or anxiety pay less than due attention to, and do not perceive, their craving sensations.

Sailors, in times of peril and shipwreck, may go from morning till night without thinking of dinner. The anxious mother, watching over a sick child, often needs to be reminded by others of the time and necessity of eating.

88. A merchant, whose business during the whole day is in the city, and whose employment often absorbs his whole attention, sometimes returns at night to his home in the country with a great appetite; for he has been so much occupied that he has forgotten his dinner. And when thus engaged, it is only at night, when business hours are passed, and his occupation has ceased, that he gives any heed to the wants of his system, or discovers that he is hungry; and then, from his previous exhaustion and want of supply, his hunger returns with double force.

89 In these and in similar cases, one may not feel appetite sufficient to warn him of the hours of eating, although, at the same time, his system is in want of nutriment, and there is real cause of hunger without the sensation. For the body is suffering from the waste of its particles and from the privation of food; the stomach is empty, and it has sent the warning of this emptiness to the brain; but if this organ gives no attention to it, no sensation is felt, nor hunger perceived. This happens for the same reason that, when we are sometimes absorbed in thought, we do not hear the church clock strike, although very near us, or even the house clock in the same room with us. In this case, the impulse was given to the air, and communicated to the tympanum of the ear, but the brain was directing its attention elsewhere, and perceived no sound.

CHAPTER X.

Appetite affected by State of Mind. — No Digestion without Appetite. — Appetite not always a Sign of digestive Power. — Appetite and Taste not identical. — Great Privation of Appetite.

90. *The appetite is affected by the state of health both of the body and of the mind.* In fever, in pain, and in certain dyspeptic states, the stomach craves little or no food. So, in mental distress, in times of great fear or sorrow, or extreme anxiety, the appetite fails. Even in a single moment the appetite may be suspended by any sudden mental affection or emotion. If any one sit down at a table with even a strong desire of food, and if, when about to eat what seems to him inviting, he should be told of the death or extreme danger of a near friend, at once all appetite is gone.*

91. Hunger is given to us as a guide to our duty in the work of sustenance; and when properly regarded, it is a safe guide. It indicates the wants of the system for more nutriment. Even in a good state of health, these wants vary

---

* Dr. Dalton says, "The secretion of the gastric juice is much influenced by nervous conditions. It was noticed by Dr. Beaumont, in his experiments upon St. Martin, that irritation of the temper, and other moral causes, would frequently diminish, or altogether suspend, the supply of the gastric fluids. Any febrile action in the system, or any unusual fatigue, was liable to exert a similar effect. Every one is aware how readily any mental disturbance, such as anxiety, anger, or vexation, will take away the appetite, and interfere with digestion. Any nervous impression of this kind, occurring at the commencement of digestion, seems moreover to produce some change which has a lasting effect upon the process; for it is very often noticed that, when any annoyance, hurry, or anxiety occurs soon after the food has been taken, though it may last only for a few moments, the digestive process is not only liable to be suspended for the time, but to be permanently disturbed during the entire day. In order that digestion, therefore, may go on properly in the stomach, food must be taken only when the appetite demands it; it should also be thoroughly masticated at the outset; and, finally, both mind and body, particularly during the commencement of the process, should be free from any unusual or disagreeable excitement." — *Physiology,* p. 149.

with many outward and inward circumstances. They differ with the manner of life, and with the quantity and energy of exercise. The laborious and active have more hunger than the idle and the slow. Children and youth who are growing in stature, convalescents who are regaining lost flesh, have more imperative appetites than others. Appetite is more keen when the body is in full vigor, and all the functions are performed with the most energy. When the blood flows freely and the muscles play smoothly, when the mind is buoyant and the spirit joyous, the appetite boldly indicates a want of food in the whole system, and a ready power of the stomach to digest it, and convert it into the nutriment of the blood.

92. *Appetite is usually the sign of digestive power.* Certainly there is no vigorous digestion without it. When the digestive organs, nerves, and brain, are apparently in good condition, and we are attentive to the warnings of the stomach, if then we feel no hunger, we may be assured that the stomach craves no food because it cannot digest it. However long it may have been without food, if it do not by its hunger declare its readiness and ability to convert it into nutriment for the blood, it is useless to eat. It is even worse than useless; for whatever is then eaten cannot be changed to pulpy chyme, nor to milky chyle, nor can the lacteals extract from it nourishment to feed the exhausted blood, or the wasted body. Food, then, eaten when we are not hungry, gives weakness and oppression, and not strength and vigor. It causes pain rather than the feeling of comfort, that follows or accompanies good digestion.

93. On the other hand, appetite is not always evidence of digestive power. In some states of dyspepsia there is a voracious desire of food, without corresponding power in the stomach to digest it. There is sometimes a diseased and continual irritation in this organ, which suggests to the brain the want of food as the only means of allaying it. If food, in these cases, be eaten according to the appetite, — or if, in some cases, any food be taken, however urgent the hunger, —

indigestion and pain will surely follow. Very frequently, during convalescence from fevers, the appetite returns before the power of digestion. The body is wasted with disease, and wants a great quantity of nourishment to restore its loss. The fever has gone; the stomach is free from nausea, but yet it is uneasy, and craves a large supply of nutritious food. But it has not regained its full strength. It can no more digest a full allowance of hearty food than the muscles or limbs can perform the full day's labor of a man in health. If the convalescent should eat a strong man's food, as appetite suggests, pain and weariness will fall upon his digestive organs as inevitably as they would upon his limbs if he should do a strong man's work.

94. *Appetite and taste are not the same.* It is a mistake in our self-management to confound the one with the other. One is a desire for food corresponding to the wants of the system; the other is mere pleasantness of the food while in the mouth. In domestic economy, when the array of successive dishes of various kinds comes before us, when all are delicious and tempting, and pleasant to the palate, there is danger of eating of one, and then another, to gratify the taste, even after appetite has been satisfied. It is, therefore, important to distinguish between the sensation of the stomach which implies the want of nutriment, and which is real hunger, and that mere sensation of the mouth which implies merely the want of something pleasant to the palate, and which is factitious hunger.

95. *There are remarkable instances of absence of appetite under disease or excitement.* Sometimes persons in a high state of mania, with the mind violently excited or absorbed, have endured entire abstinence from food or drink for three days. During this time, they could not be persuaded to take a morsel to eat or a drop of fluid. In these and similar cases, there were undoubtedly want of nourishment in the body, and power of digestion in the stomach. The appetite was suspended, because the brain had its attention intently fixed upon its delusions and distress. But when the excite-

ment was calmed and the distress alleviated, the sufferer was persuaded to eat, and ate with the usual freedom, and digested with the usual ease and comfort.

96. Some extreme instances of this are on record. "One is published in the Edinburgh Medical Essays for 1720, of a young lady about sixteen years of age, who, in consequence of the sudden death of an indulgent father, was thrown into a state of tetanus, or rigidity of all the muscles of the body, and especially those of swallowing, accompanied with a total loss of desire for food, as well as incapacity for swallowing it, for two long and successive periods of time — in the first instance for thirty-four, and in the second for fifty-four days; during all which time of her first and second fastings, she declared she had no sense of hunger or thirst, and when they were over she had not lost much flesh." *

97. The celebrated Miss Ann Moore, of Tutbury, England, lived for some years on so little food, that she was supposed to live entirely without it; and she even pretended that she was able to live without any food whatever. A woman, in consequence of lockjaw, swallowed nothing but a very little cold water for four years; and for twelve years afterward took no more food than is sufficient for a child two years old.

CHAPTER XI.

Great Eaters. — Causes of enormous Appetite and Eating. — Stomach distended by Over-eating. — Hunger recurs when Blood wants more Chyle. — Intervals between Meals vary with Circumstances — Disturbance of the usual Hours of Eating disturbs Digestion — Intervals of Meals.

98. On the contrary, there are instances of persons who from disease or perverse habit, have acquired an extraordi-

* Good's Nosology, p. 16, note.

nary and almost insatiable appetite for food. With them hunger seems to be ever present. The stomach full of food hardly allays the desire for more, or only suspends it for a short time. A case is recorded, in the "Philosophical Transactions," of "a boy of twelve years of age, who had so strong a craving that he would gnaw his own flesh when not supplied with food. When awake, he was constantly devouring, though whatever he swallowed was soon afterwards rejected. The food given him consisted of bread, meat, beer, milk, water, butter, cheese, sugar, treacle, (molasses,) puddings, pies, fruits, broth, potatoes; and of these he swallowed in six successive days three hundred and eighty-four pounds two ounces, avoirdupois — being sixty-four pounds a day on an average. The disease continued for one year." *

99. Idiots have generally an inordinate appetite, which they indulge if they have opportunity. In 1846, Dr. Samuel G. Howe, chairman of a commission appointed by the government of Massachusetts for the purpose, ascertained the measure of appetite and habits of four hundred and thirty-two idiotic children and youths. In comparison with others of their age, twenty-four of these ate less than the average; about one fifth ate the average quantity; about one fourth ate from ten to forty per cent. more; about the same number ate from fifty to ninety per cent. more than others; and rather more than one quarter ate double the usual amount, and some of these ate in still larger proportions.

100. This enormous appetite does not always depend upon the wants of the system, but in some cases, as in the boy, § 98, upon disease. In others it is caused by indulging the perverse habit of voracious eating. Great eaters feel the want of a large quantity in the stomach. They are no better nourished than those who eat less; but, without a great supply, they feel hollow, faint, and languid. The stomach, being used to this great distention, does not act easily upon a small quantity. In some persons, the stomach, being once distended, does not recover its original size. Dr. Darwin states that "a woman near Litchfield, England, who ate

* Good's Nosology, p. 16, note.

much animal and vegetable food for a wager, affirmed that since distending her stomach so much, she had never felt herself satisfied with food, and had in general taken twice as much at a meal as she had been accustomed to before she ate so much for a wager." *

101. This is an extraordinary instance of extreme distention of the stomach; but it is not unusual to find similar conditions, though in a less degree, produced by smaller errors of the same kind. This unnatural state of the stomach comes oftener from a long and gradually-increased indulgence in great eating, amounting sometimes to gluttony, than from a single gormandizing, as in the case of the woman stated in the last section. But it is an error that creeps on very insidiously, and with a seemingly good cause; and one who begins to trespass in this way is in danger of repeating the mistake, and of increasing the evil continually, without suspecting he is doing any more than obeying the natural laws of his sensations, and supplying his proper wants.

102. As soon as any of the food is digested and reduced to chyme, and sent into the duodenum, the innumerable absorbents commence their work of absorbing or taking up the chyle — the nutritious portion — and carrying it to the veins. They continue this work for several hours, more or less, according to the fulness of the storehouse of nutriment; and during this time, they replenish the waste of the blood, and enable it to supply the wants of the whole body. After a varying period of some hours, the quantity of chyle is exhausted in the alimentary canal, and can furnish no more material for the blood, and the blood can no longer meet the demands of the wasting flesh. Then there is a want of more and new nutriment — a craving for food in the stomach, and a consequent sensation of hunger, and we need to eat again.

103. *The period for the return of appetite, or the proper interval between the hours of eating*, depends upon many circumstances and conditions, such as the temperament, the

* Zoönomia, Vol. II., p. 107.

age, and the habits of the person, and the quantity and digestibility of the food previously eaten. The nutriment would be earlier exhausted, and hunger sooner return, after a light meal of innutritious food, than after a full meal of rich food. The young and growing need food oftener than the mature and full-grown, and the convalescent oftener than the permanently healthy. The expenditure of life and the waste of particles are more rapid when we are in motion than when we are still. Consequently the active and laborious are sooner exhausted, and need to be earlier recruited, and should eat more frequently than the slow and indolent. The sanguine and the nervous, for the same reason, are more impatient of hunger than the lymphatic and dull.

104.. *The return of appetite is very easily trained to regular habits,* so that it comes at about the usual time of eating; and until that hour, whatever it may be, hunger is not felt. At the usual hour of eating, appetite becomes perceptible, and, if not then gratified, it may become urgent; or sometimes it ceases till the next time of eating. The stomach, being trained to observe these hours, accommodates its wants to the periods of supply. Those who dine at twelve feel the want of food at that hour; while those who dine later, whatever may be the season, are not often disturbed with hunger until their usual time of eating comes round, and then they feel the want of food.

105. This power of the stomach to accommodate itself to the habits of life, is not only manifested in different persons, who have been differently educated from the beginning, but it is shown in the same individual at different times. We not unfrequently see an entire change of habits of the same stomach, arising from change of manner of life.

106. Some have always been accustomed to dine at twelve, and always felt hungry at that hour. Suddenly, they change their residence and their hour of eating, and wait till one or two o'clock for their dinner. The stomach does not change its habits and wants so speedily. At first, and for some time, the appetite returns at the former hour of indulgence, and

waits impatiently for its food; but gradually it accommodates its wants to the new regulation, and hunger waits quietly till the newly adopted hour. Again, the same persons have suddenly returned to their early hours, but the stomach does not go back so readily; at first, it was not in want of food at twelve; but in a short period, finding its supply come early, it manifested an early want, and became hungry at twelve.

107. Whatever the accustomed hours of eating may be, the stomach does not bear sudden changes, not even for a single meal, without some complaint. One's appetite returns at established periods; then his stomach craves food, and the gastric juice is ready to flow and dissolve it. If this want is gratified, his dinner is digested easily, and he feels comfortable, and prepared for business during the afternoon. But if, for any cause, he varies from his regular habit, and eats at a later or an earlier hour, his digestion is not so easy, and his body and mind are not so free for labor.

108. A gentleman, being one day occupied abroad, did not return to his dinner until three o'clock. He felt more hungry than usual. But, after he had eaten, his stomach reminded him that it did not perform its work with its customary ease. His body was not so light and buoyant, his brain was not so clear, as usual; he could not apply his mind with its accustomed energy to its work. And the result of the afternoon's labors was less than on other days. The same has generally happened at other times when he has postponed his dinner beyond its accustomed hour. He feels the same loss of energy and of command of his powers whenever he anticipates the hour and dines at twelve. If he had been a mechanic, he would have had the same difference in the precision and success with which he could use his tools; or, if he had been a farmer, there would have been the same failure in the energy and effect of wielding the axe, swinging the scythe, or striking with the hoe, after such a disturbance of the hours of eating.

109. Some families have no regular hours of eating. They eat whenever it suits the convenience of the cooks to

prepare the meals, or of the household to eat them. These are varied, and often very widely, to meet the plans and the accidents of business. These people dine sometimes very late, and at other times very early. There are many employed in cities at a distance from their homes. They do not return at noon, nor do they dine at any regular boarding place; but they eat at eating-houses, at any hour, when the business of their shops, their stores, or their offices, gives them leisure. Occasionally, for want of time, they omit their dinner entirely. All these irregular habits of eating disturb their regular habits of digestion, and consequently leave them with somewhat less power of application and labor for the next succeeding hours.

110. *In general, the intervals of the meals, during the active part of the day, should not be more than six or seven hours.* Dinner should follow the morning meal, and supper should follow the noon meal, within this period. The frequency of eating should follow the law of appetite, described in §§ 102, 103, p. 52; and, regarding this law, children and laborers should have shorter intervals, and eat more frequently than the mature and the inactive.

CHAPTER XII.

Breakfast should be soon after rising. — If it be late, a Lunch should be taken early in the Morning. — Health better sustained when full. — Breakfast should be before Labor or Exposure. — Hour of Dinner. — Interval between Breakfast and Dinner. — Forenoon Lunch good in some Cases. — Needed by those who breakfast early and dine late. — Night Suppers injurious. — Summary of Meals.

111. DURING the hours of sleep, there is no action of the body, and comparatively little waste; therefore the interval between the evening and the morning meal may be longer than the interval between the meals which are taken in the active part of the day. Yet the store of nutriment in the

digestive organs and the blood-vessels becomes exhausted during the night, and the system needs more food before any considerable amount of action is undertaken in the morning; for the frame is not then prepared to bear any more drafts, and it must be recruited before it can undergo any severe labor. The *breakfast* should therefore be taken soon, within an hour after rising. This is especially requisite for invalids, who have not much strength, and but little power of endurance.

112. When the morning meal is not to be eaten early, some light refreshment at the time of rising will meet the immediate wants of the system, and sustain it during the morning exercise. It is well, then, if some considerable time is to elapse between rising and breakfast, to take some food early. This is a common custom among the Creoles of Louisiana and the inhabitants of Cuba, and some classes of people in France. These have coffee, fruit, or other light food sent to their sleeping-rooms, sometimes before, and sometimes after rising, which, they think, enables them the better to sustain any fatigue before the regular breakfast is given them.

113. *The animal system sustains all action, labor, and exposure best when it is well nourished.* When the nutriment fails, it becomes sooner fatigued, and more susceptible of pain; and, besides this, it is more liable to suffer from any causes, which would impair its soundness or diminish its vitality. The contagion of disease, the infection of fever, whatever may bring on disorder, act more readily and powerfully on the hungry, and on those who are badly nourished, than on those who are well fed. We are better able to resist the influence of cold, and to maintain the natural temperature of the body, when we are full, than when we are fasting.

114. All these causes of disorder or suffering act upon the human constitution with more destructive force before breakfast than afterward. On this account, all who are about to expose themselves to any of these morbid influences, to contagion or infection of disease, or to such

exhalations of marshy countries as produce fever and ague or other malady, should eat their breakfast before going abroad. Travellers and others, who go abroad in winter, or in stormy weather, will maintain their heat better and defend themselves more effectually against the elements, if they breakfast before they go out. But if they go out in the morning hungry, they suffer much more from chills and dampness, and are in greater danger of taking cold.

115. If this precaution of early eating be requisite for the healthy and the robust, it is much more so for the feeble and the invalid. Inasmuch as those, whose strength and vitality are in any way reduced below the average standard, are more susceptible of disorder from any disturbing cause, and are more easily fatigued with labor, it is more necessary for them than for others to strengthen and defend themselves with the early morning refreshment, before they engage in laborious occupation, or expose themselves to cold or infection.

116. The time of the *dinner* differs very materially in various nations, and among people in different places of the same nation. Three hundred years ago, the king of England and his court dined at eleven. Some of the nobility, previous to that time, breakfasted at seven, dined at eleven, and supped at four. More recently, both in America and Europe, twelve at noon was the established hour; and at present, in the rural districts, almost every where, this dining hour is still observed; while in towns and cities the time varies from one to six or seven. But, in families who dine so late, breakfast is also late, and the interval between the first and second meal is not so wide as the lateness of the dining hour would seem to indicate. In about five or six hours after the morning meal, the appetite returns, and the system calls for new refreshment. This is the true guide for the time of dining. Whatever may be the hour of breakfast, not more than about six or seven hours should elapse before the system is again refreshed with food.

117. In some of the European cities, breakfast is suffi-

ciently early, but the dinner is taken as late as six or seven o'clock in the evening. This habit leaves an interval of eight to ten hours between the first and second meals, which is a much longer space for fasting than the time specified in the last section, and longer than the system can well endure without suffering from want of nutriment. Many — probably most of those who dine so late — remedy this difficulty by interposing a lunch between the first and second meal in the day. This is, with most people, a light meal, and intended merely to sustain nature through the long interval of the morning and noon; but with some it is composed of heavy and substantial food, such as would ordinarily be taken for dinner.

118. A lady went from Boston, in September, 1846, to London. Her usual dining hour had been two, at home; but, in London, it was suddenly changed to six or seven o'clock. Her morning meal was also postponed somewhat; yet the interval between these was several hours longer than she had been accustomed to. In a few weeks, she suffered materially in health, and became much debilitated, and consulted a physician, who advised a lunch to be taken in the forenoon. Following his advice, she soon recovered her wonted health and strength. A similar case occurred in New Orleans. A friend suffered in the same way, from the same cause, and was restored by a similar change in his hours of meals.

119. Growing children, and persons recovering from sickness, and men engaged in very hard labor, may do well to take this forenoon lunch, even if the interval between the morning and noon meal be not more than six or seven hours. And healthy men, in ordinary pursuits, would do it with advantage, if they breakfast early and dine very late. But for mature persons, in good health, who are not engaged in very hard labor, and whose dinner is not delayed more than six or seven hours from the breakfast, the stomach is better if at rest until the hour for the regular meal comes round at noon.

120. The *supper* is usually a lighter meal, and is needed for all who have not already eaten three times. It is the almost universal custom of the civilized nations to eat three times a day. Remembering the rule before stated, (§ 110, p. 55,) that not more than six or seven hours of active life should elapse before the refreshment of food, and that it should not usually be taken oftener than this, it is easy to determine whether any supper should be taken after dinner or not. If the dinner be as late as six or seven o'clock, and there has been a lunch taken in the forenoon or at noon, the fourth meal will be unnecessary. When the dinner is at or near night, so late that there will be not more than four or five hours between this meal and bed-time, then the supper, if taken before sleeping, would be not only needless, but injurious. It is not then wanted for nutrition, and the stomach is not in a condition to digest it. The supper, therefore, should depend upon the distance of the sleeping hours from the dinner. So that he who dines at twelve and retires at nine, and he who dines at seven and retires at four, both equally need the evening or the night meal.

121. *But supper should be eaten usually about three hours or more before sleeping.* Sleep is the rest of all the voluntary powers; then nothing but the lungs and the heart keep in motion; all the others are still. The mind, the feelings and the affections, the brain, the muscular and digestive organs, all need and enjoy this rest. If any of the organs or powers are not permitted to repose, the sleep is not profound; the rest is not entire. If, then, we eat so late that the food be not digested before we retire to our beds, the digestion is still going on while we attempt to sleep, and the sleep is disturbed by it; then dreams — sometimes distressing dreams — oppress and weary us, and the body and mind are not refreshed completely for the following day's labor. Second suppers are therefore injurious.

122. The general custom of three meals a day — a good breakfast soon after rising in the morning, a fuller and more nutritious meal near the middle of the active part of the day,

and a lighter meal a few hours before sleeping — meets the wants of the body, and corresponds with the powers of digestion. But when we add to these, lunches during the day, or take a supper of feasting for hospitality or self-indulgence at night, we overstep the demands for nourishment, and overtask the powers of digestion, and prevent the full, refreshing effects of sleep at night.

CHAPTER XIII.

Quantity of Food. — Fleshy Persons not always great Eaters. — Lean Persons not always small Eaters. — Action causes Changes of Particles. — Laborers eat more than the Sedentary.

123. THE quantity of food, as we shall see hereafter, is not to be governed by a fixed law. Men differ in their wants, and their necessities, and their powers. Dr. John C. Dalton, after making a series of experiments in diet and nutrition, says he " found that the entire quantity of food required during twenty-four hours by a man in full health, and taking free exercise in the open air, is as follows : —

Meat,	16 ounces, or	1·00 lb.
Bread,	19 "	1·19.
Butter or fat,	$3\frac{1}{2}$ "	·22.
Water,	52 fluid oz.,	3·25 pints."*

124. The seamen in the British navy are allowed 1 lb. bread, 1 lb. fresh meat, $\frac{1}{2}$ lb. vegetables; and, when fresh meat and vegetables are not given, $\frac{3}{4}$ lb. salt meat and $\frac{3}{4}$ lb. flour are allowed, being 40 oz. solid food for each day's support. The dietary for emigrants going from Great Britain to the East Indies and New Holland gives 9 oz. animal food, 12 oz. bread, 4 oz. flour, 2 oz. rice, 1 oz. raisins; in all, 28 oz. per day. The soldiers of the army of the United States are allowed to have $\frac{3}{4}$ lb. pork or bacon, or $1\frac{1}{4}$ lb. beef,

* Human Physiology, 3d ed. p. 113.

fresh or salt, 18 oz. bread or flour, or 12 oz. hard bread, or 1¼ lb. corn meal, a day besides 8 qts. of peas or beans, or 10 lbs. of rice, for every hundred days.

125. *The quantity of food* must vary with the habits of the individual, and with the energy and quantity of exercise. Some have a much greater nutritive power than others Some extract more nutritive chyle from a given amount of food than others. Fleshy persons are not always great eaters, nor are all lean persons proportionately limited in their quantity of food. On the contrary, there are many instances of great corpulence connected with an extremely small diet, and some cases where the greatest temperance in food does not prevent or diminish the fatness. On the other hand, there are some persons whose appetite is ever ready, and digestion apparently good, and who consume much more than the average quantity of aliment, and yet are miserably lean.

126. A young woman, whose body was full, round, and almost fat, came under my observation a few years ago, on account of neuralgia. The pain was, for a long time, fixed in the stomach, and then she could only eat a single cracker, or an equal amount of bread, weighing less than one ounce a day. Yet her nutritive powers were so good, that, with this small quantity of food, she maintained her full condition, and showed no sensible loss of flesh. She continued this spare diet for about six weeks, and, in all this time, retained her healthy plumpness of form. She was not strong, and yet she was not very weak. She was not confined to her bed, nor to her chamber, but was able to be about the house, and perform the light household work. In her best state of health, she was a small eater; but she was then strong and vigorous, active and fleshy.

127. The requisite quantity of nutriment varies more with quantity and energy of action which the system is called upon to sustain. All motion is connected with waste from the body; the vitalized particles exhaust their vitality in the process of action, and a change then takes place. The old

and exhausted particles, having lost their living principle, are then removed from their places in the tissues of the body, and thrown into the veins, and thence carried away. Their places must be supplied by new particles from the blood, and the blood receives these from the digested food. The body of the laborer, therefore, undergoes more rapid waste, and needs a greater and more frequent supply of food, than that of the people of sedentary habits, or idlers. It is a great mistake, then, to suppose that all men, in whatever occupation engaged, should eat the same quantity of food. The British government give to the troops, on their voyage to the East Indies, on account of the quietness of their life, 30 per cent. less of solid food than to the sailors, who are in constant action on board the same ships.

128. *For this reason, the same man should not eat the same quantity in all varieties of exercise.* At one period, he may be very laborious: he then wants more food than at another time, when he may be engaged in lighter employment. This principle ought not to be forgotten by those who make permanent changes in their occupations. Change of habits presupposes change of nutritive wants. If the action of the body be reduced, there is a reduction of waste, and of nutritive want and digestive power. But, unfortunately, the appetite is not readily reduced, nor is it perfectly easy to control it at once. But this is necessary, in order to prevent indigestion.

129. Some young men, who have been accustomed to active exercise, or even hard labor, suddenly change their occupations. They leave their farms or their workshops, and go to school or to college, or to the lighter employment of cities. Their habits of eating have been very properly adapted to their habits of labor. While they were hard workers, they were hearty eaters. Too frequently, their full diet is retained after their hard labor of body is discontinued, and they still eat the same amount of food as before. This is more than the system now requires, and more than the stomach can digest. That quantity, which was no more than sufficient to

sustain a life full of vigorous and laborious action, is too much for the inactive and sedentary life, and even becomes oppressive and injurious.

130. Hence men complain of indigestion and of loss of health in other ways, when they have become less active. The real ground of difficulty is not so much that their new occupations are necessarily injurious to digestion, as that the quantity, and often the quality, of food is not adapted to their altered habits. In the new occupation, there is less action, and consequently less waste, and of course less use and demand for food, and, necessarily connected with these, less digestive power. When these new conditions are disregarded, and the old habits of eating continued, the stomach is overburdened, and dyspepsia follows, with its usual train of evils.

131. It is not unfrequent at Cambridge — and doubtless it is the same at other colleges — for some of the most industrious students to leave on account of ill health. These unfortunate invalids are more among the older than among the younger members of the classes. And the reason is plain. Most of these were not originally destined to literary pursuits, and were engaged, in their earlier years, on their farms, or in their workshops, or other spheres of active employment. They were generally strong and healthy, but, having a decided inclination for the study of books, they changed their active habits of body for the quietness of the student's life. But their appetite and diet continued the same, and thereby they fell. Others were younger, and went through college with less suffering and fewer failures of health. These had never been laborious, nor had they acquired the habits of eating which laboring men should have. Their habits were always adapted to their present circumstances, and consequently they were spared, at least, this cause of ill health.

132. Men and women who have reached the fulness of stature, cannot safely indulge the habits of eating which were proper for them while they were growing. The quantity of food which was necessary to supply the growth in youth, is

more than is needed in mature years. And not only is it not needed, but it is a burden to the system, and imposes an injurious tax upon the powers of the stomach to digest it. As soon, therefore, as the body ceases to grow, the diet should be reduced from the fulness of youth, and accommodated to the more limited wants of the system. Convalescents, very properly, eat a greater quantity while they are recovering lost flesh, in order to meet the new conditions, and supply the new wants of the system; but the moment they have regained their usual fulness, they should return to their usual diet.

CHAPTER XIV.

Quantity of Food determined by the Wants of the System and the digestive Power. — Measured thus only when we eat slowly. — Each one must judge for himself how much he shall eat. — Excess of Food oppresses and weakens. — Due Quantity strengthens. — Time saved by hasty Eating more than lost by Oppression afterward. — Rapid Eating at Hotels and on Steamboats.

133. It has been shown, (§§ 39, 40, pp. 25, 26) that, when the whole system is in good health, the digestive powers of the stomach correspond to the nutritive wants of the body; that, when the body is in want of nourishment, the stomach prepares, or is ready to prepare, gastric juice sufficient to dissolve as much food as is needed, and no more; and that this quantity of gastric juice gives us the measure of the food which should at any time be taken. If we could then ascertain this quantity of gastric juice, we should have no difficulty in determining the requisite amount of food. In St. Martin (§ 35, p. 23) the flow of this juice could be seen through the aperture, and its quantity ascertained; but we can only obtain this knowledge by carefully watching our own sensations.

134. When the body wants nourishment, and the stomach is ready to pour out gastric juice and digest it, there is a sensation of hunger, (§ 42, p. 27;) and this sensation continues as

long as there is any of this gastric juice unoccupied by food, or until the inner coat of the stomach has poured out as much as it can give at the time. So long as this sensation continues, there is a call for more food, and more can be digested. We may safely eat, then, until this natural appetite ceases, provided we throw the food into the stomach no faster than the digesting fluid is ready to dissolve it. Mr. D.' (§ 43, p. 27) did not even eat until his hunger ceased; and yet he ate more than his gastric juice could dissolve.

135. *In order, then, to adapt the food to the wants of the system and the power of digestion, we must eat slowly;* we must masticate each morsel patiently and thoroughly in the mouth, waiting, in this manner, before we swallow this, until the previous morsel has had time to combine with the gastric juice in the stomach. So doing, we can determine whether that organ wants or is prepared for another; and, when that demand ceases, we can suspend the eating. Then we shall have eaten all that is needed for nutrition, and no more than the stomach can digest.

136. This will require us to eat, not to fulness, as is unhappily too commonly done; nor even to satiety, for that would overstep the wants of nature; but merely until the demand for nutrition ceases. Dr. Beaumont says, "There seems to be a sense of perfect intelligence conveyed to the brain, which, in health, invariably dictates what quantity of aliment, (responding to the sense of hunger and its due satisfaction) is naturally required for the purposes of life, and which, if noticed and properly attended to, could prove the most salutary monitor of health and effectual preventive of disease. It is not the sense of satiety; for this is beyond the point of healthful indulgence, and is Nature's earliest indication of an abuse and overburden of her powers to replenish the system. It occurs immediately previous to this, and may be known by its pleasurable sensations of perfect satisfaction, ease, and quiescence of body and mind. It is when the stomach says, *Enough.* It is distinguished from satiety by difference of sensation; the latter says, *Too much.*"

137. The wants of nutrition, even in men in good health, depend upon so many circumstances,—their exercise, exposure, and their temperament,—and the digestive powers differ so widely in different people, that we could scarcely find two who require exactly the same quantity of food, nor would the same man require the same quantity at all times. Therefore it is impossible to prescribe any exact weight or measure, which all should eat. But every one who learns the principles which have been stated here, who examines the circumstances of his own life, and carefully watches his own sensations, will be better able to determine how much he shall eat. If then, he faithfully obeys the law of nutrition, and applies it rigidly to his own self-management, he will not err in his diet.

138. When a man eats sufficiently, and no more,—when his stomach has received no more than it can easily digest,—he feels refreshed and easy; he soon becomes light and buoyant, and is then ready to recommence his active business. But when the stomach has more than it can easily convert into chyme, it is oppressed with labor, and feels a dead weight bearing it down. All the energies of the body are then concentrated in the effort of the stomach to perform its extraordinary labor, in the same manner as all the energies of the system are concentrated in the extraordinary muscular exertion, when we attempt to lift great weights, or to run a race. While the digestive organs, or the muscles, are making these great exertions, we can do nothing else; we can neither use the brain and think, or study, or calculate, nor can the muscles perform any other labor.

139. As much food, then, as the system needs and the stomach can digest, gives a man comfort, strength, and ability to apply his powers to business. It enables him to use his brain, and his muscles, and his bones—to work with his hands, his feet, and his mind. But all excess of food beyond this, every mouthful more than is needed or easily dissolved, gives weakness, instead of strength, for business. It is a tax upon the vital energies, and a clog upon the motions of the

body and the actions of the mind. We see this, in a remarkable degree, in the glutton, who, after his dinner, can do nothing but digest. He can neither work nor think, because all the power of body and brain are concentrated in the stomach. There are not many who indulge their appetite to this extent, and suffer so much in consequence. But there are many who err in a lesser degree. They are not gormandizers, yet they eat too much, and suffer in weakness precisely in the ratio of their error.

140. It has been before shown (§ 41, p. 26) that rapid eaters consume more than they need or can digest. Yet many eat rapidly, in order to gain in time. They imagine that twenty minutes or half an hour, spent at their table, is a waste of many minutes, which they might employ in business or labor. They masticate little, and swallow morsel after morsel in quick succession, and soon their stomachs are filled; and then they hurry back to their employment, in the mistaken confidence that they have gained by this haste, and that they shall accomplish so much the more by thus shortening the time of eating. But they carry with them a load that consumes a portion of their strength, which they might otherwise have devoted to their labor. The farmer and the mechanic — the merchant and the student — every man who wishes to accomplish the most by the use of his physical or his mental powers — will effect his purposes the most successfully, by eating slowly and cautiously, and giving ample time to the table. It therefore is bad economy to hasten at our meals.

141. An industrious merchant of Boston, when formerly engaged in business, frequently walked a quarter of a mile, and ate his dinner, and returned to his counting-room in fifteen minutes; and was then pleased that he lost so little time. While at table, he swallowed his food as fast as possible, giving insufficient time to his mouth for mastication, and as little to his stomach to mix the food with the gastric juice; and yet he ate too much, and was oppressed afterwards. He gained in time, but he lost in energy and in

power of attending to his affairs. In the afternoon he was somewhat heavy — his brain was not clear — he was indisposed to look into his accounts, or to talk with his customers upon matters of business; and he accomplished much less during the rest of the day, than he might have done if he had allowed himself sufficient time for his dinner. This was the first result; the second and remoter result is painful dyspepsia, which now, after years of error, weighs heavily upon him.

142. It is common to notice this error at hotels, where strangers gather. Travellers seem to be often compelled, by the impatience of the coach or the railroad car, to swallow their hasty meal with all possible speed; and, unfortunately, they too often continue the habit when the apparent necessity ceases. The same hurry at meals is often seen in the steamboats, and at public tables in the cities. In steamboats particularly, where the passengers, from morning till night, have nothing to do but to eat, there is commonly manifested an eager haste in swallowing food, as if all the minutes spent at the table were lost, or worse than lost, and the company were resolved to get away from it as soon as possible.

CHAPTER XV.

Appetite allowed to accompany the Duty of Eating. — Unwise to eat for this alone. — We eat too much for Appetite alone, and make this the Means of Hospitality and social Enjoyment. — Children's Appetites pampered. — All Indulgence of mere Appetite followed by Suffering.

143. It is one of the proofs of the benevolence, as well as the wisdom, of the generous Creator, that whatever duty is required of us by the necessities of our nature, is also made pleasant in the performance. Food is made necessary for the support of the body, and appetite is given to make the taking of that food a source of great pleasure. Here are

two principles to be noticed and observed : 1st, that we are commanded to eat as much food, and of such quality, as the stomach can easily convert into the material for the blood, and the system requires for its nourishment; 2d, that this food may be so selected and compounded, and cooked in such a manner, as to be agreeable to the palate. We are not only allowed by the law of our being to enjoy the pleasures of the table, but there are encouragements and inducements held out for us to obtain this enjoyment, whenever it is consistent with the first duty.

144. *But it is plain that this pleasure of the appetite is merely the accompaniment, not the main end, of eating*, and should, therefore, never be the motive for this act. To select our food, not according to its nutritive power, or its digestibility, but according to its acceptableness to the palate,— to eat when the body does not require nourishment, or, after we have taken sufficient for this purpose, to eat some more for the indulgence of the pleasure, — these are manifest perversions of the duty required of us, and abuses of a privilege granted to us. It would seem a very foolish thing in a shipmaster to load his vessel, not with the freight that it can carry best, or which is wanted at the port of destination, but with that which is the pleasantest to load; or if, for the same reason, he should, when his ship is filled, still crowd in more than the vessel can carry, or the market will justify.

145. This would be foolish indeed, but not more so than for a man to select and measure his food without regard to the wants of nutrition in his body, or to the power of the digestive organs to convert it into the material of the blood, but according to the pleasure of its passage from the table to the stomach. In the case of the ship, when the unfitting or excessive cargo is crowded into the hold, it can be taken out, and the vessel spared the danger of sinking, and the merchant saved the loss on merchandise sent to a wrong destination, and no damage need be sustained but the labor of loading and unloading. But, when the food is once in the

stomach, there is, generally, no return; whatever is once there, although out of place, must remain or go onward at the cost of comfort and strength of this organ, and of the whole body.

146. It would seem that these principles must be plain to every one, so that no man would overload or improperly load his stomach, any more than he would his ship or his wagon, for so slight a motive as the pleasure of the first step in the work. Yet this error is among the most common in society. It seems to be forgotten that the appetite and the mouth are made to subserve digestion and nutrition; and the world eats as if the whole digestive system were the mere servant of the palate, and made to carry whatever burden this may impose upon it.

147. The pleasures of good eating occupy many men's thoughts; they are the subject of much conversation; they have had their praises sung by many a poet; while the fitness of food for the purposes of life is scarcely thought of, and still more unfrequently discussed. Few there are who do not understand the flavor of the various articles of diet, or the pleasantness or unpleasantness of the different methods of cookery. Yet the nutritive power, or digestibility of these articles, and the effects of cookery upon their qualities, are almost unknown to the world.

148. With this ignorance of the true purposes and consequences of eating, and with the too common disregard of the wants of the body and the powers of the stomach, it is not surprising that appetite should very frequently be the governing law in this matter, and that men should eat for pleasure, rather than for nourishment and strength; and such is the fact. The stomach is made the receptacle of whatever the capricious and ungoverned appetite selects and sends to it. Our tables are spread, not only with substantials that nourish the body, but with delicacies that tempt the palate. We eat not only enough to support the body, but often we add to this much more for the mere enjoyment of the act, and we urge our friends to partake of the various

kinds, not because the food is nutritive, — not because they are hungry and need it, — but on account of the inviting flavor of the dishes. Our argument to our guests is not, "You are hungry, and this will strengthen you," but, "This is pleasant to the appetite, and you will for a moment enjoy it."

149. The appetite is made the means, and the stomach is compelled to bear the burden, of much of our enjoyments, and of our hospitality. Men manifest their love for their friends by offering them delicious food, by inviting, and even urging, them to partake of what their systems do not need, and their powers of digestion cannot easily bear. And, in this earnest and well-intentioned endeavor to make their hearts glad, they give their stomachs pain and disease. There is a great proneness among mankind to make many occasions of public and private festivity. There is a strong inclination, when men gather together for enjoyment, in whatever way, — whether for dancing, or conversation, or for the celebration of a public and joyful event, — to add to the social pleasure the luxuries of good and plentiful eating. And the very means they use to signify their present joy, is the source of future suffering in a greater or less degree, in proportion to their disobedience of the law of nutrition.

150. With children, the wants of nutrition appear first, before all other wants; the appetite for food predominates over other desires, and is ever seeking for gratification. This seems to be the readiest means of pleasing them; it is therefore frequently appealed to by those who wish to give them pleasure. Delicacies of many sorts — fruits, cakes, confectionery — are offered them by those kind-hearted but indiscreet friends, and too often by their parents and nurses, as a means of soothing pain or assuaging grief, or even pacifying anger, or winning approbation.

151. Men and women of every age, as well as children, generally consider it a proper and harmless privilege to indulge this appetite when delicacies invite, and when opportunities offer. Some few of them are epicures, and, finding daily opportunities, always eat to oppressive fulness. But

most of them are generally more moderate in their indulgence, yet overstep now and then, by adding to the sufficiency of their regular meals another and another portion which tempts their taste, or by taking at other times some pleasant little refreshment, which chance may throw in their way, but which afterwards becomes a source of oppression to their organs of digestion. There are others whose habits of eating are generally in accordance with the natural law, and whose daily food is usually no more and no other than their nourishment requires. Yet these will, on perhaps rare occasions, meet in parties, or go upon excursions of pleasure, which include a feast; and then they give free rein to their appetites, and indulge in the pleasures of the table.

152. All these indulgences of appetite, when for nutrition there is no call for food, or when we have already eaten as much as we can with ease convert into chyme, must necessarily lay a tax upon the stomach; and, so far as they exceed the wants of the body, they do not add to its strength, but, on the contrary, they bring upon it weakness. These are plainly violations of the law of life, and are inevitably followed by the consequences of disobedience. The consequences are not one and the same for all, whatever may be the error, but they are measured out in precise proportion to the delinquency. The constant epicure suffers more than the occasional gourmand, and the frequent gourmand more than he who but rarely indulges in eating more than he needs. But none escape. All—the least as well as the greatest offenders—have greater or less oppression. In some it is almost imperceptible; and in others it is almost intolerable It is a singular perversion of the digestive organs, to compel them to receive and to attempt to digest food of such quantities as are not needed for nutrition, and of such qualities as nature never intended they should convert into chyle for the blood. But this apparatus is sometimes perverted to stranger and more dangerous purposes than even these. Children and men put into their mouths, and masticate, and often swallow, some materials which cannot be dissolved in

the stomach, and from which no nutriment can be extracted. From diseased appetite or perverse habits, some boys and girls chew India-rubber, pitch, or slate-pencils. I once found almost an entire school in the habit of chewing one or another of these things. These unnatural things disturb and disorder the stomach, and often result in very serious disease. Some of the most inveterate and distressing cases of indigestion arose from such beginnings as these.

153. Although we are not allowed to exceed the wants of the system or the digestive power, in the least degree, without suffering, yet the opposite error is equally contrary to the law of life. Nature is very exact in her demands, as well as in her concessions. She will not give health and strength for one morsel more than her requirements; nor will she relax and give unalloyed comfort and full vigor for one morsel less. A definite quantity of nutriment, varied to suit the varieties of persons, will nourish and strengthen, and entirely meet the wants of each individual. Any smaller quantity will give less strength and power of labor of body and of mind. The strength of the laborer will fall short of its fulness, in proportion to the diminution of his nourishment. Whether this diminution be in the quantity or the quality of his food, the result of weakness is the same. Whether it be from a short allowance of good food, or the innutritious quality or the indigestible nature of bad food, — which cannot be converted into chyme or chyle, — or from whatever cause, less than the ordinary and sufficient quantity of nutriment is sent from the digestive organs to the blood-vessels, the final end is that the frame is not fully nourished or strengthened. Poor meats, thin soups, and innutritious roots, are insufficient for the laboring man; and it is bad economy to endeavor to support him on such diet.

154. Even students, and men engaged in sedentary employments, cannot maintain their full health, and their energy and clearness of brain, upon a diet lower than their natural requirements. Some students in college, for the sake of economy, endeavor to live upon very little and cheap food. But

they suffer in consequence, and are unable to pursue their studies with their original vigor. A freshman in college endeavored to support his body with eight ounces of bread a day, without other vegetable food or meat. He followed this plan four weeks, and in that time suffered from headache, nervousness, general debility, and indisposition to apply his mind to his books. But on returning to the usual but moderate diet of other students, he regained his usual health and mental vigor.

CHAPTER XVI.

Greater Flow of Blood and of nervous Energy to Parts and Organs in Motion, and to Stomach during Digestion. — Action in other Parts interferes with Digestion. — Mental Labor has the same Effect. — Rest requisite after Eating, and before Eating. — Gymnastic Exercises at Cambridge.

155. WHEN any one organ or portion of the body is in action, more blood is sent to it, through the arteries, to meet the changes, and supply its waste, and support its powers; at the same time, there is more nervous energy sent to this part, to quicken its activity. When this unusual flow of blood and nervous influence is toward one spot, there must be proportionably less sent to all the other parts, and consequently the rest of the body must be comparatively languid or inactive. No two portions or organs can, then, be kept in the fullest and most vigorous action at the same time, for the extraordinary flow of blood, and of quickening nervous power, cannot be supplied to both or all at once.

156. This is particularly applicable to digestion of food in the stomach. While this is going on, the preponderance of blood is towards this organ, to sustain this new action, and furnish the materials of the gastric juice; consequently, there must be a smaller proportion of blood in the other parts of the body. Upon the same principle, the nervous

influence flows in a larger proportion to the stomach, — the seat of action, — and in a less proportion to the rest of the frame. It is plain, therefore, that while digestion is going on, or until the gastric juice is prepared sufficiently for the digestion, the other parts have a smaller supply of blood to sustain their actions, and less nervous power to quicken their life; they must, consequently, be comparatively languid, and should be suffered to rest.

157. *Full, vigorous action cannot, then, be well sustained in two parts of the body at the same time.* If this be attempted, one or the other must fail, or both be imperfect. In order to insure perfect digestion, the stomach must be allowed to do its perfect work, and no other organ must make active exertions while this is going on. We should, therefore, let both the body and the mind rest for a short period after each meal. It is a custom in Spain to take a short nap after dinner. This is often quoted as a proof of Spanish indolence. It is no indication of indolence. It is rather a mark of wisdom; for this leaves the digestive organs an opportunity to do their work undisturbed, and to prepare for the body that new nutriment which is to give it power of action afterwards.

158. *Action of the mind, as well as action of the body, interferes with digestion;* and the digestive process interferes with mental activity. Immediately after a hearty dinner, one is indisposed to think, or thinks but lightly. He is averse to study, to business cares, to calculations, and to any matter that requires vigorous thought. Ask him then to consider a grave subject, or ask a boy, in a similar situation, to learn a difficult lesson, and either will be glad to postpone the labor until the digestive process is over. If severe mental labor be undertaken, it will not be carried on easily; and, if it be carried on at all, it will be at the cost of the digestion. Both these operations cannot be performed successfully at the same time. Yet it is not necessary that the brain be perfectly dormant. A pleasant and light action, such as accompanies cheerful conversation, or reading light works, does not interfere with this work of the stomach.

159. If any other organ or system be put in violent exercise, or in a state of high excitement, immediately or soon after the meal is eaten, the digestion is interrupted, or even suspended. A striking experiment was tried by a gentleman in England. He gave to several hounds as much food as they could eat, and then put some of them into a kennel where they had no opportunity of motion; the others were put upon the chase, and kept running in hot pursuit of game for an hour or more. At the end of this period, he killed some of both classes, and examined their stomachs. He found the food in the stomachs of those dogs, which had been running, in the same condition as when first swallowed;—it had remained unchanged. But in the stomachs of the others, which had been at rest, it was digested and converted into pulpy chyme, and had gone mostly out of the stomach into the alimentary canal.

160. It may not be necessary that we should sleep, like the Spaniard, after our dinner, but it is necessary, for perfect digestion of the food, and effectual nutrition of the body, that we do not, like the running hounds, engage in violent exercise at that time. Indeed, all active labor immediately after eating, interferes with digestion, and of course with the purposes of the meal; and this interruption must be in proportion to the activity of the motion. If violent exercise suspends entirely the work of the stomach, exercise less laborious will interfere with it in some degree. Yet absolute rest or sleep is not necessary. Dr. Beaumont says, "Gentle exercise facilitates the digestion of food." In the course of an hour from the meal, the gastric juice is sent into the stomach, sufficient for the digestion, and is completely mixed with the food. Then we may proceed to active employment, without fear of disturbing the digestive process.

161. When any of the organs or limbs have been greatly exercised, there come a fatigue in that part which has labored, and a lassitude in the whole frame; and then none of our powers are disposed to active exertion; all want rest. When the fireman's feet are fatigued with running a long

distance to a fire, he is not ready to take hold of the brakes of the engine and pump with successful vigor. Nor, after fatiguing the arms with the engine, is he inclined to make equal exertion with the feet in running a race homeward.

162. *Fatigue of the body affects the mental powers in the same way.* When the laborer has finished a very hard day's work, or when the fireman returns from his violent exertions at a fire, he is disinclined to active thought, and, perhaps, even to read, and may fall asleep over his book. Nor at any time do we readily think upon any serious subject, or attend to any business that requires grave thought, immediately after we have made great and fatiguing exertion.

163. While any limb or organ is in action, there is a greater waste of particles. At the same time, there is a greater flow of blood to supply this waste, and of nervous power to quicken the action. But, if the action be violent, the waste is greater than the new supply, and consequently the part is exhausted, and the body feels fatigued. The exhaustion of particles and the fatigue remain after the action is over. If we are then quiet, the blood and the nervous energy still continue to flow in unusual quantity, to restore the previous waste, and to revive the diminished life. By this means, we rest, and recover lost powers. While the brain and the blood-vessels are thus restoring any fatigued part, they cannot sustain a vigorous action in another; and if we then attempt to exercise the muscles, or stomach, or brain, and work, digest, or think, we shall do it but languidly, — probably unsuccessfully, — because the blood and nervous energy which are needed to sustain these actions are wanted and used elsewhere.

164. During the first process of digestion, the stomach requires a greater flow of blood and of nervous energy to sustain this action. But if these be still required in restoring the waste, and the power of other parts, exhausted and fatigued by previous exertion, they cannot be given to the stomach. If, therefore, when we are much fatigued with exertion of the muscles or of the brain, we fill our

stomachs with food, this organ cannot receive from the blood, or from the nervous system, that aid which is necessary to enable it to digest. The laborer, therefore, should not go directly from his hard work, nor the student from his severe study, nor the merchant from his oppressive anxieties, to the table. But each should allow a short interval of rest, and then he is prepared to eat and digest his food.

165. It is not uncommon for students to devote the hour before dinner to their exercise. Schools and academies, and even colleges, usually have this hour of leisure to be taken from books and devoted to recreation. That there should be rest of the brain at this time is well. When the gymnasium was established at Harvard University, in 1826, the students were invited to go to the playground at twelve, and engage in the gymnastic exercises till one o'clock. These were very active, and some of them violent, for men and boys of their strength, so that, when they left the field for dinner, they were generally fatigued, and some were almost exhausted. Those who were most fatigued, ate their dinner with less than their usual relish, and felt neither refreshed nor comfortable afterward. Their stomachs could not digest the meal with the usual ease, and consequently they were heavy, and indisposed for study in the afternoon.

CHAPTER XVII.

Cheerful Conversation at Meals aids Digestion. — Silent and solitary Meals unfavorable to Digestion. — Consequences of Abuse of digestive Organs.

166. DURING the time of eating, the body should be seated in a comfortable and easy position, and all the organs and powers, except the digestive, should be at rest. The muscles and the brain should be quiescent. The mental and the moral powers should yield, for the time, to the business of calm nutrition. The mind should therefore be free from the burden of deep reflection, care, and anxiety. None

of the evil passions — anger and envy — should ever be allowed to come to the table. All great and severe thought, all labored discussions, and matters of business, should be banished thence, and light and cheerful conversation take their places. The lively play of the social affections, the pleasant intercourse of family and friends, the enlivening flow of wit and humor, keep the brain in action, but not in labor With these, the blood moves more freely, and the nervous energies flow more joyously, and the work of digestion is more readily begun, and more easily carried on, and they should ever be present at our meals.

167. The eating hour is the time to cultivate the social nature. This harmonizes well with the lively flow of spirits that aids the digestive process. It is better, therefore, not to eat alone, nor even in silence. The solemn stillness that reigns over the table of some families, the unbroken quietness which a stern but mistaken discipline imposes upon some children, are at variance with the best interests of the time. They lay a weight upon the brain, a burden upon the spirit, and prevent that quickening which social cheerfulness would give to the stomach.

168. These several steps and conditions of digestion, of nutrition, and strengthening, were established by the Creator. They are among the very laws of our being, and cannot be changed. The only way we can gain the most strength for labor of any sort, is by perfect obedience to these laws, and fulfilment of these conditions. All failure of this must result in loss, and defeat the very purpose for which they are violated. The loss is immediate in the depreciation of power, greater or less, in proportion to the delinquency. The loss is also accumulative and remote, because the stomach itself loses power to do its ordinary duty when unnatural burdens are imposed upon it, or when it is not allowed the requisite aid to bear them.

169. The first consequence of neglect of these laws of eating is, an imperfect nutrition, a comparative weakness at the time, inability to accomplish, with the brain or the

muscles, what otherwise could have been done; therefore, a direct loss of power, and of means of production. But the later consequence is more important and lasting. The stomach being called upon to digest unmasticated food, or more than it can dissolve, is disturbed and wearied with the excess of labor, and falters. It struggles, but struggles in vain, until it exhausts much of its power in the wearisome effort. It then becomes so weak that it cannot digest even the common food, which in good health it would have easily done, and becomes so irritable as to bear only in pain the natural and proper burden. This is *dyspepsia*, which is the common result of improper use of our organs of digestion. This is a disease painful to be borne, and difficult to be relieved, and often ends only with life. In its first stage, the work of digestion is imperfect; and in the second, the digestive machine is impaired, and finally destroyed.

CHAPTER XVIII.

Animal and Vegetable Food. — Northern Nations carnivorous. — Equatorial Nations herbivorous. — Vegetable Diet. — Mixed Diet — Stimulating Food. — Climate and Season affect Diet. — Vegetable Diet in Torrid Zone. — Mixed Diet in Temperate Zone.

170. THERE is a great *variety of food*. There are many kinds that differ widely from each other; and these may be prepared, compounded, and cooked in a great variety of ways; so that the differences caused by art may be even more and greater than the differences of nature. It is important for us to know which of these afford us the easiest and the best nutrition.

171. The first natural division of food is into that of vegetable and animal origin. It is not yet a settled question, which of these is best fitted for the nutrition of the human body, — which will give to man the greatest strength and power of action, — the greatest comfort and most perfect

health, — the clearest brain and the longest duration of life. Some have contended, that man was intended to eat only of the fruits and vegetables of the earth; while others maintain, with equal confidence, that he should add to these the flesh of beasts. But none have thought that he should live exclusively upon animal food.

172. The advocates of both these doctrines find extensive examples in the various nations and the various individuals of mankind. Many nations within the tropics live upon vegetable food alone; while some tribes within the arctic circles feed almost entirely upon the flesh of animals or fish. The inhabitants of the cold regions of the earth are generally carnivorous, and the residents of the warm countries herbivorous; while those who live in the temperate climates are both carnivorous and herbivorous. Here in the United States, and in the central regions of Europe, the mixed diet is almost universal; and the people are as healthy, and have as great a duration of life, as any upon earth.

173. There are many individuals, in this and in other countries, who confine themselves to vegetable diet. They believe they enjoy better health, and maintain greater strength of body and mind, than those who live upon mixed diet. The experiment has not been tried on a sufficiently extensive range to determine its value. It has not proved a failure, nor has it demonstrated, to the satisfaction of all, that flesh is injurious. There are no advocates here for the exclusively flesh diet; but the doubt is only between the mixed on the one side, and the vegetable food on the other.

174. *It is generally believed, among civilized nations, that the mixed diet is the best for man* — that this will give him the fullest health and the longest life. The organization of the human body admits this. The form and arrangement of the teeth (§ 10, p. 12) allow him to cut and masticate both the animal and the vegetable food; and the structure of the stomach and intestinal canal enables man to digest both; so that he can use either exclusively, as the Esquimaux eat flesh, and the Hindoos eat rice; or he can use both, as most

of the inhabitants of the temperate climates eat bread and meat together. Admitting, then, this question to rest for the present, and that we are to use the mixed diet as we have done, still, there are other and subordinate questions, with regard to each individual, to be answered, before we can determine what we shall eat.

175. *The various kinds of food differ as to their effect upon the animal body.* One kind, including most meats, is stimulating, and gives a greater elasticity of life. This would excite some fever, when there is a feverish tendency. Another kind, including fish, eggs, vegetables, grain, and fruits, has no stimulating power. These would not quicken the pulse nor excite fever. The spices, and food in which they are mixed, are warm and heating. Many of the vegetables are cooling. These differences must be known, before the fitness of the various kinds to the condition of man can be determined.

176. *Climate and season affect the human body, and its wants and power of digestion, very materially.* We want a somewhat different diet in the warm and in the cold seasons. We eat more meat and stimulating food in the winter, and more vegetable and cooling food in the summer. The tribes about the arctic circle live almost exclusively upon animal food. They will eat meat in great quantities without either bread or vegetables to accompany it. They devour fish of the coarsest kinds, — whale, porpoises, &c., — such as we think unfit for our nutrition, and impossible to be digested in our stomachs. They will drink whale oil with as much apparent relish as we drink milk or water. The voyagers to these northern regions, while they are passing the winter among these people, fall into their habits of eating: they find both that they need, and that their stomachs can digest, this coarse and stimulating food, which would have been oppressive and indigestible at home in a temperate climate.

177. On the contrary, the inhabitants of the tropical regions live very much, and some nations entirely, upon vegetable food. Some of these nations never eat meat, and most

of them eat it rarely; certainly, they make it a secondary article of their diet. But in the temperate climates, in the middle regions between the extremely hot and the extremely cold, a mixed diet is generally, and almost universally, adopted. Even here, the proportions of the meat and the vegetable vary with the climate. In the warmer countries, — in the south of Europe, and as we approach the tropics, — the vegetable predominates; and among the northern nations, toward the frigid zone, the meat is the main dependence for nourishment. This is in obedience to the general law of life, that the body needs, and the stomach can bear, a more highly stimulating food, when and where the atmosphere is cold, than when and where it is warm.

CHAPTER XIX.

Temperaments. — Lymphatic. — Nervous. — Sanguine. — Bilious. — Difference of the Excitable and the Inexcitable. — Diet to be regulated according to Temperament.

178. *There are differences of individuals that should require corresponding differences of diet.* It is plain, even to the most careless observer, that men are not all alike. One is dull, difficult to be moved, and habitually inactive. Another is quick, irritable, and easily excited or depressed; he is gratified or disturbed with very small matters and is restless in his disposition and habits. These differences arise from the physical condition of the individuals, from their original organization, and constitute, in part, what are called the various *temperaments* of men. Physiologists have divided mankind into several classes, as to temperaments, according to the predominant traits in their constitution. Some have made four of these classes, calling them the *lymphatic, nervous, sanguine,* and *bilious* temperaments. Others have made more classes. But however few or many any physiolo-

gist has adopted, there have been combinations and mixtures of traits, forming other and intermediate classes, such as the *nervous-bilious*, the *nervous-sanguine*, &c.

179. In persons of *lymphatic* temperament, the form is generally full and round, fat, or tending to fatness; the skin is soft and rather full; and the flesh is loose, and somewhat flabby; the muscles are weak, and the whole body is inactive. There is an indisposition to exertion, either of body or of mind; the temper is calm and inexcitable, and the passions are not easily roused. Men of this temperament are not easily excited, and can bear much stimulation. They should therefore have food of a stimulating nature, — such as beef, mutton, coffee, — which tend to counteract their indolent disposition. On the contrary, their natural indolence is increased by a weak and unstimulating diet. Fish, oysters, eggs, and most of the vegetable roots, would aggravate the peculiarities of this temperament.

180. The *nervous* temperament is marked by predominance of the brain and nerves, and by great nervous excitability. Men of this class are easily excited, and as easily depressed. They are susceptible of high pleasures and great distress. They are very sensitive to external influence, both upon their body and upon their mind. A strong stimulating diet would increase these peculiarities; but, to counteract them, a mild and cooling, yet a nutritious diet is necessary.

181. In the *sanguine* temperament, there is greater development and activity of the apparatus for the circulation of the blood, and nutrition of the heart and blood-vessels. All the physical powers are strong, elastic, and easily excited. Men of this class are bold and resolute; they are ready to act, but not persevering; they soon become weary, especially if difficulties present themselves. Their bodily faculties predominate over their mental, and they are men of action rather than of thought. They are not scholars, nor the most cautious men of business. The boys are foremost at play, but not in school. These cannot bear excitements or stimu-

lants, without danger of disease. Stimulating food or drinks create an unnatural activity of the heart and blood-vessels, and are therefore injurious to them.

182. In the *bilious* temperament, the skin is brown, and inclining to yellow; the hair is usually dark; the form is moderately full, but not fat; the limbs are not gracefully rounded, but the muscles are well developed and very strong. Men of this class are not very quick in mind or body. They are calm and placid — not irritable in temper. They have great boldness of purpose, energy in action, and perseverance in their undertakings. They are men who succeed in their course of life, because they are cool and cautious in their plans, and indefatigable and persevering in carrying them into execution. In the higher and in the lower walks of life, they are successful.

183. Wherever persons of the bilious temperament begin life, they go up higher. There is within them a restless energy, that is not content with the present, whatever it may be. Napoleon Bonaparte is a remarkable instance of this temperament; Capt. P., who at twenty-two years of age was an hostler, and afterward became owner and commander of one of the largest steamboats on the Mississippi, is another; and most of those who in the beginning of life were day-laborers, without means or friends, but afterwards are prosperous and wealthy, and become the leading men in their towns, and the governing men in their respective business associations, all, or nearly all, belong to this class, and are of the bilious temperament. These men neither need, nor are they benefited by the stimulating diet of the lymphatic, nor by the spare and cautious diet of the nervous and sanguine. Their temper and habits of life generally imply a great amount of action of body or mind, generally of both, and consequently a great expenditure of material; they need, therefore, a full and generous diet of nutritious food, to sustain them in their activity.

184. Without supposing that any one can tell exactly the temperament of himself, or of his companions, yet one can

hardly fail to observe a difference among his associates. One is active and sprightly; another inactive and dull. One is excitable and irritable; another is slow, calm, and placid. One is hasty and impatient, quick to receive new ideas, or eager to engage in new plans; another is slow of understanding, and hesitates about new propositions. These differences are very perceptible, both in men and women. It is easy to see, among children at school, how much quicker one is than another in learning his lessons, and how much more impetuous and active at his play. He has not necessarily greater talent for learning, nor more fondness for amusement than the other, who is more patient and slow at his books, and less hasty and boisterous on the playground; but his talents and his feelings are all more active. He understands his lessons in school more readily, but he is not in the end a better scholar; he is more ready in his games of sport, but he does not play with more skill. He requires a smaller inducement to begin a task in the school-room, or a game on the playground; but he does not persevere so faithfully to the end, as the duller and slower boy.

185. These are differently affected by outward circumstances, by affairs of life, by the treatment of others, by gratifications or disappointments. They are as variously affected by the matters which they eat or drink. It is well known, that one man can drink large quantities of strong spirit without being visibly affected, while another can hardly take a small portion of wine without being intoxicated. A similar difference follows the use of stimulating and unstimulating food. What is beneficial to one is injurious to another. The sanguine and the ardent need a cooling diet. The cool and dull want meat, and other exciting food. The quick and irritable should live upon bread, vegetables, fish, and such other matters as will not excite them. If they thus regulate their diet, according to the peculiarities of their constitution, they will be better able to control themselves. If they disregard these, and eat otherwise, they will give strength to those peculiarities already too strong, and the

lymphatic will become more sluggish, and the excitable will have too high degree of action, and run on life too rapidly; and thus both will enjoy less health, and accomplish less, than if they lived faithful to the necessities of their constitution.

CHAPTER XX.

Difference of Constitution in Childhood and old Age. — Food to be varied accordingly. — The Active want more stimulating Diet than the Inactive. — When Habits are changed, Food must change.

186. THERE is a difference of excitability in the different periods of life, which should be supported by a corresponding difference in the quality of the food. In childhood, all the powers of life are more active, the blood flows more rapidly, the nervous system is more irritable, and the muscles more easily stimulated to action; the feelings and passions, and all the motions of life, more readily quickened; but there is less power of endurance, and the energies are sooner exhausted, than in maturer life. In old age, all the powers and systems are in the very opposite condition. There is a sluggishness in all the motions, and an inactivity in the limbs; and the feelings, the passions, are slow to rise.

187. There is a wide difference between these conditions of life; and, if we should attempt to support them with food of the same quality, we should fail of giving each its true life and strength. It is plain that the elastic period requires a mild and soothing diet, while the inactive period needs more stimulating food. Children then want milk, bread, and mostly vegetable food; and, if they add meat to this, it should be of the milder kinds, such as fish and fowl, rather than beef and mutton. But old men need more meat, and that of the most stimulating and nutritious kinds.

188. *The habits of the individual have an important bearing upon the quality of food.* Those can bear the greatest

stimulation who have the greatest activity, and whose exercise opens the freest outlet for their nervous energies. On the other hand, the habitually indolent and inactive, whose nervous energies are not freely expended, do not bear stimulation easily, because they have less outlet for the quickened flow of vitality. The laborious and the active should eat more stimulating food than the sedentary and the idle. Farmers, sailors, masons, carpenters, and out-of-door laborers, want more meat; while students, tailors, shoemakers, and house-employed women, want more bread. The former thrive best upon beef, mutton, and bread; while the others, when they add meat to their vegetable food, do well with chickens, fowl, turkeys, and fish.

189. *Disorders of the stomach arise from neglect of this caution.* While men are engaged in hard labor abroad they have good appetite and vigorous digestion, and very properly eat stimulating food; but if they leave their laborious occupations, and become jewellers, shoemakers, scholars, or merchants, or engage in any light employment, their lives, from being the most laborious and active, become quiet and often sedentary. They have less change in their vital particles, and, of course, less nutritive want and digestive power.

190. If, now, these men do not change their diet with their habits of exercise, their digestive powers begin to falter, and then they feel oppressed after eating. They eat with less satisfaction, and do not have the sensations of ease and comfort after their meals. They are dull, and disinclined to go to their usual employments, or their books, or their accounts.

191. When one exchanges a light for a laborious occupation, he increases his expenditure of particles, and consequently the demands for nutrition. His stomach gradually gains power, and his appetite craves more food to meet the new habit of life. A hard student at Westford Academy, in 1822, became dyspeptic and feeble. He had little appe-

tite, and digested his small portions of food with difficulty. He left his studies, and went, as a common sailor, to South America. At first, he performed little of the light labor of the ship, and ate sparingly, as he had on land; but his strength of body and power of digestion increased, and, after a few months, he was able to do a sailor's work, and eat a sailor's allowance. He ate the heavy and stimulating food of the ship,— the salt meat and hard bread,— with good relish and good digestion, and felt no oppression afterward.

CHAPTER XXI.

Digestibility and Nutritiousness of Food not identical. — Food easily digested not always best for Invalids. — Condiments excite and exhaust. — Sensibility of Stomach. — Alcohol and Wines exhaust still more. — No single Rule of Diet to govern all Men.

192. THERE are two things to be considered in regard to all kinds of food; these are, 1*st, the digestibility* — the ease or the difficulty of being converted, by the stomach and its gastric juice, into chyle; and, 2*d, the quantity of nutriment* contained in them. And these are not necessarily one and the same. One article of diet may be very easily digested, but contain very little nutriment. On the other hand, some articles are highly nutritious, yet are very difficult to be digested. Perhaps no food contains more of the nutritive principle than oil, yet few kinds require longer time to be converted into chyme.

193. That food which is most easily digested is not always the most suitable for the sick and feeble. Beef and mutton are much more readily changed by the action of the stomach than gruel. But they are also much more stimulating to the system; and, if eaten by the sick and the convalescent, they might excite fever, perhaps a return of the disease. But gruel, and bread, which may require a longer time and a greater labor of the stomach to digest them, do not excite

the circulation of the blood, nor produce fever. Dangerous and even fatal consequences sometimes ensue from a neglect of this distinction.

194. When we rub the skin with pepper, mustard, or spirit, it creates irritation; the veins and arteries enlarge; the blood flows to the place in unusual abundance; there is an increased heat in the spot, and the surface is red; there are greater action and quicker life; but these effects soon cease; and then the skin is pale, the circulation is more languid, and the life of the part is more dormant, as unusual action of the muscles leaves fatigue behind. These are the natural effects of stimulation: first, increased activity; and next, increased languor; for all unnatural excitement of the natural actions of the living system is followed by a corresponding depression.

195. The same takes place in the stomach from the use of all condiments, such as spices, pepper, mustard, with our food. The stomach is stimulated, the circulation of blood in its walls is quickened, the gastric juice flows more readily, and digestion begins more promptly. But soon this unnatural activity ceases; and then it falls below its natural standard, and digestion is finally retarded. This is the effect of once using the stimulating condiments; but, if this use be continued and often repeated, the power of the stomach, from frequent excitement and fatigue, becomes somewhat worn, and the organ is permanently enfeebled. To a healthy stomach, then, condiments and stimulants, are not only unnecessary, but injurious. They give no strength; they only quicken the action and expenditure of power already existing.

196. These enfeebling effects follow the stimulation of wines and spirits even more than that of spices. Alcohol is more speedy in its action, both of excitement and exhaustion. Dr. Beaumont saw that St. Martin's stomach was reddened after drinking spirit, and sometimes the covering of the inner coat peeled off, and left spots of canker upon the surface. The remote result of this drinking is more severe and dan-

gerous than that of condiments. The stomach gradually loses its power, until it becomes incurably dyspeptic, and is unable to digest the ordinary food.

197. The use of all stimulants, both of spices and alcohol, at first sharpens and then destroys the natural sensibility of the tongue and mouth. The healthy appetite of those who are unused to stimulants is simple, and wants simple things. They have a refined taste, and nice discrimination of the different flavor of various kinds of food, and a keen relish for what they eat, although it is neither prepared with spice, nor accompanied with spirits or wine. But those who are accustomed to highly-seasoned dishes are not satisfied without them. All simple dishes are insipid. Their taste is so blunted that these do not excite it. And, in the old and habitual drunkard, this sensibility is so deadened that nothing short of a very active or even pungent stimulant will reach and satisfy it.

198. A man who for years had drunk tea and coffee, and occasionally wine, and habitually eaten spices, suddenly ceased to use them, and drank only milk and water, and ate no other condiment than salt with his food. After eight months' practice of his simple diet, the sensibility of his tongue became so exalted, and he gained so nice a discrimination of taste, that he could distinguish and enjoy the various flavor of water from different wells, as readily as he had distinguished between the various kinds of wine, tea, and coffee. The enjoyment of the mere taste and appetite is not in proportion to the stimulating power of what we eat and drink, but in the ratio of the quickness of the sensibility. And the water-drinker, while in the mere act of drinking, enjoys his pure water more than the wine-drinker does his wines, or the spirit-drinker his stronger and more stimulating drinks. "Happy are the young and healthy," says the shrewd Dr. Kitchener, "who are wise enough to be convinced that water is the best drink, and salt the best sauce."

199. From this examination of the structure and uses of the digestive organs, and of their purposes, powers, and

liabilities, we learn that the stomach performs some of the most delicate operations, and effects some of the most wonderful changes, in nature; and that it requires the aid of the intelligent hand to supply its wants, and fit the supplies to its necessities. There is no human instinct to be our unerring guide, and to direct us what and how much we shall eat or drink. The living machinery within, and the dead material without our bodies, are prepared for our use; and the law of nature is declared to us for our government. This we are required to read and to understand, before we can perform our part in the sustenance of our frames.

200. This is not a law of appetite, that directs us always to eat when we are hungry, and take such kinds of food, and as much of it, as the palate craves. Nor is it a law of convenience, that allows such food as chance or caprice may place before us. But nature has established for every man a law which must govern his nutrition. This law is founded upon the structure of his digestive organs, the wants of his frame, his temperament, his age, and his habits of exercise. Every individual must understand these general principles, which have been described in this book, and, applying them to himself and his circumstances, he must determine what food, and how much, will meet the necessities of his own body.

201. In this matter, as there are great varieties of men and of external circumstances, so there is no one rule of diet that will apply to all mankind. All the differences of men must be supplied with corresponding differences of nutriment. Those codes of diet which are laid down for the government of all men, of every variety of temperament, and habit, and location, and which attempt to sustain all men with the same aliment, are absurd, and fail; for every man, or every class of men, have their peculiar powers, and their peculiar wants; and if they disregard these, and endeavor to support life by any other rule, they will not fully accomplish the purpose of eating.

PART II.

CIRCULATION OF THE BLOOD AND NUTRITION.

CHAPTER I.

Apparatus of Circulation. — Heart. — Structure and Divisions. — Valves. — Arteries. — Aorta. — Subclavian, carotid, facial arteries. — Branches in the lower limbs.

202. The chyle, or the nutritious part of the digested food, is carried from the digestive organs, in the abdomen, through the absorbent mouths, and the lacteal tubes, and great lacteal duct, to the great veins near the heart. There it is mixed with, and becomes a part of, the blood. This blood is to undergo certain changes in the lungs, and then it is to be distributed to all the parts of the body.

203. The apparatus for this distribution or *circulation of the blood* consists of the *heart*, or central organ of motion; the *arteries*, which carry all the blood out of the heart to the lungs and to the various parts of the body; the *capillary vessels*, in which nutrition takes place; and the *veins*, which carry the blood back to the heart.

204. *The heart* (Fig. VII.) *is a hollow, muscular organ*, or bag, composed of fibrous substance, like lean meat. It is capable of contraction and expansion, like the muscular coat of the stomach. When it contracts it diminishes its internal cavity, and presses out the fluid contents or blood that is within it. When it relaxes, its cavity is enlarged and allows other fluid or blood to flow into it.

205. The heart is placed in the centre of the chest, between the two lungs, (Fig. V. *c*.) Its larger end is upward,

and behind the breast-bone. The smaller end or apex is downward, and turned toward the left. When the heart contracts, to send the blood out, the apex is thrown forward and strikes against the ribs of the left side, near the breast-bone, where the beating is very easily felt

FIG. VII. *Heart.*

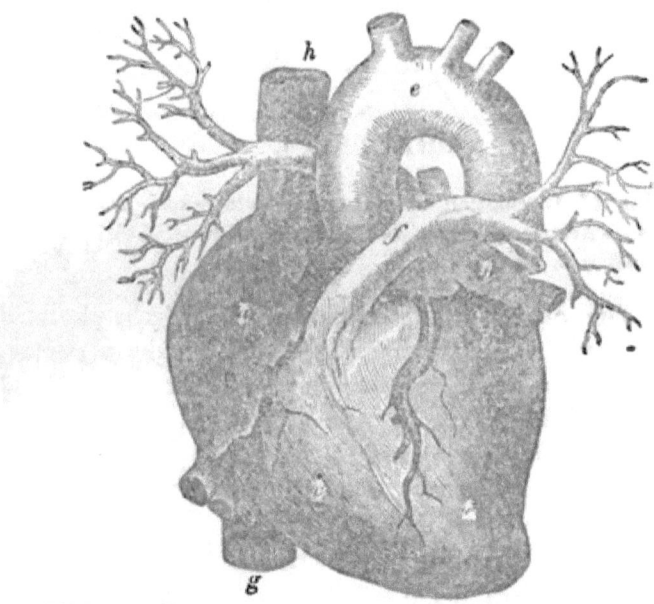

a, Right auricle.
b, Left auricle.
c, Right ventricle.
d, Left ventricle.
e, Great artery carrying the blood from the left ventricle to the body.

f, Artery carrying the blood from the right ventricle to the lungs.
g, h, Great veins carrying the blood from the body to the heart.

206. *The internal cavity of the heart* is divided by a partition wall of flesh (Fig. VIII. *c*) into two apartments, one on the right, and one on the left. This separation of the two sides of the heart is complete. There is no passage way through this wall, and consequently no direct communication between the two apartments. They are sometimes described as two hearts united together.

CIRCULATION OF THE BLOOD. 95

Fig. VIII. *Heart laid open. View of the four Chambers of the Heart.*

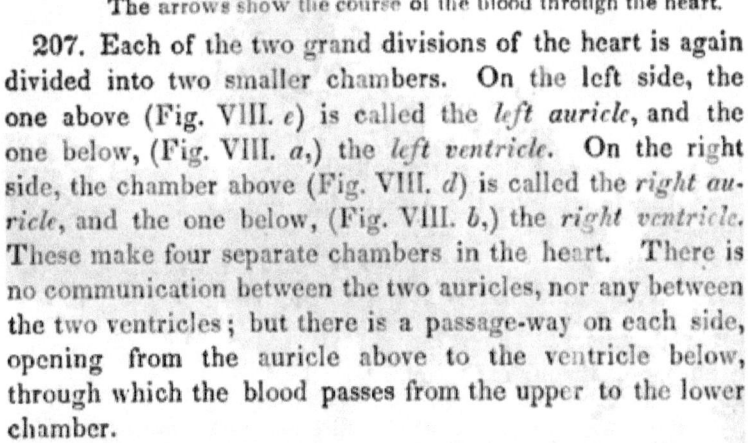

a. Left ventricle.
b Right ventricle.
c Partition wall between the two sides.
d. Right auricle.
e. Left auricle.
f, g. Partitions between the auricles and ventricles.
h, h. Great artery leading from the left ventricle to the whole body.
i, i. Great vein carrying the blood to the heart from the whole body.
k, k. Arteries leading to the lungs.
l, l. Veins leading from the lungs

The arrows show the course of the blood through the heart.

207. Each of the two grand divisions of the heart is again divided into two smaller chambers. On the left side, the one above (Fig. VIII. *e*) is called the *left auricle*, and the one below, (Fig. VIII. *a*,) the *left ventricle*. On the right side, the chamber above (Fig. VIII. *d*) is called the *right auricle*, and the one below, (Fig. VIII. *b*,) the *right ventricle*. These make four separate chambers in the heart. There is no communication between the two auricles, nor any between the two ventricles; but there is a passage-way on each side, opening from the auricle above to the ventricle below, through which the blood passes from the upper to the lower chamber.

208. The blood flows from the auricle to the ventricle, but not backward from the ventricle to the auricle. *There are valves placed at these passage-ways in the heart*, which open to allow the blood to pass downward, but they close, and prevent its passing upward. These valves act on the principle of the valve in the common pump box, which opens when the water below presses upward, and allows it to pass through;

but it closes again when the water above presses downward, and prevents its return to the well.

209. *The heart is the centre of the circulating system.* It sends all the blood to the whole body through the arteries, and receives it back again through the veins. There are two sets of these blood vessels, each consisting of arteries and veins. One set begins at the heart and extends through the whole body. The other set reaches from the heart to the lungs. The arteries open from the ventricles or lower chambers of the heart, and carry the blood out. The veins open into the auricles or the upper chambers of the heart, and carry the blood back.

210. *The arteries are round tubes.* They are firm in their structure, and retain their cylindrical form when empty. They are composed of three coats. The outer coat is dense and strong, and is the principal means of resistance to pressure. The middle coat is thick and elastic, and expands when the blood flows in, and contracts when it flows out. The inner coat is very delicate, and forms a polished surface, on which the blood flows easily. The arteries have the same structure throughout the body. There is a *valve* between the ventricle and the aorta, which allows the blood to pass from the heart into the artery, but not to flow backward from the vessel into the heart.

211. *One large artery, called the aorta,* leads out from the left ventricle upward, (Fig. VII. *e*, Fig. IX. *c, d, e,*) and then turns downward, (Fig. IX. *d*.) It passes then along the back-bone, through the chest and the abdomen. Great branches pass out from this artery to the various parts of the body. Two branches, called the *carotids*, (Fig. IX. *f, f,*) pass off from the arch and go to the head, one on each side of the neck. Two branches, called the *subclavians*, (Fig. IX. *g, g,*) go to the arms and hands. While passing through the chest, the aorta sends branches to the walls of the chest, (Fig. IX. *h, h.*) In the abdomen, the aorta sends the *cœliac artery* to the stomach, the *hepatic artery* to the liver, the *renal arteries* to the kidneys, &c., and other branches to the

various organs in that region. At the lower part of the abdomen, the aorta divides (Fig. XI.) into two branches, (Fig. XI. *e, e,*) which pass through the groins and all the limbs below.

FIG. IX. *Aorta and Branches in the Chest.*

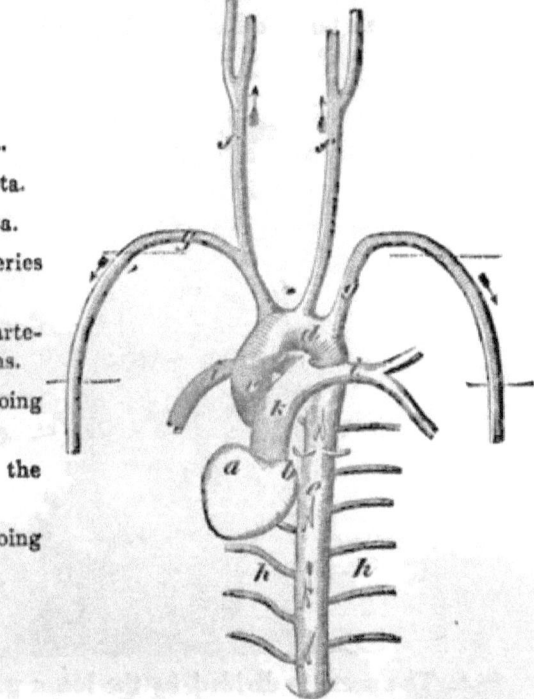

a, b, Heart.

c, Ascending aorta.

d, Arch of the aorta.

e, Descending aorta.

f, f, Carotid arteries going to the head.

g, g, Subclavian arteries going to the arms.

h, h, Branches going to the chest.

i, i, Branches to the right and left lung.

k, Great artery going to the lungs.

212. The *subclavian arteries* (Fig. IX. *g, g*) pass from the arch of the aorta, and, extending under the collar bones and through the arm-pits, they reach the arm ; at the elbows they divide into two main branches. One of these can be felt at the wrist, near the root of the thumb. These arteries send numberless little branches, which reach all the flesh of the arms and hands.

213. The *carotid arteries* passing up by the sides of the neck, can be felt near the windpipe. At the top of the neck they divide into two branches, one of which passes through the skull, and is distributed to the brain. The arteries

and branches ramify throughout all the substance of this organ.

The outer branch of the carotid goes to the outside of the head and to the face. It sends branches to the forehead, the cheeks, the chin, the lips, and the other parts of the face All these organs are thus supplied very abundantly with blood by these vessels and their numberless little branches. (Fig. X.)

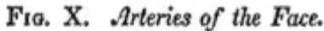

Fig. X. *Arteries of the Face.*

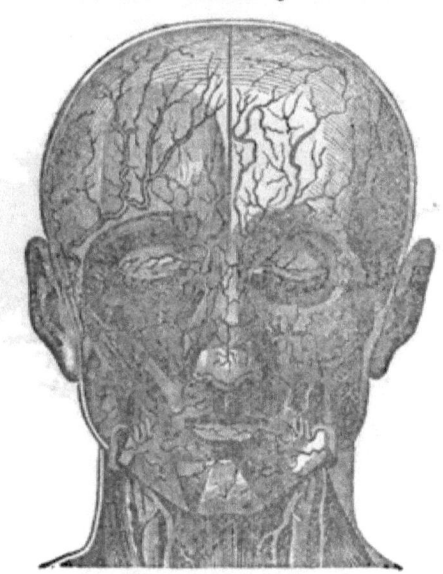

214. The aorta is divided in the lower part into two great branches, (Fig. XI. *e, e.*) These two branches pass through the groins, where they are called *inguinal arteries*, and thence into the thighs, where they are the *femoral arteries;* passing downward, they divide, and send branches to the flesh of the lower limbs, until the legs, the feet, and the toes are supplied with them.

All these arteries in every part of the body are divided and multiplied as they go from the heart, and their numberless small branches are distributed into every part of the flesh in all the organs and regions of the body. Thus all the parts, organs, and textures are supplied with blood for their nourishment.

CIRCULATION OF THE BLOOD. 99

Fig. XI. *Arteries of the whole Body.*

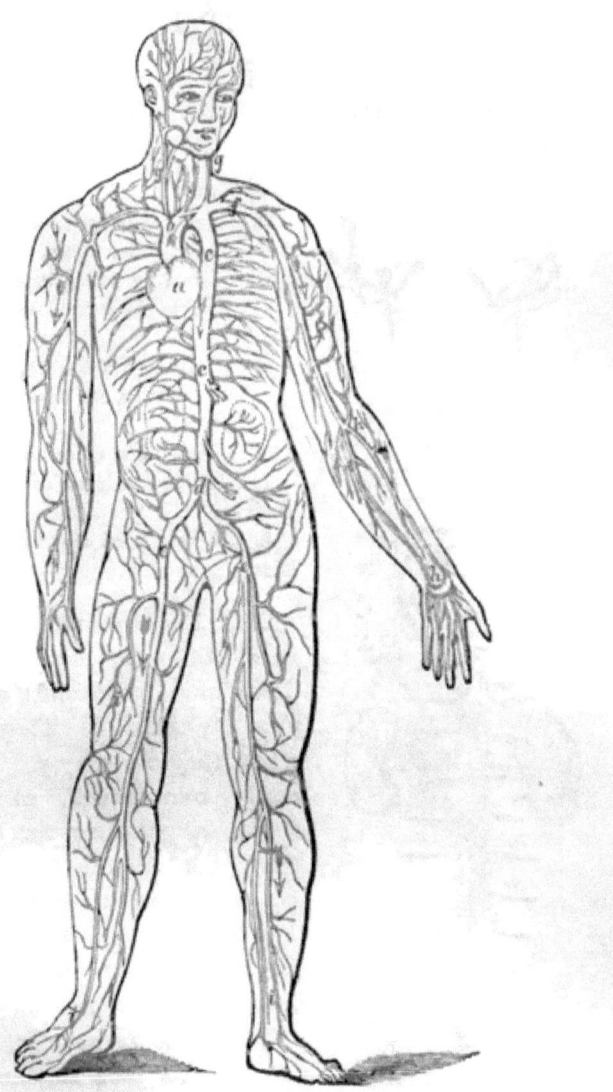

a, Heart.
b, c, c, Aorta.
d, Division of aorta.
e, e, Inguinal arteries.

f, f, Subclavian arteries.
g, g, Carotids.
h, h, Branches in the arms.
i, i, i, Branches in legs and feet.

CHAPTER II.

Veins. — Distribution. — Capillaries. — System of general Circulation. — Situation of Arteries and Veins. — Pulmonary Arteries and Veins. — Double Circulation.

FIG. XII. *Great Veins.*

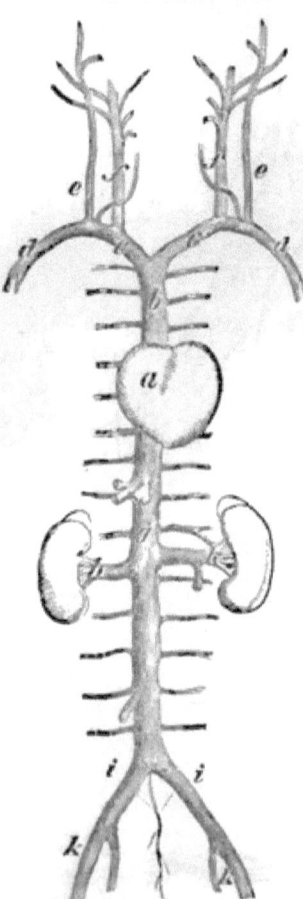

a, Heart.
b, Ascending vena cava.
c, c, Subclavian veins.
d, d, Brachial veins, in the arms.
e, e, Veins from the outside of the head.
f, f, Veins from the brain.
g, Descending, or abdominal vena cava.
h, h, Veins from the kidneys.
i, i, Great branches of veins in the groins.
k, k. Veins from the lower extremities.

215. *The veins also connect the heart with every part of the body.* Their coats are thinner and softer than the arteries. They collapse when they are empty. They are easily compressed, as can be shown by pressing the veins on the back of the hand. They have valves, which open and allow the blood to pass toward the heart, but close and prevent it flowing backward.

216. *One large vein, the vena cava, opens into the heart,. and carries all the blood of the body into it.* The upper part of this vein, the *vena cava ascendens,* extends upward, and sends branches to the head and the arms. The lower part, the *vena cava descendens,* passes

through the abdomen, along the side of the aorta, and sends branches to the organs of digestion, &c., and to the lower limbs.

Fig. XIII. *Veins of the whole Body.*

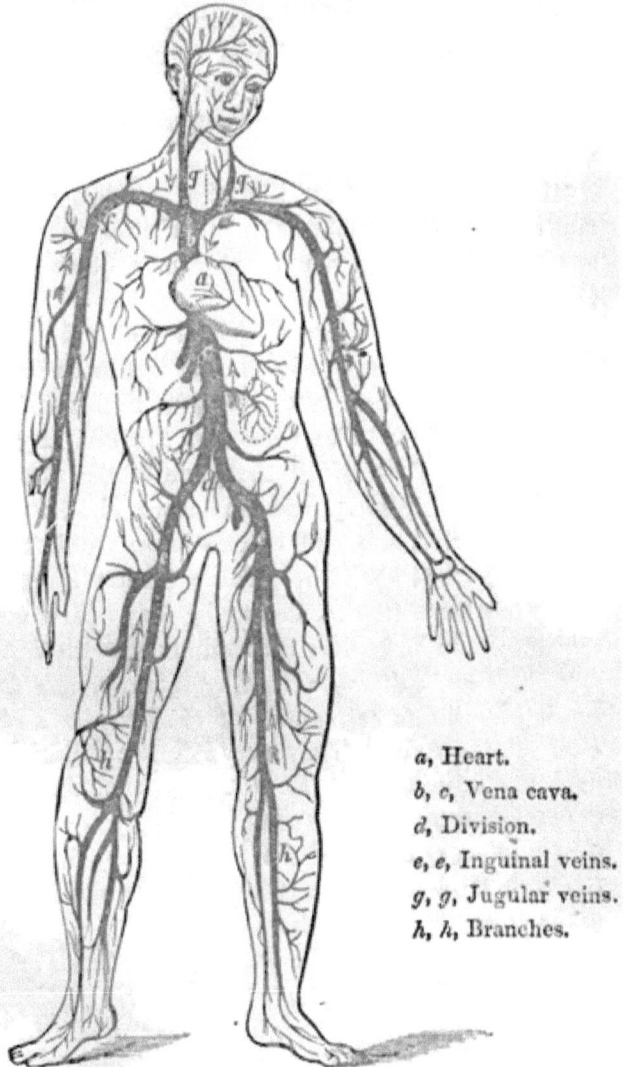

a, Heart.
b, c, Vena cava.
d, Division.
e, e, Inguinal veins.
g, g, Jugular veins.
h, h, Branches.

217. *The jugular veins* pass upward from the upper great vein, through the sides of the neck, to the head. They send

branches to the brain within, and to the scalp and the parts of the face without the skull. *The subclavian veins* (Fig. XII. *c,c*) also pass off from the upper great veins. They go through the arm-pit, and send numberless branches to the arms and hands. The great *abdominal cava* sends branches to the stomach, liver, alimentary canal, kidneys, &c., and finally it is divided (Fig. XII. *i, i*) into two great veins, which with their branches reach all the parts of the lower limbs and feet, (Fig. XIII. *h, h*.)

218. All these veins, like the arteries, are divided and multiplied into branches almost infinitely small and numerous, and thus they reach every part of the animal body, (Fig. XIII.)

219. *The arteries are said to begin at the heart* with one large trunk, the aorta, (Fig. XI. *b*,) and to end in all the near and the remote parts of the body in almost infinite numbers of minute tubes. The veins, on the contrary, are said to begin in the flesh of all the parts of the body, with tubes almost infinitely small and numerous, similar to the terminating arteries, and end in one large trunk, the vena cava, at the heart, (Fig. XIII.) These trunks meet at the heart with their large trunks, and again they nearly meet throughout the whole body with their minute extremities.

220. *The capillary system of blood-vessels is placed between the minute extremities of the arteries and the minute ends of the veins.* They are called capillaries from their hair-like minuteness. They are even smaller than this, for they cannot be seen by the eye. They are spread in every part and every organ of the body. They form the connecting link between the arteries and the veins, and carry the blood from one to the other.

221. *The system of the general circulation of the blood is thus composed of the heart, the arteries, the capillaries, and the veins.* The blood flows out from the ventricle on the left side of the heart into the aorta. It passes through this large artery into the large branches, and thence into the smaller branches, and then through the minute branches into the

capillaries, in every part of the body. From these vessels the blood flows into the minute extremities of the veins, and thence into the larger branches, and finally into the great vena cava, which pours the blood into the right auricle of the heart.

222. *The arteries carry the nutritious blood* to support the life of the textures. If they are wounded, serious consequences follow, and they are not very readily healed. They are therefore placed deeply within the flesh, where they are protected from injury. Some of them approach the surface, and their pulsations can be felt at the wrist, and at the sides of the neck, and on the temple.

223. *The veins carry the impure and wasted blood.* They suffer less, and are more easily healed, than the arteries when injured. Their great trunks are placed near the great arteries, but their branches are situated nearer to, and more of them on, the surface, than the arterial branches. They are seen on the back of the hand and on the arms; and sometimes they enlarge and become very troublesome on the skin of the lower limbs. Bleeding is usually performed by opening a vein of the arm.

224. The blood passes out from the left side of the heart through the arteries to the body, and returns through the veins to the right side of the heart. These two sides are separated by an impassable wall, (Fig. VIII. *c*.) Before the blood can reach the left side of the heart, it must pass through the lungs. This passage of the blood through the lungs constitutes what is called the *pulmonary circulation*.

225. *The pulmonary artery* passes out from the right ventricle, (Fig. VIII. *b*,) and divides into two branches, (Fig. VIII. $k, k,$) one of which goes to the right lung, and the other to the left lung. These divide, and finally spread their minute branches throughout the substance of the lungs. The pulmonary veins begin very minute in all the parts of the lungs, where the little arteries terminate. These little vessels unite again and again, until they form one large vein

in each lung, and then join together and enter the auricle on the left side of the heart.

Fig. XIV. *Double Circulation of the Blood.*

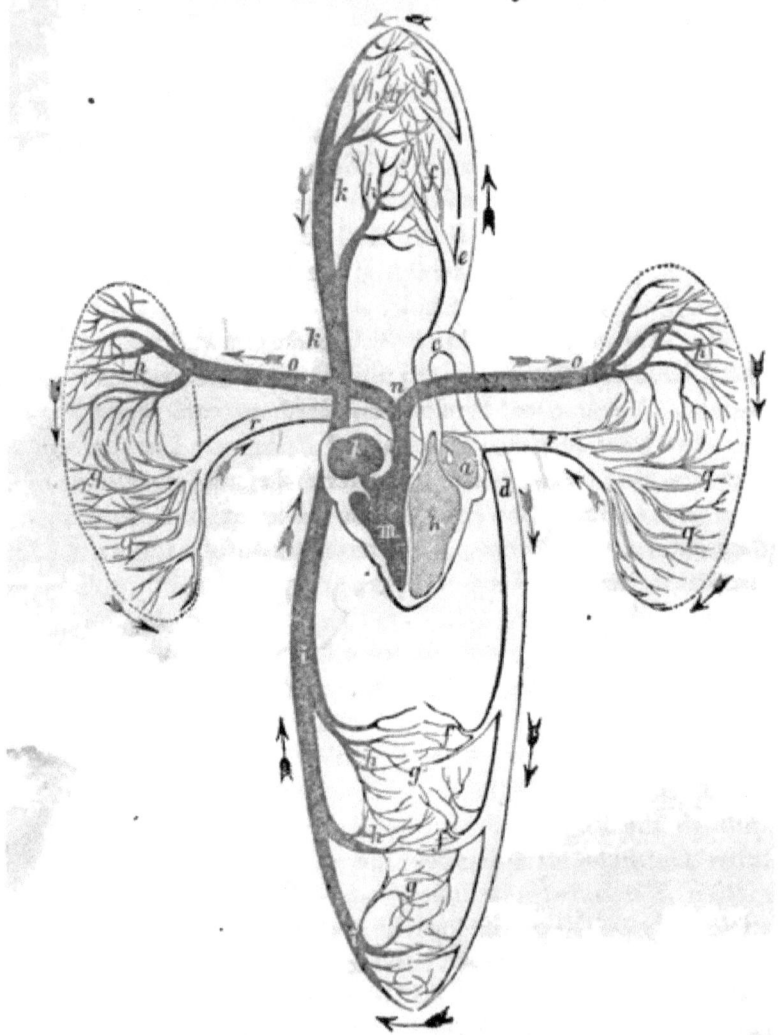

The arrows show the course of the blood.

226. *These two sets of vessels constitute what is called the double circulation.* The blood is in the left auricle, (Fig XIV. *a*,) and passes thence downward to the left ventricle,

(Fig. XIV. *b*;) thence it flows through the aorta, (Fig. XIV. *c*,) the large branches of the arteries, (Fig. XIV. *d, e*,) and the minute branches of the arteries in all the parts of the body, (Fig. XIV. *f, f, f, f*,) into the capillaries, (Fig. XIV. *g, g, g, g*.) Thence, again, the blood flows into the minute veins, (Fig. XIV. *h, h, h, h*,) and through the larger veins, (Fig. XIV. *i, k*,) back to the auricle of the right side of the heart, (Fig. XIV. *l*.)

This is the general circulation, which carries the blood from the left side of the heart through the whole body and back to the right side of the heart.

Next, the blood flows from the right auricle (Fig. XIV. *l*) to the right ventricle, (Fig. XIV. *m*.) Thence it passes through the great pulmonary artery (Fig. XIV. *n*) and the great branches of the right and left lungs (Fig. XIV. *o, o*) into the minute pulmonary branches, (Fig. XIV. *p, p*.) From these it flows into the minute pulmonary veins, (Fig. XIV. *q, q, q, q*,) and through the great pulmonary veins (Fig. XIV. *r, r*) into the auricle of the left side of the heart, (Fig. XIV. *a*.)

This is the pulmonary circulation, which carries the blood from the right side of the heart through all the lungs, and back to the left side of the heart.

CHAPTER III.

Action of the Heart. — Motion of the Blood in the Arteries. — Quantity and Flow of Blood. — Rate of Pulsation varies with Circumstances. — Exercise. — States of Mind and Feelings. Local Circulation varies. — We do not govern Circulation, but we may disturb it.

227. THESE organs of circulation are admirably contrived for their purpose. The muscular texture of the heart enables it to contract upon its contents, and expel them with

force sufficient to send them through the arteries to the farthest extremity of the frame. The power of the heart is not easily measured. Some have supposed that it could exert a force equal to that which would be necessary to raise several thousand pounds; while others have estimated it to be equal only to a few ounces.

228. The veins are continually pouring their blood into the right auricle, or upper chamber. As soon as this is full, it contracts, and empties its contents into the right ventricle, or chamber below. In the same manner, when the pulmonary veins fill the left auricle, this presses the blood into the chamber below; and then this lower cavity contracts and forces it into the arteries, and through them to the body.

229. This movement of the blood is always forward. The great vein pours its blood into the right auricle. When this upper chamber is filled, it contracts upon its contents, and "the reflux of the blood into the veins is prevented by the valves with which they are furnished;" but at the same time the valve between this cavity and the ventricle below is opened, and the blood finds free passage through it. When this lower chamber is filled and contracts, the last valve is closed, and the fluid has no way of going back to the upper chamber; but then the valve between the ventricle and the artery opens to allow the blood to enter this tube, and closes again as soon as the artery is full; so that the blood cannot go back to the heart. In the same manner, the valves on the left side of the heart open to allow the blood to pass from the veins of the lungs to the left auricle, and from the auricle to the lower chamber, and again from this chamber to the arteries of the body; but at each place they close when the next cavity is filled, and prevent the return of any fluid.

230. The arteries are capable of expansion and contraction. When any thing is forced into them, their coats stretch, and their capacity is enlarged; and when their contents are removed, they contract again, and diminish their cavity. These have no valves, except that which stands

between their great trunk and the heart; and there is nothing but this to prevent the backward flow of the blood. When the left ventricle beats, and forces its blood into the arteries, they expand suddenly to admit the increased quantity. All the arteries in the body expand and beat at very nearly the same moment that the heart beats. We can feel this beating of the arteries at the wrist, in the temples, the sides of the neck, and wherever else they come near the surface. Some can perceive it and count their pulsations in the brain.

231. While the heart is pressing the blood into the artery, the valve is opened, and the vessel expands. But as soon as this pressure ceases, the valve closes, and the artery begins to contract and force the contents out and onward through its minute extremities. The blood is then thrown into the arteries by a muscular power of contraction, and it is moved through these vessels merely by the elastic power of their coats.

232. The left ventricle of the heart will, in a man of average size, contain about two ounces, or one eighth of a pint. Every time the heart beats, this cavity is filled and emptied; therefore, two ounces of blood are forced into the arteries at every pulsation. In ordinary health, the heart of a man beats about seventy-five times in a minute, or a little more than once in a second. The quantity of blood in a man of average size is estimated to be about twenty-eight pounds, or four hundred and forty-eight ounces; and if two ounces pass through the heart at every beat, and one hundred and fifty ounces every minute, then the whole blood of the human body must pass through that organ, and through the whole system, once in three minutes. Seventy gallons of blood flow through a man's heart in the course of an hour, and sixteen hundred and eighty-eight gallons in the course of a day.

233. This is the usual rate of the circulation; but it varies with many circumstances. It is more rapid in most diseases

than in health. The heart beats faster when we are standing than when we are sitting, and faster when sitting than when lying down. The pulsation is more rapid in the morning than in the evening — in females and in children than in males and adults. All exercise increases the force and rapidity of the circulation. The rapid and sometimes violent beating of the heart when we are running, or making great exertions, is familiar to every one.

234. *While the body or any of its parts is in motion, the changes of the particles go on more rapidly,* there is more waste, consequently greater need of nutrition, (§ 127, page 61;) and, to meet this want, the heart quickens its action and sends more blood to the frame. There is a great difference between the pulse of the active boy and that of the sleeping babe. The pulsations of the laborer are strong and hard; his heart forces the blood vigorously into the arteries, and they are distended, full, and feel hard. But the pulse of the indolent man is soft and feeble. He takes no exercise; the changes of particles are few and slow; there is little waste and little need of nutrition; and, consequently, his heart sends the blood gently, and oftentimes feebly, through the arteries.

235. *The circulation is affected by the states of the mind and the feelings.* The heart beats with more force and rapidity under mental or emotional excitement; then the arteries beat with more firmness, and carry more blood. So when a man is excited with anger, or stimulated with hope, or glowing with cheerfulness, or burning with love, his blood flows more freely, his system is better nourished, and he is stronger and more capable of exertion. The depressing passions have the opposite effect of lowering the action of the heart, and the force of the circulation. While a man is suffering with fear, despair, sorrow, or gloom, his heart beats more feebly, and his blood flows more languidly; his body is less nourished, his strength is impaired, and he has less power of labor.

236. Although the heart sends the blood by the same impulse to all the arteries, and all these must then beat, almost at the same moment, and in unison, yet the expansion and contraction of these blood-vessels are not always the same in all parts of the body. Hence the blood may circulate with very different force in various parts, and some may be supplied very freely while others are but sparingly fed with this fluid. Local diseases create a greater local circulation. When one has a felon on his finger, he feels the arteries throb sometimes violently in the sides of that finger, while the beating of the arteries in the other fingers is scarcely noticed. Some suffer from cold feet in consequence of feeble circulation of blood through those extremities; others have headaches from the too great flow of blood to the brain. The arteries are more active in the parts that are in action. More blood flows to the muscles during the time of labor, to the stomach during digestion, and to the brain when the mind is actively employed.

237. This beating of the heart, and this pulsation of the arteries, are incessant during life. Day and night, asleep or awake, this movement goes on, and every part of the frame receives its supply of blood in due season, without our volition, and even without our observation. The circulation of the blood is not submitted to our care, and we are not responsible for its work, as we are for the work of digestion. Yet, though we are not called upon to aid this function, we may interfere for evil. We may, by stimulating food or drinks, excite the heart too much for health; or, by neglect of proper exercise, we may suffer it to become sluggish in its motions.

CHAPTER IV.

Object of Eating and Circulation is to nourish the Body. — All Animal Solids and Fluids formed out of the Blood. — Elementary Composition of the Blood and Flesh: nearly alike in all the Tissues, but differ in the Proportions of their Elements. — Nutrition takes Place in the extreme Vessels, and with unerring Precision. Growth and Changes of the Body are at the Cost of the Blood. — The Atoms of the Body enjoy but a temporary Life. — When one dies, it is removed. — Absorbents.

238. As the eating and digestion of the food have no other object than to supply the wants of the blood, so the motions of the heart, and the circulation of the blood, are for the sole purpose of supplying the wants of the frame. All the growth of the body in childhood and youth, all the regaining of flesh after sickness, all increase of flesh at any period of life, and all the changes of particles, are supplied by the blood. All the tissues and secretions of the body, various as they are, — the bone, muscle, brain, skin, fat, the hair and the nails, the tears, the saliva, and the perspiration, — are all taken from this same storehouse, — from this fluid that runs in the blood-vessels of the animal body.

239. The blood is not flesh, nor does it exhibit any resemblance to flesh. It is a homogeneous fluid, the same in all the arteries, wherever they may be situated. The blood that flows in the brain is of the same nature and composition as that which flows in the bones and muscles. The blood is not a simple, but a compound fluid. It contains various elements, and these are the same as those which compose the flesh. These simple elements are principally carbon, oxygen, hydrogen, and nitrogen. There are others, such as the lime, that enters the bones, some phosphorus and sulphur, that are found in the hair, the nails, and the brain. These are the most common elements in nature. Oxygen and nitrogen compose the air; oxygen and hydrogen form water. Carbon

is the principal ingredient in charcoal; it is the predominant element in vegetable substances.

240. *All these elements are found in the blood.* But they are not all found in every tissue of the animal body. There is no lime in the brain, no sulphur in the muscles, and no nitrogen in the fat. Yet, with few exceptions, all the various parts and organs are mainly composed of the same elementary atoms — carbon, oxygen, hydrogen, and nitrogen. The difference of these organs is owing, not to the difference of their component elements, but to their different combination or arrangement. Combined in one proportion, they form tendon; in another, they form muscle; in another, potato, tea, coffee. The same elements in various proportions, and with some ashes, form flesh, peas, beans, oats; and, with some sulphur, they form hair, bone, nails, and cheese. The blood is the grand storehouse which supplies all these, in their due proportion, to every organ and texture.

241. The arteries carry this blood to all the parts of the body; every point, however minute, receives its supply through these tubes. The transformation of the blood into flesh, or the separation of such elements from this fluid as will compose the kind of flesh that is needed, is done in the minute extremities of the arteries, or the capillaries, which stand between the arteries and the veins. This work of nutrition is done with unerring precision in health; just the requisite proportions of carbon, oxygen, and hydrogen, and of the other elements, are measured out; and flesh of the proper kinds is formed, each in its appropriate place; and thus the body increases in size and stature.

242. All additions to the weight of the body, the growth during early years, and the increase of flesh at any time, create a certain demand upon the blood for nutrition; but the changes of particles, during the whole of life, create a much greater demand upon the blood for new atoms. After we have reached our fulness of stature, in ordinary health, we eat and drink three to four or more pounds of

solid and liquid matter a day; and yet our weight does not usually increase. Even after making this daily addition for successive years, we weigh about the same at sixty as we did at twenty. This food is digested and converted entirely, or in part, into blood; and this blood is converted into flesh, muscle, fat, nerve, &c.; and yet these organs and parts remain of the same size.

243. The diligent arteries are almost incessantly adding atom after atom to these organs, and yet do not enlarge them. The object of their work is, not merely to make new atoms of flesh, but to make those which will supply the place of other atoms, which have served their purpose in the living body, and have been carried away. It has already been stated (§ 1, p. 9) that however long the body, as a whole, may continue to live, none of its component particles can enjoy any considerable duration of life. These particles are deposited, by the arteries or the capillaries, in the various organs and tissues; and then they are endowed with life, and the peculiar properties of the part in which they are placed. In the muscle, they have the power of contraction, and in the brain, the power of feeling and perception; and in the bone, they are hard and strong, and apparently insensible.

244. *But in a little while this vitality, this property of life, is exhausted, and the atom is dead.* Then it is removed, and another atom takes its place, to go through the same course of life, action, and death. This succession of particles, this change from life to death, and this renewal of life, are constant, and almost universal, in the animal body. We are dying, atom after atom, daily, hourly, momently; and we are renewed and revived in the same degree, and at the same time. We enjoy, therefore, a constant freshness of life. This is the united work of the arteries, which bring the new and living atoms, and of the veins and *absorbents*, which carry the old and dead atoms away. The arteries and veins have already been described. The *absorbent vessels* seem to be spread throughout all the tissues. Wherever

there is a minute artery to deposit a living atom, there is an absorbent ready to carry it away when it shall have finished its life and died.

245. Dr. Edward Johnson, in his interesting letters on "Life, Health, and Disease," thus graphically describes these vessels: "There is arising from every point of your body a countless number of little vessels, actively engaged in the pleasant task of eating you up. They may be compared to a swarming host of long, delicate, and slender leeches, attached, by their innumerable mouths, to every point of your fabric, and having their bodies gradually and progressively united together, until they all terminate in one tail, which tail perforates the side of one of the veins at the bottom of the neck, on the left side; so that whatever is taken in at their mouths is all emptied, by the other extremity, into that vein, where it becomes mixed with the blood."

CHAPTER V.

Action of the Nutrients and Absorbents. — Feeding Sheep on Madder colors Bones. — Balance of Nutrition and Absorption. — In Youth, Nutrition, and in Old Age, Absorption prevails. — Both more active in the Laborer. — Laborer should eat more Food. — Parts that are used more nourished. — Wens and Swellings. — Produced by excessive Action of Arteries, and removed by greater Action of Absorbents.

246. The arteries bring the blood, and deposit the new particles of flesh, while the veins and absorbents take and carry away the old particles. These two systems are constantly at work, antagonizing each other. One set pulls down the old fabric, taking it away atom by atom; at the same time, the other set rebuilds the fabric anew, and replaces the old and the dead with new and living portions. In this manner, we are undergoing a perpetual change; and we are not precisely the same to-day as we were

yesterday. Perhaps we have not now an atom of the flesh that we had ten years ago.

247. Though the individual atoms change, the whole, the totality, remains unchanged. The new atoms have the same character, the same sympathies, and perform the same functions, as those that went before them. The animal body, in this respect, is like a community, or a town of a definite number of people. The individual members of this community are continually changing; some go out to other towns — some die; but their places are filled by others that come from abroad, and by some that are born. The individuals are not all the same from year to year; and, in course of a single generation, they are all exchanged; and yet the body, the town, remains unchanged. The same character and habits are there, the same principles govern them. The community is, in fact, the same, even after every one of its original component elements has been removed and replaced by others.

248. The experiment has been tried of feeding pigs and sheep upon madder, which is a pink coloring matter. When some of these animals were killed, while they were living upon this food, the bones were found to be tinged with red. But some others were kept, for the same time, upon madder, and afterwards were fed, for a period, with hay and grain; then being killed, their bones were found to be as white as those of animals which had never eaten madder. There is no doubt that the bones of these last animals had been stained with the madder while they lived upon it, and that they became white when they again were fed on other food.

249. This is easily explained by the action of their nutrient and absorbent systems. The coloring matter of the madder was carried in the chyle to the blood, and in the blood to the bone, and there deposited; and, when this is absorbed, more red matter is brought and left there; and this continues as long as madder is eaten. But, when the food is changed, no more red matter is carried to the bones, and

that which was there is taken away, and its place is supplied with white material.

250. During the middle periods of life, these two sets of vessels are equally active, and usually perform a similar amount of work. The arteries carry and deposit as many atoms in the flesh as the veins and absorbents carry away. The one builds up as fast as the other pulls down; so that, though some pounds are added daily to the body, it does not gain in weight; neither does it lose, though as much is carried away. But this is not the case at all the periods of life, nor in every condition of health.

251. During the period of youth, and the time of increasing flesh, nutrition predominates; more atoms are then brought in the arteries, and deposited, than are taken away by the absorbents. When the body is wasting, absorption predominates, and carries off more than the arteries deposit. When we grow fat, nutrition is the more active; but when we grow lean, absorption prevails. The whole of the movements of life are more rapid in the earlier years, and slower in old age, than in the middle periods of life. The heart beats quicker, the flow of blood is more abundant, and both nutrition and absorption are more rapid in the former, and slower in the latter period.

252. Liebig supposes that every action of any of the parts of the body is attended with change of particles; that when a finger moves, some of the atoms in the muscle that produces the motion die and leave their places, and are replaced by others. When we move the arm, the legs, or use the muscles of any other part of the body, the same changes take place in the muscular atoms. All exercise increases the activity of both the nutrition and absorption. In order to meet the increased demands for new flesh to supply this waste during exercise, the heart beats quicker — more blood is carried to the moving parts; and thus more new flesh is formed, as long as more is absorbed. Every one knows that the heart beats rapidly, and sometimes almost palpitates, while we run or otherwise exercise violently. Those who

suffer from disease of heart cannot make great exertions, because the heart cannot carry the blood needed to supply the greater waste.

253. The waste of the old atoms of flesh, and the demand for new, being increased by exercise, of course more blood is then consumed to supply the want which is thus created; and, consequently, more food must be eaten and digested, to supply the blood with the new chyle sufficient to repair this loss. (§ 127, p. 61.) The laborer must therefore eat more than the indolent, and individuals in active youth need more food than in quiet old age. But, in inactive life, the absorption is comparatively little, and the nutrition and the consumption of blood are equally small; there is less demand for food, and a corresponding diminution of appetite. If the idle disregards this law of his nature, and eats as much as the laborious, the stomach is troubled with the burden; and, if it digests and converts the food into chyle, the blood-vessels are overfilled with the amount of blood which they cannot use, and the whole frame is heavy and sluggish. The apostle's command that, "if any would not work, neither should he eat," which was given as a moral law, is equally binding as a physical law, and cannot be disobeyed without suffering.

254. The processes of destruction and creation have usually the same comparative activity in all parts of the body, so that no part grows fat or lean more than another. But this is not always the case. If one organ or part is more active than the others, it grows more than they. Thus the arms of some laborers, and the legs of others, grow disproportionately large, because they are more used than their other limbs. But parts that are not used at all waste away, and are often withered. The arm of one of my neighbors was palsied about twenty years ago, and it is now shrunken to the size of a child's arm; the unused muscles are nearly absorbed.

255. Wens, and other fleshy tumors, are the effect of the unnatural activity of the nutrient vessels, which deposit

more fatty or other fleshy atoms in the part affected, than the absorbents take away. Physicians are often asked to scatter these tumors. This is done by stimulating the absorbents to a still greater activity than the arteries, so that they may carry away more atoms than the others bring. The glands of the neck in scrofulous persons sometimes swell, and afterwards the swelling disappears. A boil often appears upon the skin with a prominent and painful swelling; but, without coming to a head, or discharging any matter, it goes away. In both these cases, the tumor is produced by the superior activity of the arteries, and is carried away by the greater action of the absorbents.

CHAPTER VI.

The Young have fresh and new Atoms of Flesh. — The Aged have old Atoms. — Flesh of the Active is new; and of the Idle, old. — Blood and Vessels same in all Parts. — Vessels select Elements from the Blood, in due Proportion, to form the various Tissues. — Unerring Precision of Nutrition. — Difference of the Blood in the Arteries and in the Veins.

256. This double work of nutrition and absorption, producing changes of the component parts of the body, never ceases from the beginning to the end of life. But it is not equally rapid in all periods, nor in all persons. In the earlier years of childhood and youth, all the processes of animal life are more active, and the particles are more frequently changed than in middle life. Consequently, their flesh at these periods is ever new and young. But in old men, all these operations are more sluggishly and feebly carried on. Their particles are not frequently changed, and therefore the atoms of their flesh are old, as well as their whole bodies.

257. As these changes are frequent in the active and industrious, their atoms remain but a short time in the living body, before they are taken away. This gives them a per-

petual freshness of youth in their flesh; they are, therefore, lively and prompt in action. These changes take place more slowly in the inactive; their atoms remain a longer time; and their flesh is therefore always old, and indisposed to action. It is easy to see this difference between the energy and sprightliness of one who has always accustomed himself to action abroad, and the heavy sluggishness of another, who has lived delicately, and avoided exercise. Compared with his years, the one is ever young, while the other is ever old.

258. We have no means of knowing how, or by what means, the final work of nutrition is done. We only know the instruments with which it is accomplished, and the materials that are used. Anatomists have examined the blood-vessels, and chemists have analyzed the blood, and have taught us the shape of one and the composition of the other; and there our knowledge stops. We cannot penetrate any farther into the mysteries of nature. So far as the eye of man can discover, the blood-vessels are the same in structure and character, in all the organs and tissues of the body, and the same blood is found in all. And yet, with a wonderful precision, these little nutrient vessels select out of this common storehouse of nutriment just those elements, and in just their varied proportions, that are needed to form the various kinds of flesh and substance that compose the animal body.

259. In the fat, the organs of nutrition select from the blood 79 parts of carbon, $11\frac{1}{2}$ parts of hydrogen, and $9\frac{1}{2}$ parts of oxygen; and with these form fatty atoms. In the hair, they take 50 parts of carbon, 6 parts of hydrogen, 17 parts of nitrogen, and 26 parts of oxygen and sulphur, and make an atom of hair. And from the blood in the muscle, they take 51 parts of carbon, 7 parts of hydrogen, 15 parts of nitrogen, 21 parts of oxygen, and 4 parts of other matters, and form muscular particles. In a similar manner, the vessels of the brain select the brain; and in the skin, and in all other organs, they select the very kinds and proportions

of the elements that compose each, and no other. They take just enough of each element, neither more nor less, and combine them in the due proportion of each kind, to form the part which is wanted.

260. Although the different tissues of the animal body are so nearly alike in their composition, and so slight a variation would produce another kind of flesh, yet, in health, no mistake is made. Each organ and tissue receives flesh of its own kind. Muscle is not deposited in the brain, nor bone in the muscle, nor tendon in the liver. In every part, the blood-vessels act with such unvarying and beautiful precision, and perform their work with such faithfulness to their purpose, that they might almost seem to be endowed with a special intelligence, if we were not assured that they, even the minutest of them, are under the constant supervision and direction of that paternal Providence, without whose notice not an atom moves nor a sparrow falls to the ground.

261. Thus all the atoms of flesh, all the parts of the animal body, were first in the stomach, and next in the arteries; and then they became living flesh, and acted a while, and died. Then, again, all these, with the exception of the hair, the nails, and the outer skin, which grow out and fall, are once more taken into the vessels, and are found in the veins. The blood in the arteries differs from that in the veins, in its nature and its composition. In one, it is scarlet, rich, nutritious, loaded with new particles of digested food, and is therefore capable of giving life and strength to any of the tissues. In the other, it is dark purple; it has lost its rich particles, and is therefore innutritious; it is also loaded with the dead and wasted particles that have lived and died in the body. If the venous blood be thrown into the arteries, and circulated through the system, it not only must fail to nourish and give new particles of flesh to the tissues, but, with its wasted and offensive burden, it must carry disease or death to the body.

262. The veins are incessantly receiving additions of the particles of the exhausted flesh, and would soon be so over-

loaded as to be incapable of action, if there were not some means provided to carry these out of the body. This might seem a difficult matter. These dead atoms are in the veins, and those are buried in the deepest recesses of the body, apparently beyond the reach of any external influence, and without any outlet to the world abroad. But Nature has no difficulties. Her means are always adequate to her wants. Her process of relieving the living body of those useless and burdensome matters is made simple and easy, by means of the lungs and respiration, and of the skin and perspiration.

PART III.

RESPIRATION.

CHAPTER I.

Wasted Particles carried out of the Body. — Composition of Blood in right Side of the Heart. — Lungs protected by Bones of Chest. — Spine. — Breast-Bone. — Ribs. — Position of Ribs.

263. The wasted particles of the animal body — those that have lived and have been a part of the living system — have been removed from their places in the various organs, and carried into the veins, and through them to the right side of the heart. As nutrition and absorption are continually going on in the body, these old particles must accumulate in the veins and heart; and, as they amount to several ounces a day, they would soon overload and destroy the living system, if they were not carried out from it. This is done; and they find an outlet through the lungs as fast as they are removed from their original places of life and action in the various tissues.

264. The venous blood — that which is gathered in the

right side of the heart, from all the various parts of the body—consists of three kinds of materials: 1st, the residue of the arterial blood after nourishing the body, or that which is left after the particles have been selected for the nourishment of the textures; 2d, the old and dead atoms of flesh; and, 3d, the chyle, or digested food brought through the lacteals. Neither of these three elements can nourish the body The remnant of the arterial blood has lost much, if not all, of its life-giving qualities. The dead particles would be poisonous if carried round again; and the new chyle is not yet prepared to furnish nutriment. They must, therefore, all be submitted to some process by which the first shall be strengthened, the second carried out of the body, and the third perfected, before this blood can be used again to nourish the body. All this is done by means of the air in the lungs.

265. The lungs of man are placed within the chest, at the upper part of the trunk. They are organs of exceeding delicacy in their structure, and would not bear with impunity any exposure to external injury. They are therefore protected with great care. They are covered on all sides with a bony framework, which prevents all contact with external objects. The bones which compose the walls of this chest are so arranged, and fixed with joints and muscles, that they admit of very free motion, and allow to the internal cavity great range of expansion and contraction, for admitting and expelling air.

FIG. XV. *Bones of the Chest.*

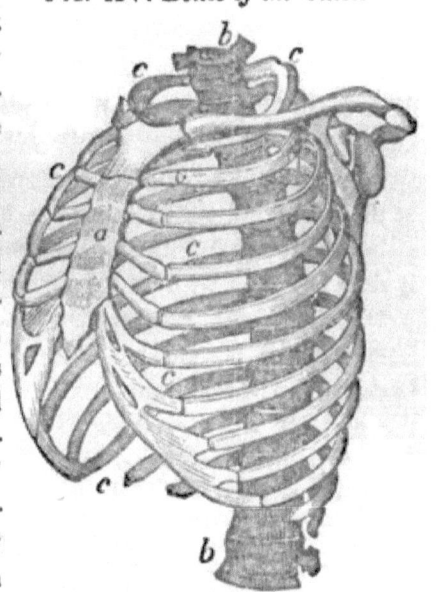

a. Breast-bone. *b, b.* Back-bone.
c, c, c, c. Ribs.

266. The chest (Fig. XV.) extends from the neck to the abdomen. It is conical in shape, being small at its upper end, and larger at the lower part. It is enclosed by bone at the top and on its sides, and by muscle at the bottom. The breast-bone, *a*, (Fig. XV. p. 121,) is in front; the spine, *b, b*, or back-bone, is behind; and the ribs, *c, c, c*, cover the sides of the chest.

267. The *spine*, or back-bone, is composed of twenty-four bones, called *vertebræ*, which are connected by thick layers of very strong and elastic cartilage, or gristle, between them. These give to the column great flexibility and freedom of motion, and such strength that, through all the chances of accidents and violence, these bones are very rarely broken or displaced. Yet it may be bent in any direction, and is capable of sustaining great burdens that may be placed upon the head. Twelve of these bones form part of the chest, and to these twelve vertebræ or bones of the back are attached twenty-four ribs, twelve on each side.

268. The *breast-bone*, *a*, is thin and flat, reaching from the neck to the region of the stomach. It is covered with so little flesh as to be perceptible to the touch. At the lower end is attached a cartilage or gristly substance, that extends about two or two and a half inches downward, and ends in a point at the pit of the stomach. This breast-bone forms the front pillar of the chest, though by no means an immovable one, for it rises and falls with all the motions of the ribs.

269. The *ribs*, *c, c, c*, compose the principal part of the framework of the chest. They surround the cavity from the back to the breast-bone, covering all the sides and most of the anterior and posterior portions of the cavity. All are joined to the back-bone by their posterior end, and ten of them are connected with the breast-bone by their anterior ends. Some of these are fixed directly to the breast-bone; others terminate in cartilages of an inch or more in length, that extend to the breast-bone; by which arrangement these ribs have a great freedom of motion.

270. The ribs nearly surround the chest, somewhat as

hoops surround a barrel. But their course is not horizontal. They incline downward from the back-bone to the breast-bone in front; consequently, the diameter of the chest is so much lessened by this obliquity of position; but when the ribs are raised to a horizontal position, at right angles with the axis of the chest, this diameter is increased, and the capacity of this cavity is enlarged. This is easily shown by the experiment of putting a large hoop obliquely upon a barrel of smaller diameter. The hoop, a, c, in its oblique position, touches the barrel; but, if the hoop be raised horizontally to b, it would extend beyond the cask, and allow it a much greater expansion.

Fig. XVI.

CHAPTER II.

Movements of Ribs. — Diaphragm. — Expansion of Chest in Inspiration. — Contraction in Expiration. — Size of expanded and contracted Chest.

271. THE posterior ends of the ribs are attached to the back-bone, and fixed. The motions are all made with the anterior ends, which are free. They are joined to the spine in such a manner that they can only move upward and downward, not from side to side. The spine being the pillar upon which the frame of the chest rests, it is fixed and immovable in breathing; but all the movements of the ribs and breast-bone are made upon it. These are *lifted up* and *fall down* at every respiration.

272. The first or upper rib is fixed and motionless; the second has very little motion; the third has more motion than the second; and the fourth more than the third. This motion goes on increasing to the eleventh and twelfth, which move very freely. There are several muscles which are at-

tached to the back-bone and to the ribs, and fill all the spaces between them. Some of these are attached by one end to the spine, and, running obliquely forward and downward, are attached by the other end to the ribs below. Others are attached by one end to one rib above, and by the other end to another rib below. When these muscles contract, they lift the ribs. The posterior end of each rib rolls in its socket in the spine; but the main portion of the bone is raised and carried outward, and the whole cavity of the chest is then expanded, in the same manner as the hoop (Fig. XVI. a, c,) would allow the cavity of the cask to be expanded, if the side c were lifted to the level of the side a, which is supposed to be fixed.

273. The ribs, spine, breast-bone, and muscles (Fig. XV. p. 121) bound the chest on all its sides. As this cavity is conical, there is hardly any surface at the top. But there is a broad and extensive surface at the bottom of the cone, which is covered by a flat muscle, called the *diaphragm*. (Fig. II. d, p. 19.) This performs a part of the greatest importance in the work of respiration. It is the flexible partition that divides the chest from the abdomen, and separates the respiratory from the digestive organs. Its edges are attached to the back-bone, and to the lower edge of the lower ribs, to the breast-bone, and to all the lower part of the chest. It forms an arch, upon the upper or convex surface of which the lungs rest; and in the hollow below some of the organs of the abdomen — the liver, stomach, &c. — are placed.

274. When the diaphragm is at rest, its arch points upward into the chest, as the bottom of a common glass bottle is turned into its cavity; and its upper point reaches as high as the fourth rib, and, consequently, must very materially lessen the capacity of the chest, and press upon the lungs. But when it is in action and contracted, the arch is drawn down, and, leaving a space behind, enlarges the capacity of the chest, and allows more room for the lungs to expand. The diaphragm is the dividing-wall between the lungs and diges-

tive apparatus. (Fig. V. *d*, p. 19.) The lungs lie in contact with it above, (Fig. V. *a, b*, p. 19,) and the digestive apparatus lies in contact with it below. (Fig. V. *e, f*, p. 19.) When it is expanded, it rises into the chest, and the lungs are pressed up, and the abdominal organs follow immediately behind. And, on the other hand, when it contracts and lessens the arch to give expansion to the lungs, it must press the abdomen and its contents downward and outward. A simple and easy illustration of the operation of the diaphragm in breathing, may be found in the common India-rubber bottle. If we hold this in one hand and press the bottom inward with the finger, the air is forced out through the neck. If, then, we remove the finger, the bottom returns to its natural position, and then the air flows through the neck to fill the increased cavity.

275. By these two combined actions of the muscles of the ribs and of the diaphragm, the chest is enlarged. The muscles on the sides of the chest raise the ribs, and extend their circle forward and outward. The diaphragm draws down its arch from the fourth to below the seventh rib, and thus enlarges the chest; and the lungs having room for expansion, the air is pressed into them to fill the vacuum left by the enlarging chest. This is the mechanical part of the process of *inspiration*.

276. After the chest is thus sufficiently expanded, the muscles of the ribs and the diaphragm relax and lose their firmness. Then the action of other muscles, aided by the elasticity of the cartilages, carries the ribs downward; and, in going down, they lessen the diameter, and consequently the capacity of the chest, by bringing the sides nearer to each other, and the breast-bone nearer to the back-bone. At the same time, the muscles that cover the abdomen press upon its contents, and force them against the diaphragm. This yields to the pressure, and rises upward and presses upon the lungs, which retreat before it, and the air is expelled. This is the process of *expiration*.

277. Fig. XVII. represents an outline of the front view of

the expanded and contracted chest. The full lines *d, d, d,* show the size of the cavity when the ribs are drawn down and the diaphragm is expanded upward. The dotted lines

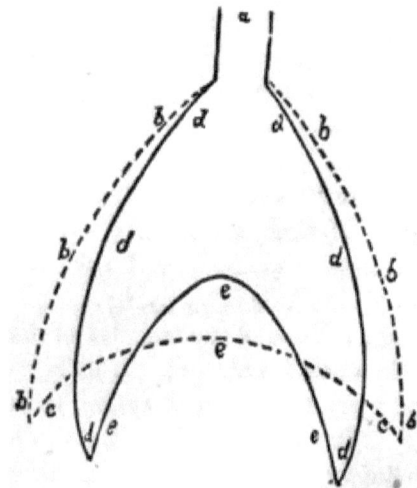

FIG. XVII. *Outline of the Expanded and Contracted Chest. Front View.*

a. Neck.

b, b, b, b, b, b. Size of the expanded chest.

c, c, c. Position of the diaphragm when drawn down.

d, d, d, d, d, d. Size of the contracted chest.

e, e, e. Position of the diaphragm when expanded.

show its size when the ribs are lifted and the diaphragm drawn down. When the ribs are at rest, they lie downward and inward, and their surface is represented by the lines *d, d, d, d, d, d;* but, when they are raised, they are carried outward, and their surface is represented by the dotted lines *b, b, b, b, b, b.* When the diaphragm is at rest, it projects upward in form of the line *e, e, e;* but, when it is in action, it is drawn downward in form of the line *c, c, c.* The upper point *e* reaches as high as the fourth rib. The upper point *c* reaches as high as the seventh rib. It is now very plain that the cavity *b, b, b, c, c, c, b, b, b,* is larger than the cavity *d, d, d, e, e, e, d, d, d,* and the difference in size must be in proportion to the extent of the motions of the ribs and the diaphragm.

CHAPTER III.

Lungs. — Situation. — Tissues. — Windpipe — Glottis. — Organ of Voice. — Air-Tubes and Cells. — Mucous Membrane. — Coughing. — Interweaving of Air-Vessels and Blood-Vessels. — Inspiration — Expiration. — Respiration. — Coöperation of Parts.

278. The *lungs* are situated within the chest, which has been now described. There are two lungs, (Fig. XVIII. *a*, *b*,) placed one on the right, and the other on the left side of the chest. The heart lies between them, (Fig. XVIII. *d*,) and these completely fill this cavity. (Fig. V. *a*, *b*, *c*.) The lungs are very soft and spongy. They contain little or no flesh, but are composed almost exclusively of tubes and cells, which are to be filled, some with blood, and others with air.

Fig. XVIII. *Lungs and Heart.*

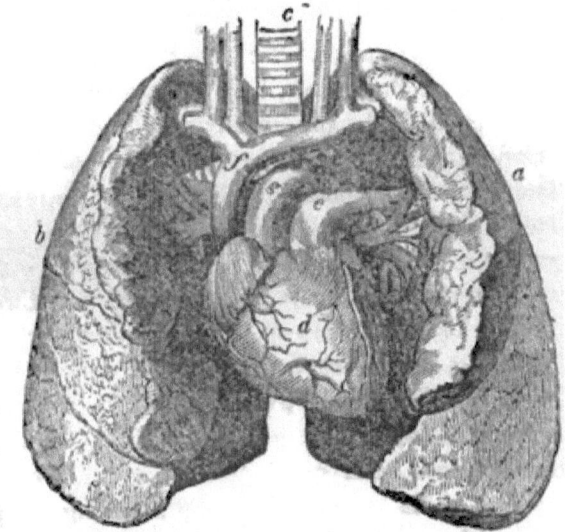

a, Left lung.
b, Right lung.
c, Windpipe.
d, Heart.
e, Great artery carrying blood to lungs.
f, Great vein.
g, Great artery carrying blood to the body.

279. The *air-tubes* begin at the back part of the mouth and nostrils with a single cylinder, which leads through the neck to the chest; but in the lungs they are divided and subdivided into smaller and smaller tubes, which are distrib-

uted throughout the whole respiratory organs. The blood-vessels (§ 225, p. 103) proceed from the heart in large trunks, and, like the air-tubes, are divided and multiplied until they terminate in minute branches of imperceptible size, which lie in contact with, and spread over, the air-cells. Then these little blood-vessels are again gathered together into larger and larger tubes, until, at last, they form two large trunks, that enter the heart. These trunks are of the same size as those which left the heart to carry the blood to the lungs.

280. The *windpipe*, or *trachea*, is composed of somewhat stiff rings of cartilage or gristle, so that it is easily felt in the front of the throat. The upper end of this tube is usually open; but the epiglottis (§§ 20, 21, 22, p. 16) is placed there to cover over and protect this passage whenever the food passes over it, on its way from the tongue to the œsophagus. But, when we are not swallowing, this valve stands open.

281. The upper end of the windpipe opens into the back chamber of the mouth, by a narrow chink, called the *glottis* This is made by the approximation of the two sides of the tube, so that the air may produce sounds when it passes between them. Some muscles, which are attached to these sides, draw them more closely together, or allow them to separate. By thus opening or narrowing this chink, the sounds which are made by the air passing through it, are varied. *This is the organ of voice;* and these sounds are the various tones which we utter in language, in singing, or in crying. When this part of the windpipe is diseased by what is called a *cold* in the throat or otherwise, these tones are changed, and often destroyed. A man then loses the control of his voice, and can neither sing nor talk as he does in health. Sometimes the voice is entirely destroyed or suspended, and then the sufferer can only speak with the mouth in a whisper. A gentleman, whom I have seen while writing this chapter, has not been able to speak a loud word for three months, in consequence of ulceration about the glottis.

282. At the junction of the neck with the chest, and just behind the top of the breast-bone, the great air-tube, the *windpipe*, is divided into two tubes, or bronchi. One of these goes to the right, the other to the left lung. After entering their respective lungs, they divide into smaller branches or tubes, as represented in Fig. XIX. These

Fig. XIX. *Windpipe and Air-Vessels of the Lungs.*

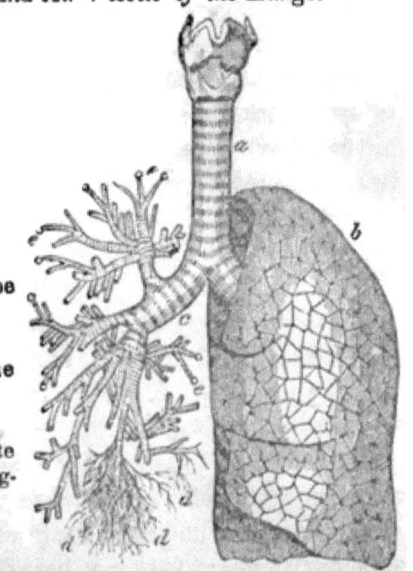

a. Windpipe.

b. Left lung.

c. Great branch of the air-tube going to the right lung.

d, d, d. Minute branches of the air-tubes.

e, e. Air-cells at the minute termination of the air-tubes, magnified.

tubes do not divide so minutely as the blood-vessels, but terminate somewhat abruptly in a great number of minute cells. (Fig. XIX. *e, e.*) These cells are estimated to be one hundredth of an inch in diameter. They are so numerous as to be distributed to every part of the lungs; and the extent of the inner surface of the whole, collectively, is estimated variously by physiologists. Some suppose it to be 20,000 square inches; others thirty times the whole surface of the body. But all agree that there is a very extensive surface presented to the action of the air.

283. *These air-tubes and air-cells are lined with a mucous membrane* of exceeding delicacy, which, during the whole of

life, will bear the presence of pure air, but will not tolerate, for a single moment, the presence of any other substance, not even a drop of water, as most of us have had occasion to know in some part of our lives. This lining is thin and sensitive, and very liable to be disordered. Most of our colds and catarrhs are but affections of this membrane. And our coughs are mostly caused by some irritation applied to it, or derangement in it. Whenever it is not in a comfortable condition, — when it is dry for want of mucus, or covered with too much of it, — or when any particle of food, or any other foreign substance, is lodged in any part of the air passages or cells, or even gets within the glottis, — then nature sets up a violent expulsory effort to press the air out suddenly, and to blow and force away the intruder, or relieve the irritation. This is *coughing*.

284. We perceive this sometimes when we breathe dust, or offensive gases, or pungent matters, all of which irritate and offend this sensitive texture. But, delicate as it is, it may, by use, lose its sensibility, and become accustomed to bear very injurious substances, as the sole of the barefoot boy loses its delicacy, and will bear the rough surface of the street without suffering; and as the hands of the smith and of the dyer lose much of their sensitiveness to heat, so that they can, without apparent pain, handle iron and plunge into dyes so hot as to burn others; so the membrane of the lungs becomes used even to tobacco-smoke, and bears its frequent and almost perpetual presence, without appearing to suffer any harm.

285. The heart is placed about the middle of the chest, and between the lobes of the lungs, (§ 205,) (Fig. XVIII. p. 127,) and sends its blood-vessels to the right and to the left, through each of these lobes. These vessels ramify through every part of the organ, and are interwoven with the air-tubes (Fig. XX.) These myriads of minute arteries come in contact with the air-cells, and are separated from them only by an exceedingly thin membrane, so thin that gases

can pass through it from the air-cells to the blood-vessels, and from the blood-vessels to the air-cells; but it is sufficiently thick to prevent the passage of fluids.

FIG. XX. *Interweaving of the Air-tubes and Blood-vessels in the Lungs.*

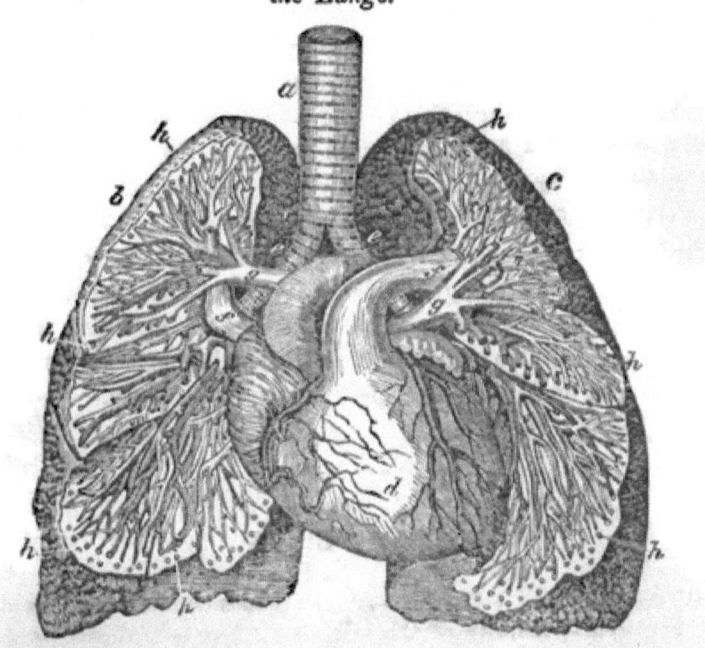

a. Windpipe.
b, c. Right and left lung.
d. Heart.
e, e. Divisions of the great air-tubes going to the right and left lungs.
f, f. Pulmonary arteries carrying the blood from the heart to the lungs.
g, g. Pulmonary veins, carrying the blood from the lungs to the heart.
h, h, h, h. Air-cells at the terminations of the air-tubes.

286. When the ribs are lifted and the chest expanded at the sides, and the diameter thereby increased, and when the diaphragm is drawn down, and the chest enlarged below, there must be a vacuum within this cavity to be supplied by air. The only passage into the chest is through the mouth and windpipe, and into the lungs; consequently, when the cavity of the chest is enlarged, the air rushes into the air-

tubes, and fills all the air-cells. On the other hand, when the ribs fall, and the abdominal muscles press upon the digestive organs, and force the diaphragm to rise, the lungs are compressed, the air is expelled, and the air-cells closed.

287. During life, there is a constant succession of these actions. The air is at one moment drawn into the lungs, by the contraction of the diaphragm and the lifting of the ribs, and at the next moment it is expelled, by the falling of the ribs and the contraction of the abdominal muscles. These two operations constitute what is called *respiration*. Each respiration supplies the lungs with a new quantity of air.

288. This is the operation of respiration. All the parts of its apparatus, — the framework of bone, the muscles of the ribs, and the diaphragm, — all coöperate in the work, and are necessary to effect its purpose, which is to bring the air and the blood together, and to relieve the latter of its impurities, and fit it for the support of the living body.

CHAPTER IV.

Waste Particles. — Carbon. — Air, Composition of. — Oxygen. — Nitrogen. — Affinity of Oxygen for Carbon. — Carbonic Acid. — Carbon meets Oxygen in the Nutrient Vessels. — Blood absorbs Oxygen from the Air, and gives out Carbonic Acid.

289. The respiratory apparatus and its operation, which carry out from the body the dead and waste particles, afford striking evidences of Nature's skill, and beautiful illustrations of her handiwork. The principal elements of this waste matter, now mixed with the venous blood, are carbon and hydrogen. These have a stronger affinity or attraction for oxygen, one of the elements of the air, than they have for the fluid of the blood in which they move.

290. *Carbon* is one of the most important elements of the animal body. It enters into and forms a part of all flesh, and of all vegetable matter. The brain, the muscle, the

fat, and the bile, the solid fibre of the wood, the pulp of the cherry, and the flour of the grains, are all composed, in a great proportion, of this substance. *Hydrogen*, also, is a very essential element in the composition of flesh. Combined with oxygen, it forms water; and in this state it is found in the fluids, and in the more solid textures of the animal body. But it is also found, in different combinations, with carbon, oxygen, and nitrogen, in the various kinds of flesh.

291. It will now be necessary to examine the composition of the *air*. This is apparently a simple element; nothing seems purer or less compounded than air, as we breathe it. But chemical analysis shows it to be composed of two elements, — oxygen and nitrogen, — in about the proportions of twenty-one parts of oxygen to seventy-nine parts of nitrogen. Beside these, there are generally some other gases, — a little carbonic acid gas, and a little vapor, — amounting to one or two per cent., in the atmospheric air.

292. *Oxygen* is one of the most prevalent substances in nature. It enters into the composition of all animal and vegetable matter, and is the perpetually necessary element of life, in all its forms, and in all its stages. It is the essential ingredient of most acids. With sulphur it forms sulphuric acid; with nitrogen, in large proportion, it forms nitric acid, or aqua fortis; and, in smaller proportion, atmospheric air; with hydrogen, it forms water; and with carbon, carbonic acid. In air it is a gas; in water it is a liquid. When separated and alone it is a gas; in the rust of iron it is solid.

293. *Nitrogen* forms about four fifths of the volume of the air we breathe. It is a little lighter than air, and, of course, lighter than oxygen. It unites with oxygen in several proportions, forming very different substances, according to the proportions of their mixture. Nothing is more mild and bland than air, and few things are more caustic and harsh than aqua fortis, which is a combination of the same elements.

294 Although nitrogen and oxygen are apparently so

closely united, yet the oxygen has a stronger affinity for carbon and hydrogen than for nitrogen; and whenever, under appropriate circumstances, the carbon or hydrogen is presented to the air, the oxygen leaves the nitrogen and unites with the carbon, and forms carbonic acid, or with the hydrogen, and forms water. In other words, the air is decomposed, its simple elements are separated from each other, and a new compound is formed by the union of carbon and oxygen, or by the union of hydrogen and oxygen.

295. *Carbonic acid* is a composition of oxygen and carbon. This is a gas heavier than air, and lighter than water. If it be in a vessel with water, it rises to the top; and if in a vessel with air, it sinks to the bottom. It is so much heavier than air that it can be poured from one tumbler to another, like water. It is found in some caves, and at the bottom of some wells. It is the fixed air which is the product of fermenting bread, beer, wine, and cider, and fills the bubbles that rise to the top of these liquids at the time of their fermentation. When the beer is drawn out from the vats, in the great breweries, this gas often falls to the bottom, and partially fills these reservoirs. It is also the product of combustion of charcoal; and often, where this fuel is burning without any outlet near the floor for this gas to run off, or a chimney of sufficient draft to carry it upward, it partially or entirely fills the room. Wherever this gas is, there can be no pure air, for this is excluded by it as certainly as it would be by water; and it is as unsafe for a man to enter a cavern, well, vat, or a room containing it, and carry his head below the surface of this gas, as it would be if these contained water.

296. When the chest expands, the air rushes in and fills all the air-tubes and the air-cells throughout the lungs. There it comes almost in contact with the venous blood, which is distributed in the numberless little vessels, and separated from the air-cells only by a thin film of membrane, through which the gases can pass. There an interchange takes place between the fluid and the gas. The blood absorbs from the air some of its oxygen, and the air takes from

the blood some of its carbonic acid and its water. By this change the blood is relieved of its exhausted and dead particles, and receives new and life-giving particles in their stead. The color is changed from a dark purple to a bright scarlet. After this, the blood is ready again for the sustenance of life, and is sent back, through the pulmonary veins, to the left side of the heart, to be sent again, through the arteries, to the whole of the body, carrying nutriment to support it, and oxygen to combine with its dead carbon.

297. *Carbon and hydrogen compose the principal portion of the wasted and exhausted particles of the living body, and these are thrown into the veins.* There they meet with the oxygen that has been absorbed from the atmosphere in the lungs, and carried in the blood, through the arteries, to the capillaries and the minute veins. There these, the hydrogen and the carbonic particles, and the oxygen, meeting together, unite and form carbonic acid gas and water. These new compounds are then sent, with the venous blood, through the veins, to the heart, and thence to the lungs.

CHAPTER V.

Venous or purple Blood changed to arterial or scarlet Blood. — Color of venous Blood seen in Veins of Hand, and of arterial Blood in flushed Cheek. — Oxygen consumed in Respiration. — Carbonic Acid given out. — Water given out through Lungs. — Other Matters. — Foul Odors in Breath. — Offensive Breath.

298. THE blood enters the lungs a compound of three kinds of matter — the old blood, which had not been used for the purpose of nutrition, the old wasted particles, which are now seeking an outlet, and the new chyle from the digestive organs. (§ 264, p. 120.) All this heterogeneous mass is unfit for the nutrition of the animal body. Its color is purple. In this compound no free oxygen is present, but carbonic acid and water are abundant. When the blood

returns back to the heart from the lungs, it is one homogeneous compound; it has lost its carbonic acid and water, and received a supply of oxygen, which now pervades the fluid.

299. The difference of the color of the blood is seen in the veins of the hand and arm, which appear to be blue, while the flushed cheek is of scarlet red, from the presence of the blood in the arteries of the skin. When a person is bled from the arm, the vein is opened; the blood that flows is venous, and of course purple. The inexperienced mistake this natural color for the effect of disease, and often remark, "that the blood is very black; the patient needed bleeding to be relieved of such dark impurities."

300. *This change of the blood is effected in the lungs*, of course, there must be a corresponding change in the air. The oxygen which the blood receives is the oxygen of the air; and the carbonic acid and water which are thrown off from the blood are mingled with the air. The air is therefore changed by this process, and to this extent. During a state of repose, the air at each respiration has a little more than four per cent. of its volume of carbonic acid gas added to it, and the oxygen is diminished in proportion necessary to form this acid. If the same air be respired over and over several times, all the oxygen is consumed, and the air becomes loaded with carbonic acid gas.

301. Sir Humphry Davy enclosed one hundred and sixty cubic inches of air in an oiled silk bag, and breathed this for the space of one minute. In this time, he made nineteen respirations. On examination of the air, he found that nearly one half ($\frac{19.4}{4.4}$) of the oxygen was consumed, and its place supplied by 15.2 inches of carbonic acid, which had been generated in the blood-vessels, and given out from the lungs in one minute.†

302. The quantity of carbonic acid gas which is found in the air that has been breathed, varies in different circumstances, and in different conditions of the human body.

---

* Lehmann's Physiological Chemistry, Vol. II. 436.
† Muller's Physiology, p. 295.

It is greater in the waking than in the sleeping hours. It is more when the body is in action than when at rest; in winter than in summer. It is greater when the body is well nourished than when fasting. Vierordt found that the quantity discharged was fourteen per cent. greater two hours after he had dined than it was on other days at the same hour, when the dinner had been omitted.*

303. Besides the carbon of the blood, which is to be carried away through the lungs, there is *water* which is not needed in the body, and which finds its passage through the same outlet. This water goes out in the form of vapor, and ordinarily is not perceptible. But every one is familiar with the visible cloud of vapor that accompanies his breath in a cold winter's morning. This is but the condensation of the vapor that is invisible in a warm day. The same may be ascertained at any time by breathing on a looking-glass, when the vapor is condensed, and becomes visible in the form of water.

304. There are other matters carried off from the body through the lungs by the air. Their perceptible qualities differ in various men. The breath from one is sweet, from another sour, from a third foul and offensive, and from a fourth, it is without perceptible odor. These disagreeable odors may, in some cases, proceed from decayed teeth, or from disease in the mouth, the air passages, or the lungs; but more commonly they come from direct secretions in the lungs of certain matters, which existed previously in the animal body; as, when one has eaten onions, his breath smells of garlic. The odor of wine or spirits which have been taken into the stomach, is perceptible in the breath, long after the mouth has been thoroughly cleansed of these matters. In other cases, these unpleasant odors proceed directly from some foul secretion in the lungs. The habit of chewing or smoking tobacco gives to one's breath an odor peculiarly offensive to others who may inhale the same air. In some persons, this odor is so powerful as to taint the air of a whole room as soon as they enter it.

* Lehmann's Physiological Chemistry, Vol. II.

CHAPTER VI.

Air changed by Respiration unfit to be breathed again. — Dyer colors with Dye of full Strength. — Air should have full Proportion of Oxygen. — Respired Air, loaded with Carbonic Acid Gas, can take away no more. — Air saturated with Water can take no more from Lungs. — Air spoiled for Respiration in three Ways.

305. An examination of the air, after it has passed out of the lungs, shows that it is very different from the same air before it went in. At first, it had about twenty-one per cent. of oxygen, seventy-eight per cent. of nitrogen, and one per cent. of carbonic acid; but, when it has been respired, it has lost about one fourth of its oxygen, and has gained carbonic acid and vapor, in proportions varying with many circumstances connected with the state and health of the animal system. If, then, it is necessary for the blood to consume at each respiration one fourth of the oxygen of the air which is inhaled, it would follow that, if the same air be breathed twice, one half of the oxygen would be consumed; and if breathed three times, three quarters; and if breathed four times, all would be consumed. If so much oxygen is not consumed, there must be so much less of the carbon and the waste of the blood carried away.

306. But, if the lungs consume only one fourth of its oxygen at each respiration, it by no means follows that the air can be breathed four times over, and at each time have the same effect in purifying the blood. In respect to relieving the blood of the carbon, the oxygen may be considered as the strength of the air. When this constitutes twenty-one per cent., it is of full strength; when it is only fifteen per cent., it is only three quarters of full strength; and, at most, it can have only three quarters of the due power. So, when, after being once respired, it is reduced by another and another respiration to one half and one quarter its full strength, it, of course, has so much less power of performing that which is required of it.

307. When the dyer has determined what strength of dye will give the due color to his cloths, he adds fresh coloring matter as often as one piece has weakened it, in order to keep up the dye to its full strength, and to give to each successive piece of cloth the same hue; for, if the dye be weakened, it will give a weaker color. He would not, therefore, continue to dip his cloths in it, after it is reduced, because there was some coloring matter left; nor would he think of exhausting all the power of the dye, unless he was satisfied to produce a duller shade.

308. Precisely analogous to this is the effect of the air in purifying the blood of its corrupting carbon. The stronger the air, — that is, the greater the proportion of its oxygen, — the more effectually will this carbon be carried away; the weakened air must produce a weak effect, and take away less of the impurities. Air, therefore, which has been breathed once or more, having lost a certain part of its oxygen, must be, in that proportion, unfit to do the work of respiration. If we breathe pure oxygen, or air too strongly oxygenated, — that is, air containing more than twenty-two per cent. of this gas, — the carbon would be taken from the blood faster than it could be spared, and the body would be wasted. If we breathe air containing less than twenty-one or twenty-two per cent. of oxygen, it will not carry the carbon off so fast as is required. It is only by breathing air of the natural strength that this work is best performed, and the carbon carried away neither too rapidly nor too slowly. Air, therefore, should be breathed once, and once only. We need a fresh draft of air at every inspiration, as much as the dyer needs a fresh dye at every coloring.

309. Air, when it has been breathed, not only loses its oxygen, which is its active principle, but it is loaded more or less with carbonic acid gas, which is increased by every respiration. Therefore, when we breathe air over and over, we not only breathe a weaker gas, but a fouler one; we receive back into the lungs, and into the vital system, that

dead and corrupting matter which nature had so carefully removed.

310. There is another consideration in this matter. Supposing the air were merely a passive vehicle to carry off the carbonic acid gas, and had no active duty in the work, it would be a natural question to ask, How much of this gas can it bear away? Is there any limit to its capacity of taking up and bearing off this offending matter? Now, it is well established that the air will not receive and hold an indefinite quantity; but, after having received a certain proportion, it will receive no more. When it has arrived at this point of saturation, — that is, when it is so full that it can receive no more, — it then is useless as a vehicle to carry off any more from the lungs. Bernan says* that, when the air holds in solution only about three and a half per cent. of its bulk of carbonic acid gas, it is unfit for respiration.

311. The air, for this purpose, may be considered as the water which the dyer would use to wash his colored cloths. It is plain that, when the water is once befouled or saturated with the loose coloring matter, it would take no more from the cloths; and therefore the judicious cleanser changes his waters as often as they become foul; and, whenever he can, he selects a running stream, so that the water is carried away as fast as it is befouled, and its place is supplied with fresh and clean.

312. Upon the same principle, the lungs cannot be thoroughly cleansed of the impurities which come to them through the blood, unless the air is supplied to them fresh and untainted at every respiration. For foul air, loaded with carbonic acid gas, can no more cleanse the lungs, than foul water can cleanse the colored cloths.

313. *The blood is relieved of its superabundant water through the lungs.* If this does not find an outlet here, and if it is not carried off by the air, it must be carried back in the blood to the heart and the arteries, to overload the sys-

---

* Art of Warming and Ventilation.

tem and impede its operations. There is a definite quantity which must be carried out, and the air has a limited capacity for holding water, and of taking it away. When this limit is reached, and the air is saturated with water, it can take up no more. The air would be saturated with moisture from the lungs in about the same number of respirations that would consume its oxygen; after this, it would be useless for the removal of either carbon or water.

314. Thus we see that, in three ways, the air becomes vitiated, and unfit for continued respiration: 1st, by the consumption of its oxygen, so that it is unable to remove the carbon from the blood; 2d, by being loaded with carbonic acid gas, so that it cannot take it up any longer from the lungs; 3d, by being saturated with moisture, so that it cannot aid in relieving the system of its superabundance of water.

CHAPTER VII.

More Oxygen consumed, and Carbonic Acid given out, in cold and dense Air, and less in warm. — Air on Mountains does not support Life as in Valleys. — Impurities in Air diminish Oxygen. — Feeble, consumptive, and melancholy Persons give out less Carbon and Hydrogen.

315. THE amount of oxygen received, and the quantity of carbonic acid gas and watery vapor carried off, differ at different times, and vary with varying circumstances. A dense atmosphere is more concentrated, and, consequently, contains more oxygen in a given space, than a rare one. The air is more expanded in a warm than in a cool climate, and in hot than in cold weather. We therefore do not inhale so great a weight of air, and, consequently, so great an amount of oxygen, in summer as in winter. The oxygen received being less, the carbonic acid given out is diminished in the same proportion.

316. We experience faintness and languor in the warm season, because the air does not purify a sufficiency of blood for the vigorous sustenance of the system. Then people complain that "the air is heavy," which is directly opposite to the truth, for the air is really light, and it does not contain sufficient oxygen to invigorate them, and give sufficient strength and elasticity to bear the burdens and operations of life easily, and hence all these are heavy to them. We observe animals puff and breathe rapidly after running in summer, and we do the same on any active exertion, more in warm than in cold weather. We do this to bring a more frequent supply of air to the blood, and thus to compensate for the lightness of the air, and deficiency of oxygen, by the rapid renewal of both.

317. The air is more dense in the lower regions of the earth, and on the level of the sea, than on the heights of mountains. As we ascend from below to the higher elevations, we find the air lighter and more expanded, and we are compelled to breathe more rapidly. Travellers all complain of the increasing languor and faintness as they ascend, and enjoy the bracing and invigorating effect of the air as they come down the mountains.

318. All kinds of impurities in the air, and every thing that diminishes the proportion of oxygen, have the same effect of weakening or diminishing the vital properties of the air. In some mines, a gas is given out, called the *fire damp*, which is carburetted hydrogen. Whenever this is present, it lessens the vivifying power of the air, by excluding its oxygen. If it is breathed in small quantities, it occasions giddiness, sickness, and diminished nervous power; and, when it is in great proportion, the miners are unable to breathe, and often fall a sacrifice to it.

319. The general state of the system affects the quantity of matter which is carried out through the lungs. When the whole frame is well, and all the functions are carried on vigorously, — when the circulation of the blood is easy and the respiration well sustained, — then the old particles of the

body are freely separated, and the new and vitalized ones take their places, and the former are carried rapidly away, and life is frequently renewed, and vigorously sustained. But when the system is feeble and languid, — when it is feverish and generally disturbed, — when it is exhausted by fatigue or want of sleep, — the reverse happens; the circulation is languid, the nutrition feeble, absorption slow, and a smaller proportion of carbon and hydrogen is carried off.

320. *The system is relieved of these dead matters, more or less, according to the condition of the lungs,* in proportion as they are healthy or unhealthy. In some diseases, their texture is changed from an exceedingly porous and spongy body, to one partially or entirely solid. In consumption, a part of these organs is filled with tubercles, or abscesses. Sometimes these occupy almost the whole substance of the lungs, and leave so little room for air that respiration cannot be carried on. Then the sufferer literally dies for want of breath, because the lungs cannot receive sufficient air to purify as much blood as is necessary to sustain life. The impure blood which comes to the lungs, to exchange its carbon for oxygen, does not find air to give it relief, and goes back to the heart nearly as corrupt as when it came out.

321. In lung fever, and some other diseases, a portion of the lungs becomes solid, like liver, and the air-cells are closed. If the whole of the lungs becomes consolidated, death must follow; but this state more frequently prevails in a part only of these organs; then the blood is sent back in an imperfect condition, and the frame is then only partially nourished. Some have sustained life for a considerable period with only one sound lung; but theirs was a feeble and lower life, and they could not perform all the work, nor enjoy all the comforts, of ordinary well-sustained existence.

322. The states of the mind and feeling, as well as those of the body, affect the discharge of waste matters through the lungs. Cheerfulness and exhilaration, and the exciting passions, increase the separation of carbon; while the depressing emotions — fear, grief, and anxiety — diminish it.

CHAPTER VIII.

Lungs must have Capacity to receive sufficient Air. — Action of Respiration performed by the Muscles of Chest and Diaphragm. — Ribs spread outward in Inspiration. — Action of Diaphragm presses the Abdomen downward and outward.

323. It is not only necessary that the lungs should be in good health, and be supplied with pure air, but they should be able to receive it in sufficient quantity. This implies that the chest should be of the natural size, and that it should have the due power and opportunity of expansion and contraction.

324. Although it is absolutely necessary that the air reach the blood in the lungs, yet it has no active power to get there. It is merely passive. It does not enter of its own accord, but it is pressed into these organs, when, by the enlargement of the cavity, a vacuum, or rather, more room, is made for it. Nor have the lungs any active power of expansion. They, too, are merely passive. Their air-cells do not extend themselves, and thus press the walls of the chest outward. But, when these walls are extended, the air rushes, or rather it is pressed, into these air tubes and cells, and compels the lungs, thus filled with air, to swell and completely fill the cavity of the chest.

325. The structure of the chest is arranged like the common bellows for expansion and contraction. (§§ 270—277, pp. 122, 125.) The bony framework is furnished with joints, on which the ribs move. The muscular covering contracts and sets this framework in motion, while the diaphragm draws down, and both coöperate to enlarge the internal capacity of the pulmonary cavity; then the ribs fall, and the abdominal muscles press the diaphragm up, and both combine to diminish this cavity.

326. In Fig. XXI., the full black lines represent the outline of the chest and the abdomen when the lungs are empty, and the dotted lines represent the same when the lungs are

filled with air. When the air is inhaled, the walls of the chest are expanded from a, e, to a, b, and the diaphragm drawn down from e, d, to b, d, and consequently the walls of the abdomen are carried from e, c, to b, c, and the diameters of both are increased.

FIG. XXI. *Side View of expanded and contracted Chest.*

e, d, Diaphragm drawn up.

b, d, Diaphragm drawn down.

a, e, c, Front wall of contracted chest and abdomen.

a, b, c, Front wall of expanded chest and abdomen.

327. This expansion of the chest and abdomen cannot take place unless there is room outwardly. If the body is enclosed in any inelastic girdle or dress which fits it closely when the chest is empty, it must be confined within that limit, and its expansion prevented.

328. Some of the fashions of the dress of females of modern times, and in civilized nations, have precisely the effect of bandages to confine the ribs, and limit the expansion of the chest, and prevent the inhaling of the due quantity of air. The corsets are made of inelastic materials, and usually so constructed and laced as to exactly fit the shape of the bust, and lie as closely to the surface as possible. When these are worn, and the other garments are arranged upon

the same principle, and with their fastenings bound closely to the body, so as to mould the form, they confine the ribs, and prevent their movements upward and outward.

329. In this confinement of garments, whenever the muscles attempt to raise the ribs and extend them outward, they meet with resistance. These muscles are not very strong; they are made for a definite purpose — merely to raise the free ribs, and to expand the unobstructed chest, but not to break bands, force lacings, or stretch layers of compact cloth. Hence, finding all labor ineffectual, they after a while cease their attempts to move the ribs, or at least diminish their exertions very materially, and leave the main business of respiration to be done by the diaphragm.

330. When the diaphragm descends out of the chest, it must press the digestive organs downward before it. But these organs cannot be compressed; they are not made to occupy less room than before; they are merely removed from their upper position to a lower and a broader one. Therefore they must find room for extension below and outward. And if this is prevented, — if the abdomen is so bound or compressed that it cannot expand, — the stomach and liver cannot give way before the diaphragm, and then this muscle cannot descend to make room for the lungs, nor can we breathe by this part of the respiratory apparatus.

231. The consequence is, whenever the fashion of the female dress extends the pressure of the waist beyond the ribs, and encloses a good portion of the abdomen, unless quite loose, or whenever the costume of the male presses upon this part of the body, it must interfere with the freedom of motion of that part of the system, and so far prevent or restrict respiration by the diaphragm.

CHAPTER IX.

Common Notion of Beauty of Chest unnatural. — Chest Seat of most important Organs. — Size of Chest corresponds to Size of Body. — Natural Chest not conical. — Shape of Bones changed by Pressure. — Comparative Form of Chests.

332. THERE is a common and mistaken notion of beauty of the female chest. The *beau ideal* of many requires that it should be of a small and taper form, diminishing from the shoulders downwards to the waist. This opinion is encouraged and strengthened by the fashion of female garments. But however general this form may be, and however long established in the world, it is artificial, and not natural. It is opposed to that principle of beauty which nature has clearly and every where established — that grace is secondary, and not primary; that it is the proper and becoming arrangement of those parts that are necessary and useful.

333. The chest is not a mere connecting link between the upper and lower portions of the animal frame; but it is the depository and the workshop of some of the most important of the vital organs, without the action of which life cannot be for a moment sustained, and without whose free and perfect operation life must be impaired and enfeebled.

334. As the chest is made for the use of the body, and not for ornament, — as it was created to contain the heart, and to give room and motion to the lungs, so that respiration could be carried on in the best manner, — it would follow that that form and size of this part of the animal frame is the most beautiful, which would best answer these purposes, and allow the lungs to perform their functions most effectually.

335. The size of the chest should bear a proportion to the size of the body, so that it may receive a quantity of air proportioned to the quantity of blood that must be purified in the system. Therefore, a small waist becomes only a small person, and a large waist is necessary to the grace of a large

person, precisely as a large or small head is becoming to a frame proportionally large or small.

336. This is the plainly established principle of Nature. We see it in all her works. If we examine the little child, who has never worn any close dress, we find the circumference of its chest about as great as that of the body at the hips, and a line from the arm-pit to the hip would be nearly straight. If the waist is never subjected to the pressure of clothing, which would interfere with the motions of the ribs, the chest will be continued through life in nearly the same shape as that of the child, (Fig. XXIII.,) or of the Indian female, whose garments have never been bound about the chest. We see the same in many laborers, more especially those from the middle and the north of the continent of Europe. The ancient statues show the full chest, the expanded waist, and the broad freedom of the lungs for motion.

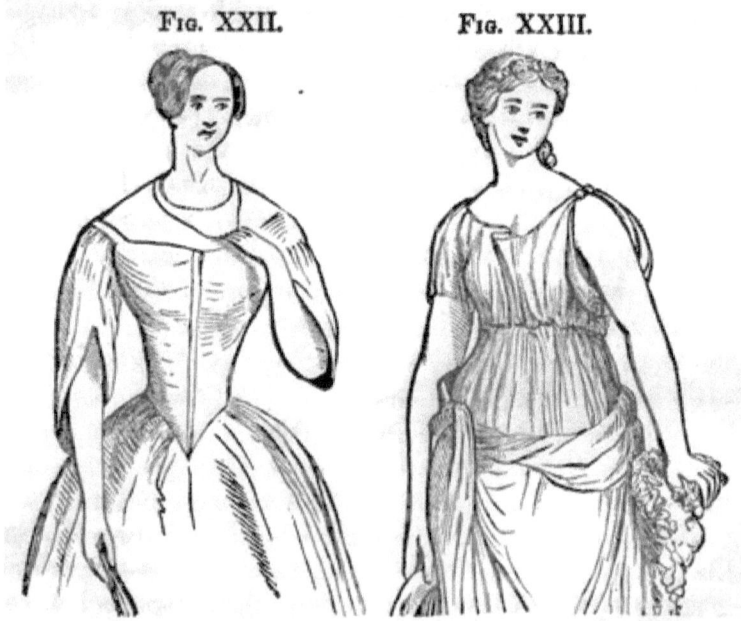

Fig. XXII.   Fig. XXIII.

337. But the chests of many who are incased in a close costume are small, and taper downwards from the shoulder

to the waist, (Fig. XXII.) In some, this is the temporary effect of present pressure; and when the close garments are taken off, the ribs rise to their natural position, and the chest expands to its natural size. But in others, whose chests have been long subjected to this close confinement, this distortion of ribs and contraction of chest become fixed and permanent; and then they need no outward covering to confine the respiratory organs within these narrow dimensions.

338. Figure XXIII. represents the chest of those who have always worn loose dresses. Fig. XXII. is that of one used to tight dresses. The former receive much more air into their lungs, and carry off the impurities of their blood more freely, and hence their changes of particles must be more rapid, their vigor and elasticity of body must be much greater than the others enjoy.

Fig. XXIV.
Bones of a natural Chest.

Fig. XXV.
Bones of a distorted Chest.

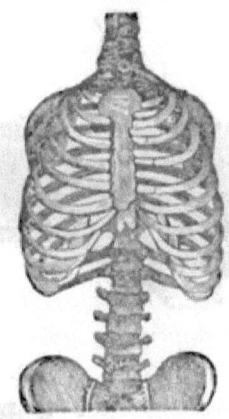
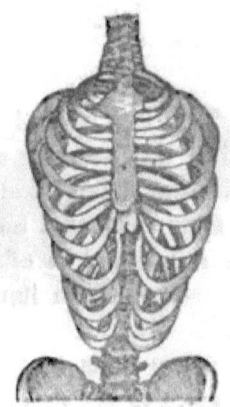

339. *The bony frame by pressure may be altered*, and made to assume forms very different from that which nature intended. In the process of nutrition, (§§ 244–247, pp. 112, 114,) the old particles of the animal body, in all its parts, are continually going away, and new ones are taking their places. But if any pressure bear upon the depositing vessels

on one side and close them, the new particles are not placed there, but the blood is poured more freely into the other side, and there the growth is increased; and thus the shape of the organ, the bone, or the flesh, is changed.

340. This distortion necessarily follows in the form and size of the ribs, from the pressure of corsets or any tight clothing upon them. They gradually yield to the external form, and, bending inward, assume the shape which the outer mould makes for them; and the chest, which was originally of a size in due proportion to the rest of the body, now becomes permanently small, and the internal capacity of the lungs corresponds to the external measurement. Fig. XXV.

CHAPTER X.

Action of Diaphragm is affected by State of Stomach. — Frequency of Respiration, Capacity of Lungs, Amount of Air inhaled, should correspond with the Carbon and Water that are to be carried away. — Quantity of Blood in the Body. — Quantity of Blood and Air flowing through Lungs.

341. The free operations of the diaphragm are sometimes impeded by the disorders of the stomach. In some forms of dyspepsia, the sufferer feels as if the cavity of the chest were already filled, and that no more air could be inhaled. He breathes short, and is often convinced that there must be serious disease of the lungs. In some of these cases, the stomach is distended with gas, and presses upward upon the diaphragm so as to prevent its motions downward. In other cases, the peculiar kind, rather than the quantity, of gas affects this organ and impairs its power of motion.

342. *A man in good health will breathe about eighteen times a minute.* Some breathe more rapidly than this; others not so frequently. Children and women breathe more rapidly than men. Exercise, especially fast running, quickens the respiratory movements. So, also, the exhilarating affections — cheerfulness — laughter — have the same effect.

On the other hand, fatigue, depression of spirit, grief, and anxiety diminish the frequency of respiration.

343. The lungs of a man of average size, and in usual health, when at rest, when neither expanded nor contracted, will hold two hundred and ninety cubic inches, or a little less than a gallon of air. But, when distended by ordinary inspiration, they receive twenty inches more. This will make three hundred and ten inches when full. This twenty inches is the usual extent of respiration. This is the amount of air which the lungs need, and which they receive at every inspiration, when allowed freedom of motion, eighteen times a minute, and one thousand and eighty times an hour.

344. This quantity of pure air is not merely wanted to fill the capacity of the chest and lungs, but it is needed for the purification of the blood. Bearing in mind that the blood receives from the system carbon and hydrogen of which it must be relieved; and knowing that it receives with the chyle more water than is wanted; and that, when these are combined, they go to the lungs to be disburdened of their superfluous and noxious elements; it is natural to suppose that the amount of air should correspond with the quantity of these matters, which are thus to be removed.

345. Then in order that the air may meet the wants of the blood, the size of the chest corresponds to that of the body, and the motions of the ribs and the expansion of the lungs correspond to the flow of the blood. This one would suppose to be the case from a mere general view of the harmonies of nature; for the Creator makes all his works consistent one with another.

346. The quantity of blood in the whole system of a man of average size, amounts to about twenty-eight pounds. (§ 232, p. 107.) The heart beats in a man about seventy-five times a minute, and forces out of itself about two ounces, or half a gill, at each pulsation or contraction; and, consequently, in one minute, more than nine pints of blood are sent to the lungs to be acted upon by the air. In the same time, twelve pints of fresh air are brought into the lungs; and

the amount of carbon in the nine pints of blood corresponds very nearly with the capacity of the twelve pints of air to carry it away.

CHAPTER XI.

Air spoiled by Loss of Oxygen and by Carbonic Acid Gas. — Capacity of Air to receive Vapor. — More Air saturated by Vapor of Breath in cold than in warm Day. — Vapor from Skin saturates some Air. — Amount of insensible Perspiration. — Quantity of Air spoiled by Loss of Oxygen, by Carbonic Acid, and by Water.

347. BREATHING air once destroys or weakens, and, partially at least, spoils, for the purpose of respiration, 720 cubic inches of air a minute, by the mere consumption or use of its oxygen. On this account, we need about one half a cubic foot of air every minute. When the air goes out from the lungs, it contains about four or five per cent. of carbonic acid gas, (§ 302, p. 136;) but if it contains more than three and a half per cent. of this gas, it is unfit to be breathed again.

348. It necessarily follows, then, that the air which has been once breathed, contains about two and a half times this proportion, and therefore the quantity of this injurious gas that is the product of one respiration, is sufficient to corrupt nearly once and a half as much more. The quantity which is exhaled in one minute would give three and a half per cent. to 1800 inches, and render so much unfit to be breathed again.

349. There is a limit to the power of the air to take up and carry away the watery vapor from the lungs, and this limit differs with the temperature. When the air is cooled down to 32°, or freezing point, a cubic foot of it will hold about two and a half grains of water in solution. When it is raised up to 65°, which is usually the proper temperature of sitting-rooms, it will hold a little more than seven grains; and at 90°, which is very nearly the temperature of the air

when it goes out from the lungs it will hold fifteen grains. This vapor is invisible, and generally imperceptible. But if the air at 90°, containing fifteen grains in a cubic foot, be cooled down to 32°, it then can hold only two and a half grains; and the difference between these quantities — twelve and a half grains — will be condensed and become visible in the form of water.

350. The cloud of vapor which one seems to expire in a cold day, is caused by this condensation. If, in winter, one or more persons sit in a room sufficiently warm to be comfortable, the air becomes filled with pulmonary vapor. If the temperature of the room is the same throughout, this vapor is imperceptible to the eye; but the air near the windows, if these are not double, becomes cooled by the action of the outward air, and then this vapor is condensed, and lodged upon the glass in the form of water. If the air abroad is cooled below freezing point, this condensed vapor freezes upon the windows, and the glass becomes coated with a layer of ice. So we usually find the windows of our sleeping-chambers covered with ice from this cause, in the cold mornings of winter.

351. The whole of the water thrown off from the lungs in this state of vapor amounts to about seventeen ounces in a day,* which will make five grains and two thirds a minute. This will saturate nearly one half a foot of air at 90°; but, as air usually contains about one grain of vapor in each cubic foot, it can absorb so much less, and more air will be saturated with the pulmonary vapor.

352. *The insensible perspiration is another and very fruitful source of moisture in the air.* The skin is a very active agent, and is incessantly throwing off watery vapor from its surface. When this runs freely in drops, it is called *sweat*, and seems to be very abundant. But this is only a small part of the whole of this fluid, which is thrown off through the external surface; for a much greater quantity is sent off in an invisible form.

* Valentin and Dalton.

353. The quantity of this insensible perspiration varies from thirty ounces a day in the northern, to forty ounces a day in the southern, countries of Europe. Carpenter estimates it to be thirty-three ounces in England; this is eleven grains a minute; others give a much higher estimate. Cruikshank's experiments demonstrated it to vary from twelve to forty-five grains a minute; and he assumes the mean, from persons of both sexes, of average size, and in good health, to be twenty-three grains a minute. Taking the last as the standard, and adding these twenty-three grains to the five grains and two thirds of vapor thrown out from the lungs, we have enough to saturate somewhat more than two feet of air with moisture; and, so much air being saturated, it can take no more vapor from the lungs.

354. Thus we see that, in these three ways, the air loses its power of relieving the blood of its superfluous carbon and water, and is thereby rendered unfit for the work of respiration; — first, by the loss of its oxygen, in each minute, 720 inches; secondly, by saturation with carbonic acid gas, 1800 inches; thirdly, by saturation with vapor from the lungs and skin, 3590 inches.

CHAPTER XII.

Seven to ten Feet of Air spoiled each Minute. — Want of fresh Air in Houses, but not provided. — Size of Parlors, and Number of Occupants. — Small Sleeping-Chambers. — Lodging-Rooms in Boarding Houses, and in temporary Houses. — Cabins of Steam and Canal Boats.

355. *About four cubic feet of air being rendered by each person partially or entirely useless for the purpose of purifying the blood and giving it new life*, it will, of course, be necessary that we have so much new and fresh air supplied every minute for each one. If this, after having been once breathed, or saturated with vapor, were carried immediately away, this

quantity would be sufficient; but, as the corrupted air mingles with the pure, this is partially corrupted; therefore we need a larger supply to support respiration. The best authorities on the subject of ventilation consider seven feet as the least that should be supplied to each person; and Dr. Reid allows ten feet. Taking the lowest estimate, seven feet will be considered as necessary for the maintenance of the healthy respiration of each person in each minute of life.

356. If we always dwelt in the fields, we should have fresh air enough, without any effort on our part. But when we live in closed houses, it becomes a question whether we are thus supplied, and the wants of nature are satisfied; and, if our rooms are made air-tight, then it is necessary to determine whether they contain air sufficient for the consumption of all that inhabit them, as long as they stay there. If this be not the case, then it is necessary to find some means to carry off the foul air as fast as it is rendered so by respiration, and to bring in a new supply from abroad to take its place.

357. *A continued supply of fresh air for all inhabited rooms is as necessary as a continued supply of heat in cold weather.* And yet provision is not usually and intentionally made to meet this necessity in the arrangements of our dwellings and our public rooms. The architect and the builder provide carefully for warmth, but they generally make little or no provision for respiration. Fortunately, the imperfection of the builder's work obviates, in some small degree, and generally prevents, the immediately destructive consequence of the defects of the architect's plans. It is difficult — almost impossible — to make a room so tight that no air can force itself into it, when the internal atmosphere is heated, or vitiated by respiration.

358. A room sixteen feet square, and nine feet high, will contain 2304 cubic feet. This will be sufficient for four persons less than an hour and a half for ordinary day purposes. This room, though not so large as some that are inhabited by day or by night, is yet as large as most, and

much larger than many rooms so occupied. It is esteemed a proper economy to have small and tight parlors and sitting-rooms, for the occupation of the families during the day and evening. On an average these do not probably contain more than 1700 feet. If only four persons inhabit one of these, they would have air sufficient for less than one hour.

359. It is considered, by many, a prudent architectural design, to have many and small sleeping-chambers. Room for the bed and wardrobe, and for convenient dressing, is all that is thought absolutely necessary. At least, the plan of a good dwelling generally includes a portion of these narrow chambers. Many of these will not contain more than 500 cubic feet; and in such, two grown persons, often more than two children, sleep during the night. Here is air enough to last two persons a little more than half an hour.

360. In public boarding-houses, in some taverns, and in the houses where the operatives of factories are boarded, it is an object to lodge the family as cheaply as possible. Consequently, the lodging-rooms are often made as small, or to hold as many sleepers, as they can. Oftentimes these lodgers are so closely crowded, as to have hardly air enough for half an hour's respiration. At one of our large manufacturing establishments, eight sleep in one chamber containing 2574 feet; in several other chambers, two have $262\frac{1}{2}$ feet, four have 1800 feet, six have $973\frac{3}{4}$ feet, four have $686\frac{1}{2}$ feet, for a night's respiration. These rooms contain air enough to supply their occupants from twenty-five to sixty-six minutes.

361. This close crowding of sleeping-chambers is carried to the greatest extent in some of the lodging-houses built for temporary use on some of the railroads, and other public works. I have the measure of one of these chambers. The room was in the attic, — sixteen feet long, and fourteen feet wide. The height was six feet ten inches in the middle, but the roof met the floor at the sides, so that the average height of the room was three feet five inches, and the whole cubic contents of this chamber were 765 feet. There were no means whatever provided for the ventilation of this room.

There was neither window nor door. The only opening made into the chamber was a small hole in the floor, through which the sleepers ascended from the room below. This lower room was not ventilated much better. It had less than 1400 cubic feet of space; and there nineteen persons boarded or lived in the day, and five slept at night, and there all the operations of cooking, eating, and washing were carried on.

362. In this chamber, with less than 800 feet of air, fourteen men slept through the night; and for eight hours these men breathed over and over the air from each other's lungs, in the vain attempt to purify their blood, and refresh their frames, and invigorate themselves for the next day's labor. Here was air provided sufficient to last them less than nine minutes, and yet it was required to last them 480 minutes; and in this, as well as in other crowded chambers, nothing but the undesigned ventilation through the crevices of the imperfect carpentry, saved these sleeping occupants from suffocation.

363. The crowded state of the cabins of steamboats, in which the sleeping apartments are below decks, and of the canal boats at night, leaves less air for respiration than even these rooms. Not unfrequently, fifty or even sixty persons sleep in the narrow cabin of a canal boat, which contains no more space than some of the airy chambers where only two cautious people would usually spend the night.

CHAPTER XIII.

Crowded Workshops. — Chambers. — Public Halls. — Churches. — School-Rooms. — School-Rooms filled with foul Air. — Habit of breathing each other's Expirations. — Foul Air offensive. — Ventilation.

364. SOME of the trades require a very small space for their operations. The shops in which these are carried on

are therefore constructed in reference rather to the convenience of the work, and the economy of heating them, than to the health of the workmen. Consequently, these men are sometimes so crowded and confined as to have insufficient air for respiration. In a room ten feet square, and eight feet high, with 800 feet of air, six and sometimes eight men can work, without interfering with each other; and this is thought good accommodation.

365. Family rooms, lodging-chambers, cabins, and shops are not the only places where men and women gather in numbers beyond the capacity of the air to support their healthy respiration. Public rooms, lecture-rooms, churches, concert halls, and, above all, school-rooms generally, are badly ventilated. They are not supplied with air sufficient for the ordinary numbers, and still less for the occasional crowds, that meet in them. The general plan of these is to hold many persons; and the idea of the architect is to so arrange the seats, that the greatest possible number may be gathered into a given space.

366. A part of these pages on respiration were read as a lecture before an associated audience, which assembled in a hall forty-seven feet long, thirty-seven feet wide, and nine feet high, measuring 15,650 cubic feet. This room is made to hold five hundred persons when full; and usually from three to four hundred meet there. In the former case, there are thirty-one feet, and in the latter thirty-nine to fifty-two feet of air for a person. The sittings of this society vary from one to three hours.

367. A church which has been recently built has one hundred and eighty-three feet of air for a person, in an average audience, and one hundred and thirty-six feet when crowded. Another has from seventy-five to ninety-nine feet of air for each of the people, according to their numbers. Many other churches afford about the same proportion of air to their occupants. I have not their exact measurement, as of these above stated; but the foulness of the air which one perceives on entering them late in the forenoon, or in the

afternoon, too plainly shows that they have not sufficient ventilation.

368. In times of great excitement, the crowds in the churches or halls are more dense even than these. It is estimated by those observant of the matter, that, in the closest crowds, a man standing will not occupy more than two square feet of surface; and therefore a room can hold half as many as there are square feet of floor. This would give twice as many cubic feet of air as the height of the room above the heads of the people. I have stood in Faneuil Hall when each man had very little more space for air than that which was over his head to the ceiling above.

369. School-houses seem to be as imperfectly supplied with air as public halls. It is rare that one enters a school-room from the fresh air abroad, after the scholars have been in a few minutes, without perceiving the foulness of the atmosphere within. A room thirty feet square is ordinarily supposed to be large enough for eighty or ninety children; and, if the room be nine feet high, this will allow eighty or ninety feet of air for every child, which is sufficient for their respiration twelve or thirteen minutes. The air of these rooms becomes loaded with carbonic acid gas, with the foul secretions of the lungs, and the excretions of the skin. It is offensive, so much so as sometimes to produce sickness and faintness in those who enter from the external air.

370. But "not the least remarkable example of the power of habit is its reconciling us to practices which, but for its influence, would be considered noxious and disgusting. We instinctively shun approach to the dirty, the squalid, and the diseased, and use no garment that may have been worn by another. We open sewers for matters that offend the sight or the smell, and contaminate the air. We carefully remove impurities from what we eat and drink, filter turbid water, and fastidiously avoid drinking from a cup that may have been pressed to the lips of a friend. On the other hand, we resort to places of assembly, and draw into our mouths air loaded with effluvia from the lungs, skin, and clothing of

every individual in the promiscuous crowd — exhalations offensive, to a certain extent, from the most healthy individuals; but when arising from a living mass of skin and lungs, in all stages of evaporation, disease, and putridity, prevented by the walls and ceiling from escaping, they are, when thus concentrated, in the highest degree deleterious and loathsome." *

371. When one enters any rooms thus crowded, and inhales the air thus exhausted and corrupted, he perceives, at once, an offensive and oppressive smell, and there comes a feeling of suffocation about his throat and chest, followed by some degree of faintness. But those who live in it, having by degrees become accustomed to it, do not perceive the smell; the sensibility of their lungs and nostrils is blunted; they are not offended with the foul odor of the atmosphere; yet their lungs do not find the oxygen to purify the blood, and cannot perform their work successfully. They are not relieved of the waste of dead atoms of flesh within them.

372. It is evident that unless there is some way of removing the respired and foul air from these rooms, and of replacing it with new and fresh air from abroad, the work of respiration cannot be carried on as it should be, the blood cannot be purified of its dead particles, and the system cannot be nourished with life and energy; and then the conditions which nature established for our existence cannot be fulfilled.

373. *Ventilation, or the means of supplying fresh air to every inhabited room, every parlor, sleeping chamber, schoolhouse, public hall, church, or shop, in which people live, is, then, as necessary as the supply of food.* After the air already in the room is consumed or vitiated, it must be removed, and as much brought in every minute as is used or spoiled. There must, then, be two constant currents; one outward, carrying off the foul air, and the other inward, bringing in pure air. The outward current may pass upward through the chimney,

---

* Bernan, Art and History of Warming and Ventilation, Vol. II. p. 313.

or through the crevices in the upper part of the ceiling, or through a passage-way provided for the purpose. The inward current more commonly comes through the unintentional crevices which the skill of the architect and mechanic seldom entirely prevents, and which admit air sufficient to save the occupants from the death of the Black Hole, but not enough to save them from some sickness, or faintness, or certainly some depression of life. As those crevices are inadequate to supply the air that is needed to sustain the fulness of life, every room that is inhabited, and especially every hall that is filled with people, and every school-room, should be provided with means of ventilation sufficient to admit and to carry away at least seven feet of air a minute for each occupant. For this purpose, a school-room, with forty persons, should have a ventilator a foot square, through which the air should move upward at the rate of two hundred and eighty feet a minute, and as much fresh air should be received.* In ordinary circumstances, air cannot be compressed; no more can be received into a room than is carried out. It is therefore useless to provide means for the admission of fresh air by a furnace or otherwise, unless there be some avenue, either accidental or designed, for the foul air to escape. Nor can a room be emptied of air; none will go out unless as much comes in. A ventilator will not, then, carry away the foul air, unless there be some place accidentally left, or especially provided, for the admission of other air to take its place.†

* This current of air upward is accelerated by placing a large burning lamp in the flue of the ventilator. In most school-houses, a larger ventilator — one measuring four or more square feet — will be better, and will carry off the foul air sufficiently with a slower current.

† "Experiments have been made, in a room prepared expressly for the purpose; and in the House of Commons, every day of the session, for two years; and the results show that it was rare to meet with a person who was not sensible of the deterioration of the air when supplied with less than ten cubic feet per minute."— *Wyman on Ventilation.*

For the best practicable methods of ventilation of dwelling-houses, school-rooms, and public halls, Dr. Wyman's valuable work can be advantageously consulted.

CHAPTER XIV.

Connection between Fulness of Life and Respiration. — Hybernating Animals stupid. — Man is lively or dull in Ratio of Respiration. — Consumptive Persons have less Energy of Life. — Diminution of Air and Respiration lowers Life. — Lodgers in unventilated Chambers unrefreshed in Morning.

374. *Nature has connected a fulness, buoyancy, and energy of life with the amount of respiration.* Reptiles, snakes, frogs, have a small respiratory apparatus; they breathe but little, and are dull, heavy, and inactive. They have comparatively little muscular energy, and little nervous power. As we ascend in the scale of animals, we find the correspondence between the activity of their respiratory functions and their general vital energy to be more and more manifest. Birds have more life and muscular power than other animals, and they have a fuller development of their respiratory apparatus, and breathe a freer air. Man, also, has a larger preparation for breathing, and more energy of nervous and of muscular life.

375. The hybernating animals retire to holes and caverns in winter; and there they spend the cold season in a torpid, insensible, almost lifeless state. But, in the spring, they come out with new life and activity. In the dormant state, the hedgehog breathes only four or five times, and the dormouse eight or nine times, a minute, and both with so little motion as to be scarcely perceptible. While their respiration is thus feeble, all their voluntary functions, their power of motion, and their sensations, seem entirely suspended, and their vital energies reduced to the lowest point consistent with the bare continuance of life.

376. But the warm weather of spring gives them a new life; then the lungs again expand, and work with their accustomed activity; the blood circulates freely; the old particles are taken away, and new ones supply their places, and respiration carries off the offensive matters, and the whole

animal is revived into buoyancy and energy. In these animals, while respiration is low, life is low; and, on the contrary, while respiration is active, life is in the same condition. A similar relation between the amount of respiration and the fulness and activity of life is shown in the various races of animals. "The development of their locomotive powers, and the degree of heat maintained in their systems, will be found peculiarly connected with the activity of respiration." * Those which breathe most are the most vigorous, lively, and active, while those which breathe least are the most sluggish, stupid, and feeble.

377. The same law holds good for the different individuals of any class, as well as for the various races of animals. There is a manifest connection between any man's fulness and energy of life and the development and free use of his respiratory organs. Wherever the lungs are imperfect, or air insufficiently supplied, there is a lower life, a feebler power of locomotion, less muscular energy, a duller nervous system, a more inactive brain.

378. These effects are not always noticed and referred to their true causes, yet they are none the less certain. In persons suffering from consumption, the lungs are more or less filled with tubercles and abscesses; the air-vessels are, to the same extent, closed, so that the air cannot penetrate them, and reach the blood, to purify it. These men are not well nourished, for want of pure blood, and therefore they waste away; their muscles grow thin and weak, and their buoyancy of life is extinguished. Their lungs become filled more and more as the disease progresses; and, at last, when respiration can no longer be carried on with sufficient power to effect its due purposes, they sink in death.

379. Whatever may be the cause that prevents the lungs from receiving a full and requisite quantity of air, the result is the same — a lower degree of life. Whether the chest be originally small in proportion to the size of the body, or

---

* Carpenter's Comparative Physiology.

made so by artificial means, or whether it be encased so as to prevent its natural expansion for the admission of air, there necessarily follows the same diminution of energy in the performance of the function of respiration.

380. The effect of imperfect respiration upon the blood, and upon the energy of life, is the same, whether it come from want of room in the lungs to receive the air, or from want of oxygen in it. Those who breathe impure and corrupted air, and those who live in small and ill-ventilated rooms, show the same languor and feebleness, the same want of muscular power and buoyancy of spirit, as those who are suffering from consumption, or who have deformed or diminutive chests.

381. The object of sleep is to restore the exhausted energies, and give us new life for labor in the morning. But for want of sufficiency of air, this balmy restorer often fails in some measure of fulfilling its purposes; and, in some instances, it comes very far short of it. In small and crowded chambers, the sleep is not sound and refreshing, and the sleeper awakes in the morning unrefreshed, indisposed to get up, and irresolute in regard to labor.

382. The laborers who slept in the narrow attic of the shanty (§§ 361, 362, pp. 156, 157) assured me that they awoke in the morning almost as weary as when they went to their chamber; they felt no vigor nor elasticity; they were not refreshed by their sleep; they felt a slight nausea, and a sinking about the heart, and some headache, after they rose; and they ran, as soon as possible, out of doors, to breathe the fresh air. After being in the open air a while, they recovered their comfortable feelings, and then had some appetite for their breakfast. Even then they had not the muscular vigor, nor the power for labor, which they would have had if they had been well supplied with air during their sleep. It is a mistaken economy to give laborers, or others who are expected to use their powers, such small lodging apartments. I have felt the same languor and sickness after sleeping in the cabins of boats, and have seen the passengers rush to the

deck in the morning, even in cold and stormy weather, to inhale the fresh air, and remove the oppression, and recover themselves from the weariness of their night's lodging.

CHAPTER XV.

Crowded Audiences uneasy and impatient. — Children in unventilated School-rooms uneasy and dull. — Deficiency of pure Air depreciates, and total Want of it extinguishes Life. — Breathing Carbonic Acid Gas. — Drowning. — Breathing impure Air impairs Constitution. — Consumption among Females.

383. A CROWDED audience in a lecture-room or concert-hall, after a while, become weary and uneasy, and indifferent to the lecture or the music before them, although the one may still be as interesting, and the other as exquisite, as in the beginning. Their senses grow dull; they neither understand the arguments of the speaker so readily, nor enjoy the harmonies of the music so keenly; and yet they are more impatient of mistakes and imperfections. Some complain that they never return from such assemblies without a headache. The weariness, the restlessness, the impatience, and the pain, all arise from one and the same cause — the foulness of the air. For want of oxygen, the blood is not purified; then impure blood is sent to the muscles, and cannot strengthen them to support the body; the same is sent to the brain, and irritates it, and disturbs the nervous system.

384. *After children have sat in crowded school rooms for some time they grow dull and heavy.* Their blood is not then relieved of its carbon and hydrogen: impure blood is sent back to the heart; and thence it is sent again, with all its imperfections, to the whole body. The brain, being fed with this corrupted and corrupting blood, instead of being enlivened, is made inactive and heavy. It then works languidly, or refuses to work at all. The children become uneasy, restless, and oftentimes sleepy; they are averse to

mental labor, for it is difficult for them to fix their attention upon their studies; and they are fatigued with the ineffectual attempts to learn that which at other times is easy. But the moment they are dismissed, they run eagerly from the impure air of the room to the pure atmosphere abroad, and then feel a return of life, and even a glow of exhilaration.

385. Whenever, in these and other ways, the lungs are not supplied with a sufficiency of pure air or oxygen, life is depreciated, and this depreciation is in proportion to the foulness of the air. If men dwell in rooms that are perfectly air-tight, so that no fresh air can be admitted, all the oxygen is soon consumed, and then their blood can be relieved of no more of its burden of dead atoms, and the vital powers, not being sustained, sink, and life is as effectually extinguished as it would be if they were buried in the water. In this manner, one hundred and twenty-three men died in the Black Hole of Calcutta.* The difference between the faintness and languor of a crowded room and the death in the Black Hole is a difference only in degree, but not in kind; and it is only by step after step, in the same course of corrupting atmosphere and depreciating life, that our children in the unventilated school-rooms, and our sleepers in the small chambers, and our audiences in crowded lecture-rooms, might go from the inconvenience they there feel to the death from which they shall awake no more.

386. If breathing air loaded with more than three and a half per cent. of carbonic acid gas be injurious, the breathing this gas in its pure state is destructive. This gas is heavier than the air, (§ 295, p. 134,) and therefore it falls to the bottom of a vessel or room, like water. Hence it is unsafe for a living creature to go to the bottom of wells and vats that contain it. Fire will not burn in this gas. Workmen,

---

* One hundred and forty-six persons were shut up in a room, called the *Black Hole of Calcutta*, on the night of the 20th of June, 1756. This room was eighteen feet square, and eighteen feet high, "open only by two windows, strongly barred, from which they could scarcely receive the least circulation of air." "At the dawn of day, only twenty-three persons remained alive out of one hundred and forty-six."

when they wish to enter a well or vat where they suspect its presence, first sink a lighted candle down. If it burns, there is air, and it is safe for them to descend; but, if the candle is extinguished, there is no air but carbonic acid gas; they cannot go down in safety. For want of this precaution, some have been suffocated, and even lost their lives.

387. Probably more have perished from breathing the fumes of charcoal than from breathing any other gas. A pan of coals is sometimes left burning in a small bed-room, which has no open fireplace, while some one sleeps on the bed. The gas given out falls to the floor, and fills the bottom of the room, rising as fast as it is produced, until it reaches the sleeper's head. At first, he suffers difficulty of breathing, violent pulsations of the heart, which are soon followed by a partial and almost entire suspension of the respiration and of the circulation of the blood. Then the organs of sense lose their power, the sensibility is destroyed, the prostration is extreme, and the want of power of motion so complete that the sufferer seems dead. If removed, he may possibly be restored; but, if he remains in this gas, destruction of life follows as surely as if the sleeper were overwhelmed with water.

388. Drowning produces death, not, as is commonly supposed, by filling the lungs with water, but because the water prevents the access of air to the respiratory organs, and the sufferer dies from suffocation.

389. The effects of limited respiration, and of breathing impure air, have thus far been considered only in the immediate depreciation of life, or the production of death, by the mere deficiency of pure, well-oxygenated air.* But often injurious and even fatal consequences afterward come. Some of the survivors of the Black Hole were seized with putrid fever, and subsequently died. Those who breathe charcoal gas are for some time drowsy, and are apt to fall into a deep sleep, or lethargy, from which it is difficult to rouse them; and those who live in close rooms have less mental and bodily activity, less sprightliness and energy,

less power to sustain themselves under the exposures of life, and less strength to resist the causes or the attacks of disease.

390. *Consumption is more frequent among females than among males.* The deaths from this disease in Massachusetts, during the registered years 1845 to 1863, were, of males, 32,512; females, 44,091. And in England, in the ten years ending with 1860, there were, of males, 239,305; females, 269,318.* This shows that the mortality from this cause was, in these years, in Massachusetts, 25·5 per cent., and in England, 7·5 per cent., greater among the females than among the males in proportion to the population of each sex in these countries. Dr. Farr says, "The higher mortality of English women by consumption may be ascribed partly to the in-door life which they lead, and partly to the compression, preventing the expansion of the chest, by costume. In both ways they are deprived of free draughts of vital air, and the altered blood deposits tuberculous matter with a fatal, unnatural facility." †

CHAPTER XVI.

Lower Animals can bear Privation of Air longer than higher. — Some Men, by Practice, can bear this longer than others. — All Animals need Air. — Air covers all the Earth. — Animals consume Oxygen, and give out Carbonic Acid Gas. — Vegetables use Carbonic Acid Gas, and give out Oxygen.

391. If a mouse or rabbit be placed in the exhausted receiver of an air-pump, it will die in less than a minute; and a bird, which needs more air and that more frequently, could not survive this privation more than half a minute. But the lower animals, which have less energy of life, endure this much longer. Reptiles, serpents, frogs, &c., will live a considerable time in a vacuum, or in such gases as cannot

---

* Supplement to Registrar-General's 25th Report, p. 2.
† Registrar-General's 2d Report, p. 73.

be respired; a tortoise lived twenty-four to thirty-six hours, and frogs lived near an hour when placed in oil, while insects died immediately, if placed in the same fluid. Fishes die if the water be boiled and the air excluded; yet gold fishes have lived in water thus prepared one hour and forty minutes.

392. But if a man be deprived of air, or of the power of admitting it to the chest, the circulation of his blood will generally cease within ten minutes, and his power of motion within five, often within three minutes. Yet some men, by long practice, acquire a power of suspending their breath for this period, without suffering any apparent loss of power. The divers of Ceylon are in the habit of remaining under water three, four, or even five minutes, in search of pearls; and, when they come up, they seem wearied, but not exhausted.*

393. This necessity of good air is imposed upon all the animated creation, though in an unequal degree. Yet every animal, the highest and the lowest, the man and the worm, and all intermediate grades of creatures, must sustain life by their breath. All of these, from the first to the last moment of their existence, are continually absorbing and consuming the life-giving oxygen of the air, and sending back in its stead the poisonous carbonic acid gas.

394. The air covers the whole globe, and reaches to forty-five or fifty miles from it. It is so subtile, that it penetrates the smallest crevice; and, if not excluded by other matter, it fills all space within forty-five or fifty miles of the earth. Yet, abundant as this air is, it might be feared that the respiration of so many millions of creatures, carried on for thousands of years since the world began, would consume all its oxygen, and leave nothing but nitrogen, and carbonic acid gas, and vapor, in its place.

395. To one who looks no farther into the order of nature, this, perhaps, might be a reasonable fear. But a

---

* Carpenter's Physiology, p. 393.

more thorough examination of the plans of the benevolent Author of all things, shows that there is no natural want without a due supply. And if that want be permanent, the means of gratifying it are equally so, and coëxtensive with it. The works of the Creator are all arranged in infinite wisdom. There is no deficiency — there is no want of harmony. The oxygen, which is so continually and universally consumed by the animal creation, is restored by agents equally universal and permanent.

396. *Animals and vegetables meet each other's wants, and supply each other's necessities.* The animal uses oxygen, and gives out carbonic acid gas; while, on the other hand, the plant uses carbonic acid gas, and gives out oxygen. The vegetables, like animals, breathe air; but, unlike them, they breathe it for the carbonic acid, and not for the oxygen. Through the leaves of some, which are provided with them, and through the bark of others, the carbonic acid is absorbed from the air, and then, within this vegetable respiratory apparatus, it is decomposed — the carbon is retained to nourish the plant, while the oxygen is thrown out for the use of the animated creation. Thus the equilibrium is maintained; and, as long as both live together, there need be no fear of their suffering for want of air suited to their necessities.

397. This process of respiration of vegetables is conducted only in the presence of light. In its absence, in darkness, precisely the reverse takes place, and the vegetable respiration is similar to that of animals — oxygen is absorbed, and carbonic acid given out.

398. Plants, then, as they aid animal respiration when they have the light of the sun, are proper and healthy accompaniments of any inhabited room in the daytime. But, on the other hand, as, in darkness, they consume the oxygen that animal respiration needs, they are unhealthy and injurious to be kept in rooms which are occupied in the night for sleeping or other purposes.

PART IV.

ANIMAL HEAT.

CHAPTER I.

Internal Heat of living Bodies usually greater than the Heat of surrounding dead Matter. — Whales and Porpoises in the Arctic Ocean as warm as at the Equator. — Man's Heat does not vary in Extremes of Temperature. — Blagden's Experiment. — Natural Tendency to Equilibrium of Heat in all dead Matter. — Living Matter sustains its own Heat.

399. It is easy to see that the temperature of most animals is higher than the surrounding medium. Our own bodies are usually warmer than the air about us. In winter, especially, when water freezes and the air is colder than ice, this fact is to be noticed. If we then lay our hands upon the body of a horse or a kitten, or upon our own flesh, we find them to be warmer than the air. If we take ice into our hands, it melts, from the natural heat of our flesh; and yet this flesh is not cooled down to the coldness of ice, and, although it loses a little heat while it is in contact with the ice, it soon recovers it after the ice is taken away.

400. The porpoise and the whale dwell under the ice, in the waters of the Northern Ocean. Above them, the air may be cooled down to 50° below freezing point; the temperature of the ice is at least as low as 32°, and the water is nearly as cold, and yet they are warm. Their temperature is sustained at about 100°, as high as that of other animals of the same kind, in the burning regions of the equator.

401. Man dwells in all climates; he finds a home in

every country, from the equator almost to the poles. Under the equator, the temperature of the atmosphere is elevated to 100°, bodies exposed to the sun are heated to 130°, and the inhabitants are there subjected to a perpetual heat. In the northern regions, Captain Parry found the thermometer as low as 55° below zero; and Captain Back found it 15° lower than this, or 70° below zero. These were 87° and 102° below freezing point. And yet, in these extremes of external temperature, the internal heat of the human body varies very little. There are greater differences than even this. In France, some bakers entered their ovens heated up to 278°, or 66° warmer than boiling water, without increasing their own heat. And some philosophers of London tried the experiment, to ascertain how great heat could be borne without injuring or increasing the temperature of the living body.

402. Sir Charles Blagden entered a room, prepared for the purpose, in which the thermometer stood at 260°; and there he staid for eight minutes. Eggs were put into the same room, and were soon roasted quite hard. "Beefsteak was not only dressed, but almost dry." And yet here, in this great heat, in which water boiled and meat was cooked, the thermometer, when placed under the tongue, was raised only to 100°, two degrees above the usual standard. There have been many other experiments and observations of this kind, which show the same principle — that the heat of the living body does not change, or changes very slightly, with the temperature of the air or water which surrounds it. A dyer will hold his hands in water at the temperature of 130°, and the ice-cutter has his hands in contact with ice at 32°; and, in both instances, the temperature of the body is about the same, neither raised in one case, nor depressed in the other, materially.

403. There is a natural and almost universal tendency to equilibrium of heat. When a warm and a cold dead body are brought in contact, their heat is shared in common between them. One loses, and the other gains, heat, so that in a

short period, they have equal temperatures; neither is warmer or colder than the other.

404. If a piece of wood or of dead flesh be put into hot, or even boiling water, it soon is as warm as the fluid. If it be put into cold water or snow, it soon becomes as cold as that. If ice be put into hot water, it receives a part of the heat of the fluid. It first melts, and its water is then warmed up to the temperature of the original water; while this, losing its heat, is cooled down to the temperature of the water from the ice, and, finally, both have the same degree of heat. The same effect is seen when any substances are placed in air of different temperature. When the atmosphere is at $32°$, water freezes, and solids become as cold as ice. On the other hand, water boiled, and the eggs and the beef were heated up to the temperature of the room which Sir Charles Blagden entered.

405. But it is not so with living beings. Their temperature does not follow that of the surrounding and contiguous objects. The temperature of the warm-blooded animals,—of man, horses, and birds, for instance,—scarcely varies with any extremes of cold or heat to which they may be exposed. The usual temperature of man is $98°$. If a thermometer be placed in his mouth, in the East Indies or in the arctic regions, it will be found the same. The body sustains its own temperature in the cold medium, and is no warmer in the heated room.

406. If this were not so, if the temperature of our bodies should follow that of the surrounding medium, the most fatal consequences would ensue. The blood and the flesh would be frozen, and all our motions stayed, and life extinguished, in the severe weather of winter, even in the temperate climates; and, on the other hand, the fatty portions of our frame would sometimes, in the tropical climates, melt, and the blood would boil in such experiments as Blagden tried.

CHAPTER II.

Law of Equilibrium of Heat different in Regard to living and dead Matter. — Animals maintain their own Temperature, and give Heat to other Bodies. — Animal Heat generated within. — Warm and cold blooded Animals. — Power of sustaining Heat varies with respiratory Apparatus. — Fishes breathe by Gills, and have little Heat. — Whales breathe by Lungs, and have much Heat. — Animals have internal Apparatus for generating Heat.

407. BESIDE this maintenance of its own warmth, the living animal body is continually giving out heat to other substances which are cooler than itself; and yet it does not apparently lose its own heat; at least its temperature remains undiminished. If we hold a piece of ice in our hand, it is melted, but the hand is not much cooled; or, if cooled, it soon regains its heat after being separated from the ice. But, if we place the ice upon a piece of iron heated to the temperature of the hand, the ice melts there, as in the other case, and the iron is cooled down to a lower temperature, and the water of the ice is raised to the same degree. The ice cools the iron, and the iron warms the water of the ice; and then the temperature of both remains the same, until some external influence changes it.

408. Here, then, is a manifest difference in the law that governs the heat of living and that of dead substances. One class seems to have heat only in common with contiguous and surrounding objects. If they are warm, the dead matter becomes warm; if they are cold, this is cooled to the same degree. It neither warms itself nor cools itself, but depends upon others for its heat. But the living body is neither cooled nor heated materially by surrounding matters. Its own heat seems to be independent of them.

409. Our heat is not borrowed from external objects; certainly not from the atmosphere, for we have seen that the human bodies are warm when the air is extremely cold; nor from the sun or fire, for we are warm in the absence of

both. Nor is our heat derived from clothing, for this has no active power of giving heat — it has no warmth in itself; it only tends to prevent changes of temperature. If we wrap a piece of dead flesh in flannel, it is not warmed; it remains the same as before. If, in the winter, this flesh be heated by fire, the flannel wrapped about it keeps it warm. If, in summer, we put ice in flannel, it prevents the melting. Clothing, then, only prevents the passage of heat. It keeps a warm body warm, and a cool body cool; but it creates and gives no heat. If, then, animal heat is not given from without, it must originate within the body. There must be some internal means or apparatus by which we and other living beings create and sustain our temperature.

410. There are two grand classes of animals, divided according to their temperature. One is called the *warm-blooded*, and includes man, birds, quadrupeds, &c. Their heat is ever of the same degree, and does not vary with the temperature of the atmosphere or the water in which they live. The other class is called *cold-blooded*, and includes snakes, oysters, fishes, worms, toads, turtles, &c. Their heat is but little higher than that of the medium in which they live. The earth-worm, leech, and shell-fish are usually $1\frac{1}{2}°$ warmer than the air, or earth, or water which surrounds them. Fishes are $2°$ to $5°$ warmer than the water. Reptiles, frogs, lizards, have a still higher heat relative to the air or water, yet not so high and permanent as that of the warm-blooded animals.

411. There is, in these two great classes, a great difference of power of maintaining their own heat. Man maintains his usual temperature in the midst of air varying $320°$ from extreme heat to extreme cold; and therefore he may be at least $160°$ warmer, or $160°$ cooler, than the surrounding medium; while a fish is only $2°$ or $3°$ warmer or cooler than the water in which it lives. It is natural, then, to ask, What is the difference in the structure of these classes, from which arises this difference of internal heat? On examination, we find that the principal difference that runs through

the whole of these classes is in the apparatus of respiration. The warm-blooded animals breathe more and purer air than the cold-blooded. Fishes breathe only by gills, and receive only the little air that is in the water, and they are cold; we breathe with full lungs, and receive a more plentiful supply of air, and are heated to 98°; while birds have the largest means of respiration, and breathe the purest air, and are consequently from 2° to 13° warmer than even man. Insects have generally larger means of respiration, and a higher temperature.

412. There is a remarkable difference, in this respect, among the inhabitants of the sea. Fishes — such as the pike, cod, haddock, sturgeon, smelt, &c. — which breathe by gills are dependent solely upon the air in the water. They can therefore neither obtain nor consume more than a very small portion of air, and consequently they are cold. On the contrary, whales, porpoises, and dolphins breathe by lungs. They rise to the surface of the water, and inhale the free air above it. They find this abundant, and consume it plentifully, and consequently their temperature is about 100°, and independent of the heat of the water. It is neither depressed in winter nor raised in summer. They are therefore classed with the warm-blooded animals.

413. In order to maintain this heat within the animal body, constantly and independently of the influence of surrounding and contiguous matters, two conditions are necessary: 1st, each animal must possess some internal apparatus for generating or creating this heat; 2d, the skin, or the external covering, must be endowed with such a power of regulating the transmission of heat, that it may prevent its too rapid passage out in winter, or when the air is colder than the body, and also its passage into the body in summer, or when the air is warmer than the body.

CHAPTER III.

Latent and sensible Heat. — Heat applied to Ice forms Water, and to Water, makes Steam.

414. THE warm-blooded animals breathe more than the cold-blooded. The same difference prevails among the subdivisions of these classes, for the warmest kinds breathe more than the coldest. It would seem, then, that the internal heat arises out of, or is in some way connected with, respiration. And this we find to be strictly true, upon examination of the nature and properties of the elements of air, and of the chemical effects produced by this gas upon the blood and particles of the animal body.

415. When ice is melted and changed to water, it is easy to see that heat is given to it, and absorbed by it. The heat necessary to produce this change has united with the ice, and both together have become water. Again, if much more heat is applied to this water, it boils and is changed to vapor or steam. By continuance of the same process that produced the first change, the second one is produced; and, by the union of heat with water, steam is formed. It is obvious that steam contains more heat than water, and water more heat than ice. Heat added to ice produces water, and heat added to water produces steam; and, in both cases, most of the heat becomes latent or hidden in the new substance. If, now, we reverse the process, and return the steam back to water, heat must be given out. Just so much is given out as was originally required to convert the water into steam. If we continue this process further, and change the water to ice, there must be a further discharge of heat; and as much heat will be given out, in this process of freezing the water, as was before required to melt the ice.

416. There is a general law of matter, that rare or light substances require more heat than dense or heavy matters. Liquids commonly require more heat than solids, and gases more than liquids, to keep them in their respective states;

and abstracting heat renders them more solid, while adding heat renders them more fluid. By a great reduction of temperature, airs or gases can be condensed to fluids; and, during this change, heat is given out.

417. The apparent heat of a body, as measured by the thermometer, or as perceived by the touch, is not always an exact measurement of the quantity of heat in that body. If, for instance, we mix a pound of water heated to 100°, with a pound of spermaceti oil at 50°, it might be supposed that the temperature of the mixture would be 75°, the exact medium between them. This would be the case, if each of these bodies were raised to the same temperature by the same quantity of heat. But the temperature of the mixture is actually $83\frac{1}{3}$°. If, again, the experiment be reversed, and water at 50°, and oil at 100°, be mixed, the result is a temperature of $66\frac{2}{3}$°. Instead of the warmest substance losing 25°, and the coolest gaining as much heat, we find that, in the first instance, the water loses only $16\frac{2}{3}$°, while the oil gains $33\frac{1}{3}$°; and, in the other case, the oil loses $33\frac{1}{3}$°, and the water gains $16\frac{2}{3}$°. That is, the quantity of heat that will warm water $16\frac{2}{3}$° will warm oil $33\frac{1}{3}$°; or, the water requires twice as much heat as oil does, to raise it to any definite temperature. It will, then, be clearly understood, that the same substance has different quantities of heat in its different states; and also that one substance requires more heat than another to give it the same apparent heat.

418. The burning of wood and all other fuel shows both of these principles. Oxygen exists in the air in the state of gas. (§ 292, p. 133.) When wood or coal is heated, this oxygen combines with the carbon of the fuel, and forms carbonic acid. In this process, the oxygen enters into a new state, and becomes a part of a compound more dense than it was before. On two accounts it loses heat; 1st, oxygen has greater capacity for, or holds more, heat than carbonic acid gas, and therefore, when this new gas is formed, the surplus heat, or that excess of heat which oxygen can hold over that which the other gas can hold, must be given out; 2d, the oxygen is in a denser state when it

composes a part of the carbonic acid gas than when it is pure and uncombined, and therefore holds less heat, (§ 416, p. 177,) and must give out some when it enters the compound. From both of these causes, heat is derived from fire of every kind. Whatever may be the theory or explanation, the fact is evident, that heat is evolved from the union of oxygen with fuel — carbon or hydrogen. This is what we call *combustion* or *fire*. The amount of heat thrown out from this union or combustion is always in proportion to the amount of material consumed. A pound of wood in a solid block gives out the same quantity of heat, in burning slowly, as a pound of shavings of the same wood, in burning rapidly.

419. Upon these principles, it will now be easy to understand how the warmth of the animal body is obtained. The particles of our flesh are continually changing. (§§ 242—244, pp. 111, 112.) The old ones are going away, and new ones taking their places. The principal components of these old particles are carbon, nitrogen, and hydrogen. (§ 290, p. 132.) When the air is received into the lungs, and brought into contact with the old and venous blood, it is decomposed, or divided into its two elements; the oxygen is separated from the nitrogen, and united with the blood. (§ 297, p. 135.) The blood, at the same time, throws out into the air carbonic acid gas and vapor, (§§ 297, 303, pp. 135, 137;) and then, being relieved of these impurities, it is returned to the heart, and thence it is circulated throughout the body, carrying the newly-acquired oxygen with it.

420. It was once generally believed, by chemical physiologists, that the oxygen does not enter the blood, but that these dead particles are brought unchanged to the lungs, and there the carbon and the hydrogen, meeting the oxygen, combine with it, and form carbonic acid gas and water, which are given out with the returning air.

421. But it is now, with better reason, supposed, that the oxygen of the air enters the minute arteries in the lungs, and is there mingled with the blood. It is then carried with this blood to the heart, and thence sent through the arteries all over the body. When this blood, and the oxygen

which it carries along with it, reach the minute arteries and the capillaries, where the work of nutrition is carried on, the interchange of the old for the new particles of flesh takes place. The old — those which have finished their work and are dead — give way, and the new ones, fresh with living vigor, take their stations and perform their part in the work of life. As these old and dead atoms of flesh pass from their stations to the vessels, the oxygen in the blood meets them, and they unite together and form new compounds. The union of the oxygen with the carbon produces carbonic acid, and its union with hydrogen produces water. These unions take place in the same manner, and the same results follow, as when the carbon of the wood in the fireplace, and the hydrogen of the gas lamp, unite with oxygen. The carbon and the hydrogen are burned, fire is produced, and heat is evolved; the carbonic acid and the water are then carried through the veins to the heart and the lungs, and the heat is left in the textures of the living body. During this process, precisely the same amount of heat is given out from this internal fire — this slow combustion of the wasted particles of flesh — as would result from the combustion of the same amount of fuel, carbon, and hydrogen elsewhere.

CHAPTER IV.

Exercise increases Combustion of Carbon and Evolution of Heat. — Whatever increases Flow of Blood increases Heat, and whatever diminishes Flow of Blood lessens Heat. — Oxygen and Fuel necessary to support internal Fire. — Whatever interrupts Supply of Oxygen or Air to Lungs, prevents Development of Heat. — Tight-Lacing lessens Heat. — Bad Air — foul Air — lessen it. — Food supplies Fuel. — Well-fed warmer than the Ill-fed. — Alcohol does not increase the Heat of the Body. — Meat supplies more Fuel than Bread. — More Meat eaten in cold than in warm Weather.

422. This work of interchange of particles, and of burning the old flesh, is carried on throughout the whole body;

consequently, every part of the body is warmed. The more rapid is the circulation and the more frequent are the changes of living for dead particles, the more carbon and hydrogen are burnt, and the greater is the heat given out. Whatever increases the interchange of particles, the work of absorption and nutrition, and consequently the flow of blood, increases the internal fire and the evolution of heat. Motion is attended with greater waste of particles, and, of course, with greater absorption of carbon and hydrogen, (§ 127, p. 61,) and greater development of heat. Labor, therefore, warms the body, and, if violent, may heat it uncomfortably. The watchman keeps himself warm with exercise, and the passenger leaves the vehicle to warm his feet with running. The hardy laborer heats himself with his exercise, and sits down quietly to cool his body.

423. On the contrary, whatever interrupts the circulation and the interchange of particles prevents the development of heat. If we bind up the arm or finger with a tight cord, and prevent the flow of the blood through it, the limb becomes cold. It is a common and a true observation, among shoemakers, that a loose boot is warmer in winter than a tight one, because the latter presses upon the blood-vessels, and interrupts the full flow of blood. So we find, if one side or one limb be palsied, that side or that limb becomes cold, for the same reason.

424. This animal heat then, is sustained by the combustion of the dead atoms of the flesh in all the parts of our frames where the blood circulates. In order to maintain this combustion, the same things are requisite that are needed to support fire elsewhere; these are fuel and air. To deprive the body of either would be as fatal to its internal heat as taking away the fuel or the air would be to the fire of the stove. As the wood in the fireplace burns by aid of the oxygen which it derives from the air, and as this fire burns freely in proportion to the quantity of air which it receives, so the internal fire of the animal body, deriving its oxygen from the air, burns in proportion to the fulness of its

supply. Consequently, whatever impedes the flow of air into the lungs, and its access to the blood, must so far prevent the development of internal heat, as certainly as any interruption of the draught or diminution of air would lessen the fire and the heat of the fireplace.

425. Whatever, then, restricts the expansion of the chest, or limits the capacity of the lungs for the admission of air, — any pressure of clothing without that prevents the motions of the ribs or the diaphragm, or any disease of the lungs that closes the air-cells within, — any of these obstructions, by lessening the amount of oxygen that the blood receives, diminishes the combustion of the atoms of dead flesh and the evolution of the internal heat, as certainly as shutting the draught of a stove would lower or extinguish its fire. For this reason, asthmatic persons, and those whose air-cells are partially closed with disease, are only partially warmed, and cannot endure so severe a cold as men in health. A poor woman, whom I saw sick with consumption in a very cold room, was frozen to death, one night, in her bed, in the winter of 1830, while some other women, who slept in the same room, and under the same quantity of clothing, awoke in vigor, though suffering with cold.

426. Nothing but oxygen can support this internal combustion of fuel. We must not only receive a sufficiency of air into the lungs, but that air must contain its due proportion of this gas. If, then, the air contains less than the due quantity, if it has been breathed over, and its oxygen has been consumed, or if, in consequence of mixture with other gases, the twenty inches which we inhale contain less than twenty per cent. of oxygen, then the internal combustion is impeded, heat is sparingly evolved, and those who breathe this impure or weakened air are comparatively cool. After a crowd has been long in session in a close hall, or children in an unventilated school-room, in winter, they begin to complain of being cold. Notwithstanding the fire may glow in the stove, and the thermometer indicate no reduction of temperature, still, for want of oxygen in the vitiated air, the

fire burns languidly in the bodies of the people, they are not well warmed, and their sensations persuade them that the room is growing cooler.

427. The narrow-chested are colder than the broad-chested; the tight-bound, than the loosely-dressed; and those who breathe the impure air of close and unventilated rooms, than those who breathe the free air of the fields. The former need more external protection of houses or clothing, or more outward heat from fires, than the others. A free expansion of the chest, with a good supply of pure air, is, therefore, an economy of clothing and of fuel. When the air is dense and heavy, as when cooled, it contains a greater weight of oxygen to the cubic inch, than when it is rare and light, as when heated. We therefore breathe more oxygen in winter than in summer, and the fire, consequently, burns most actively when it is the most needed.

428. Fuel, as well as air, is necessary to keep up this internal combustion in the animal body. This is supplied by the atoms of wasted flesh that have died in the various textures, and are ready to be burned or combine with oxygen, and need to be carried away. The combustible matters of the flesh — its carbon and hydrogen — are originally supplied by the food that contains the same materials. These elements of the food, being converted first into chyle, then into blood, and next into flesh, are at last burnt by their union with oxygen, and carried out through the veins and the lungs. As the food is the only source from which this fuel is supplied, of course, all other things being equal, the internal fire must burn, and the body be warmed, in proportion to the amount of carbon and hydrogen which is eaten, and incorporated into and becomes a part of the tissues.

429. The well-fed and well-nourished, — those who live upon good and generous food, — having a better supply of fuel, are therefore better warmed than the hungry, or those who live upon a poor and meagre diet. The traveller who has been long exposed to the severe weather without eating, in winter, complains that he is both hungry and cold. His

hunger and his low temperature may seem to him to be merely coincident circumstances, accidentally coming together: but, in truth, one is the cause of the other. For want of food, his body is not supplied with fuel, and its internal fire burns feebly, and therefore does not warm him. One of the best means of protection against the effects of exposure to the cold air of winter is proper nutriment.

430. Alcoholic spirit is sometimes taken for this purpose, but with a mistaken view of its effects upon the heat of the body. It stimulates the stomach, excites the nervous system, and quickens the action of the heart, and the flow of the blood. It supplies to the flame carbon and hydrogen, the most combustible of materials; but these soon burn out, and their fire is then exhausted, and the body is afterward cooler than it otherwise would have been. Food, alone, can sustain a permanent fire. Two travellers met, in a very cold day of January, 1810, at a tavern in Groton, Massachusetts. One of them called for a mug of hot flip, and advised the other to do the same; for, he said, "When I am going out in the cold, I always drink hot spirit." The other refused, but said, "When I am going out in the cold, I eat a good dinner." The temperate traveller acted from his own experience, and also, without knowing it, upon the truest physiological principles.

431. Flesh, containing more carbon and hydrogen, supplies more fuel to the fire than vegetable matter. Meat, therefore, warms a man more than bread, and we eat it more freely in the winter than in the summer. For this reason, the coachman, the sailor, and the teamster, who are exposed to the coldest air abroad, need more meat than the mechanics, who work in warm shops, or those persons whose life and occupations are in warm houses. In the northern regions, where winter reigns with great severity, there is a more rapid loss of heat through the skin, and of course a necessity of creating more within the body, than in the warmer regions, at and near the equator. To keep the body warm, there must be more fuel, or food containing more carbon and hydrogen, in the cold than in the hot climate. Nature sup-

plies this necessity by the difference of food, and of digestion, of the inhabitants of these diverse regions. The principal diet of the people within the torrid zone is of vegetable origin, while the inhabitant of the frigid zone lives mostly upon flesh; and the people who inhabit the countries in the temperate zones, between these, have a mixed diet, in which the meat predominates as they approach the arctic circle, and the vegetable predominates towards the tropics.

CHAPTER V.

Other Influences may affect Supply of Heat. — Some Diseases increase, some diminish it. — Fatigue and Exhaustion lessen Evolution of Heat. — Infants and old Men have less Heat. — Less Heat evolved in Sleep. — Carbon consumed and Heat evolved in a Day. — Heat must be carried out of the Body through the Skin. — Evaporation of Perspiration carries off Heat. — Greater Internal Fire in cold than in warm Climates. — Winter and Summer Constitution. — Animals cool more rapidly in Summer than Winter at same Temperature.

432. This chemical explanation of the origin of animal heat is shown at length in Liebig's Animal Chemistry. There are doubtless other influences that affect the development of internal heat, beside the supply of carbon, hydrogen, and pure air. It is ascertained by the later physiological chemists that this process of combustion or combination of oxygen with carbon and hydrogen accounts for only a part of the heat that is developed in the animal body. "Animal heat is a phenomenon which results from the simultaneous activity of many different processes, taking place in many different organs, and dependent, undoubtedly, on different chemical changes in each one." *

433. Even when the body is well supplied with both good food and pure air, there is not an equal development of heat in all states of the system. In some diseases, such as fever,

* Dalton, Human Physiology, p. 263.

inflammations, &c., the heart beats quicker, and the flow of blood is more rapid than natural, and there is a greater production of heat. But in some other diseases, as asthma, cholera, &c., there is, on the contrary, a greater coldness The internal warmth is affected by the condition of the nervous system, by excitements and depressions, by the emotions, the passions, and the states of mind. One is burning with anger or with love. The exciting and the ardent passions quicken the flow of the blood, and increase the internal heat, while the depressing passions diminish it. Cheerfulness and merriment promote the evolution of heat, while fear, sorrow, and despondency impede it.

434. Fatigue, exhaustion, hunger, night-watching, sleeplessness, indigestion, or any thing that depresses the system and diminishes the energies of life, lessens the production of heat, and the power of resisting cold. In this condition, one cannot bear exposure to a low temperature as well as when he is fresh and vigorous. He is then more liable to take cold. Visiting a friend, a public officer, in the afternoon of a pleasant day of March, I found him shivering over a fire, though otherwise in good health. He said that he had been out to walk, and was chilled. It was a warm day, and other men complained of the heat; but they were vigorous, for they had been refreshed by their night's sleep; but he, having an important report to finish, had sat up, and labored upon it with all his mental energy, until two o'clock in the morning; then, being exhausted, he retired, but awoke in the morning still fatigued and unrefreshed; consequently, he had not sufficient power to maintain his proper heat, even in a temperature which was comfortably warm to men in the enjoyment of their usual vigor.

435. In the different periods of life, there is a difference of power of producing internal heat. It is more feeble in infancy and in old age than in the vigorous years of youth and manhood. Dr. Edwards exposed some young and old sparrows to a temperature of 64° with the same amount of protection. At the end of a definite period, the young were cooled down to 66°, while the older birds maintained their

temperature at 102°. Full-grown magpies lost 5° of heat in the same atmosphere, and in the same time that young birds of the same species lost 25° of heat. The same law applies to children and men. Infants and old men cannot, therefore, endure the cold so well as men of middle life, and need more careful protection of clothing when exposed.

436. "The state of natural sleep is in general accompanied by a diminution of the power of producing heat." The body is then more susceptible of the influence of cold. Thus the consumptive woman ($\S$ 425, p. 182) was frozen during her sleep. Night travellers are in much more danger of suffering from the cold if they allow themselves to sleep than if they keep awake.

437. The amount of heat given out from the combustion of a definite quantity of carbon or hydrogen, or the union of either of these with oxygen, has been determined by experiments. It is found, also, to be the same wherever this combustion takes place, whether in or out of the living body, and whether it happens rapidly and with a flame, as in the fire of a furnace, or slowly, atom by atom, as in the textures of the animal body. If, then, we can ascertain the amount of these elements which are consumed in the living system, and in any given time, we can determine the amount of heat which will be then evolved.

438. In the course of twenty-four hours, there are, on an average, 13.9 ounces of carbon converted into carbonic acid gas, and given out from the lungs of every adult in good health. Every ounce of carbon, during the process of combustion, evolves as much heat as would raise 78.15 ounces, or almost five pounds, of water, at 32°, or the temperature of ice, to 212°, or the boiling point; and, consequently, the 13.9 ounces of carbon, which are consumed in the human body daily, must give out heat enough to raise 67.9 pounds of water from 32° to boiling heat. So much heat from this cause is, then, generated in the body of a person of the average size and in good health, in each day.*

* Liebig's Animal Chemistry, Part I. $\S$ V.

The amount of hydrogen consumed is not so easily determined; but it is supposed by learned chemists that a great proportion of the animal heat is given out by it.

439. If so much heat be daily added to the body, the same amount must be carried off in some way, otherwise it will warm the body too much, and cause distress. But it does not increase; when the body is at its usual temperature, it does not become any warmer, although so much heat is continually added to it. This quantity must, then, find its way out, through the surface and through the passages. Whatever goes from the body, carries some of its heat. However cold may be the air which we inhale, it becomes warm within the lungs, and is then exhaled at the temperature of the body.

440. The skin is the main avenue of the heat outward; and through this it is continually passing away, both winter and summer. When the air is considerably colder than the body, it is very plain to every one that heat goes off by transpiration through the outward surface, in order to maintain an equilibrium with the surrounding atmosphere; and thus the internal temperature does not rise. But when the air is as warm as, or even warmer than, the body, this transit of heat outward is not so manifest, yet it is equally certain.

441. The skin not only affords a passage-way for the heat to go out, as through any dead substance, but it has an active power to furnish the means of carrying off the surplus heat, when it would otherwise accumulate in the body. The skin is constantly preparing and throwing the perspiration upon its surface, where it is usually converted into vapor and absorbed by the atmosphere. This change of the perspiration from a fluid to a gaseous form — from water to vapor — is effected by the addition of heat, (§ 416, p. 177,) which is absorbed from the body, and therefore cools it. This perspiration is most abundant in warm weather, when the air can absorb the most, and causes the greatest cooling when it is most needed. It is a common, and by no means an unfounded notion, that one is cooled and refreshed, in summer,

by drinking a cup of hot tea. The tea excites the perspiration, which creates the necessity of evaporation; and this is done very much at the expense of the heat of the body, which is thereby cooled. All the insensible, and most of the sensible perspiration is converted, on the skin, into vapor; and, by this conversion of liquid into vapor, a large portion of the excess of heat is carried out of the body, and the standard of the internal temperature is preserved.

442. This evaporation was very rapid in Sir C. Blagden's experiment, (§ 402, p. 172;) consequently, the temperature of his body was kept down to about its usual standard, which was 162° below that of the surrounding air which he breathed. He received dry air into his lungs heated to 260°, but when it went out, it was cooled down nearly to 98°; and, when he breathed this air upon his skin, it felt cold, instead of warm, as it usually does.

443. By the beautiful adaptation of Nature's supply to her wants, the animal body is kept cool in the summer and warm in the winter. The greater appetite and greater desire for animal food, supply more carbon and hydrogen, and the density of the air supplies more oxygen, and consequently a greater fire is maintained, in the cold season, and in cold climates, than in warm seasons, and in hot climates, when and where the appetite craves, and the stomach digests, vegetable diet, which gives less fuel, and the atmosphere affords less oxygen for the support of the combustion.

444. It is this greater supply of internal heat, and the lesser cooling by evaporation from the surface, that give us what is called the *winter constitution;* and the diminished internal fire, and increased evaporation of the perspired fluids, give us the *summer constitution.* By these means, the body is able to endure a greater degree of cold in the winter, and in the climate of the polar regions, than in the summer, and in the tropical countries; and we can bear a greater degree of heat in the summer and in warm climates, than in the winter and in cold climates. That degree of cold which we bear without discomfort in January,

is almost intolerable in August. We are sometimes nearly overcome, at least languid, with the heat of a thawy day of February, when the thermometer is no higher than 40°. But we are chilled with the air of the same temperature in July. For this reason, we need to have our sitting-rooms somewhat warmer in the summer than in the winter. An ice-house is a sufficiently warm and comfortable place for a man to work in while storing ice in the winter, but it is chilly and often dangerous to those who enter it in the summer to take ice away.

445. To demonstrate how much more rapidly the heat passes away, and how much less power of resistance to cold the animal body possesses, when it is under the influence of its summer, than when under the winter constitution, Dr. Edwards, of Paris, took several sparrows from their warm rooms, in the month of February, and put them in a cage surrounded by ice, where the temperature was, at the highest, 32°; after remaining there three hours, they had cooled less than 2°. He tried the same experiment in the month of July, with the same conditions and in the same time; the sparrows lost 21° of heat.*

446. We gradually pass from the intensity of summer's heat through the autumn to the severity of winter's cold, and back again through the spring; and as each of these opposite seasons comes upon us, we receive the constitution adapted to it, and endure the extremes of temperature without suffering. But we cannot leap from the one to the other with impunity. A resident of Massachusetts would be enervated by suddenly arriving in the West Indies in the winter; and if, after residing under the equator for a season, he should as suddenly return to Boston in January, he would suffer from the cold.

447. The dwellers in warm houses, and the workmen in warm shops, retain the summer constitution through the winter more than the out-of-door laborers, and cannot bear

---

* Influence of Physical Agents on Life, Part III. Chap. III.

cold as well as they do without suffering, and therefore need more clothing when exposed to the same temperature. A shoemaker or student, going from his warm shop or room and taking the outside seat of the stage-coach, by the side of the driver, in winter, must wear thicker garments than his companion who is daily exposed to the weather; if he does not do so, he will suffer more than the coachman. Those who live in houses heated by furnaces, in which all the entries and rooms are more or less warmed, and who seldom go abroad, hardly receive the winter constitution in the proper season, and cannot bear exposure to the open air without much additional clothing.

PART V.
THE SKIN.

CHAPTER I.

The internal Structure needs Protection. — Skin. — Cuticle; thickened by Friction if gradually applied. — Blisters. — Corns.

448. The inner framework and vital machinery of animals — their lungs, heart, and blood-vessels — their muscles, nerves, and digestive apparatus — are all very delicate, and would ill bear exposure to the action of the elements, or even the contact with other bodies. They are, therefore, protected with some outward covering, which is different in different animals. Yet, in all, it stands between these organs of life and the external world. In man, and in many other animals, this outward covering is the skin, which is a soft and pliable, and yet a strong membrane, that is not easily injured or torn, does not suffer from contact with other substances, and will bear wide variations of heat and cold.

449. *The outer skin* (Fig. XXVI. *a, a*) *or cuticle, covers the body.* It is lifeless and insensible. The hangnails of the fingers, the peeling of the lips when we have a cold, are parts of this skin. If we pinch them, we do not feel it. So, also, we may run a pin through this skin at the corners of the fingers or thumb, or trim the thickened skin of the heel, and suffer no pain. It does not ache with the cold nor suffer with the heat. It has no nerves to feel nor blood-vessels to give it life.

450. The cuticle, sometimes called the *scarf-skin*, is formed by, and grows from, the true skin beneath it, and is constantly casting off its surface in the form of powdery scales. But it is as constantly renewed. This process of change never ceases in health. Sometimes this outer layer peels off from the lips in case of a cold, or from the roots of the nails; but soon another takes its place. When it is peeled off, it leaves the true and sensitive skin bare and tender. But, when it is cast off naturally in dead scales, it leaves a layer behind, which protects the more delicate parts beneath.

451. Over the whole of the child, and on the parts of the

Fig. XXVI. *Skin and perspiratory Apparatus highly magnified.*

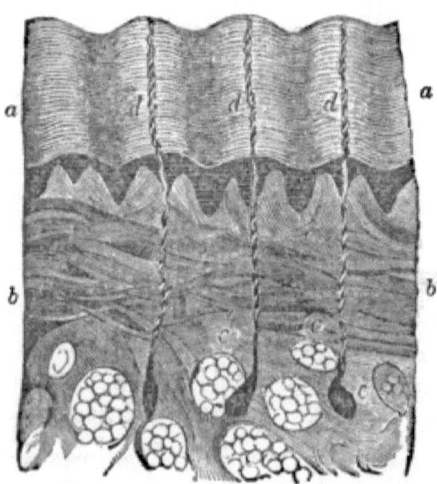

a, a, Cuticle.
b, b, True skin.

c, c, c, Perspiratory glands.
d, d, d, Perspiratory tubes.

adult which are not exposed to contact with other bodies, and especially on the lips, this cuticle is thin and delicate. But, whenever it is exposed to the elements or friction, it becomes thicker and tougher; for any friction, if moderately applied, instead of wearing it out, causes it to grow more and more; the under skin throws out more of the matter that forms the cuticle, and this latter is thickened and strengthened. This is most observable in the sole of the foot and palm of the hand; the more they are used, the thicker and harder their cuticle becomes, so that the bare-foot boy treads on the rough pavement without injury.

452. Though the cuticle becomes thick and hard from friction and labor, yet these must be applied cautiously and gradually, otherwise the reverse will happen. When the student or clerk undertakes to cut wood, or rake hay, or row a boat, for several successive hours, the cuticle of his palms, instead of growing thick and hard, becomes thin and sore. The outer skin may be worn off, or it may separate from the other, and the under skin, instead of throwing out more matter to be formed into cuticle, throws out a watery matter, which fills a little sack between them, and forms a blister.

453. But if this friction had been applied gradually, and continued for a long time, it would have stimulated the inner skin to form more and more of the outer or scarf-skin, to meet the want and the pressure, instead of throwing out water, and causing pain and soreness.

454. By the gradual application of friction, the skin becomes so fortified with this thickened outer layer, that it will bear very rough usage without suffering; so that the hands of the stone-layer and of the mason are neither scratched nor inflamed by the rough stones, nor irritated by the lime in the mortar. In the same way, the hands of the blacksmith and the founder become accustomed and prepared to handle very hot and rough metals without being burned or suffering pain.

455. But if one unused to labor with his hands attempts at once to become a stone-layer or brick-mason, he would soon find the tender skin of his hands blistered and torn. The

new apprentice in a blacksmith shop or a foundery burns his hands in doing the very work which the older workmen do without any suffering.

456. When the feet are pinched by new and very tight shoes, painful pressure is made upon the skin, and sometimes blisters are raised in walking. But, if this pressure be more gently and gradually made, and long continued, the cuticle becomes thickened on the prominent joints of some of the toes, by the formation of new underlayers. These layers are broad at the top and narrow at the bottom, and the whole thickening is somewhat wedge-shaped or conical, with its point inward. This is a *corn;* and the shoe, bearing upon it, presses upon the tender flesh beneath, often producing acute distress.

CHAPTER II.

Cuticle defends true Skin from external Injury. — Nails, Hoof, and Horn. — Seat of Color. — True Skin has many Blood-Vessels and Nerves.

457. The cuticle, placed between external objects and the true skin, protects it from their contact. It bears their hard usage, but suffers no pain. By means of this protection, we are enabled to handle, not only rough and hard substances, but many matters which would be poisonous to the more delicate skin beneath. The dyer or the chemist holds his hands, if the outer skin is unbroken, in strong mixtures, without pain or irritation; but if the cuticle is broken and the inner skin bare, great pain and sometimes disease are the consequence. Physicians often examine the bodies of those who have died of putrid diseases, and, if the scarf-skin of the operator is entire, no bad consequence follows; but, if there be the least cut or scratch of this cuticle, through which the poison can gain access to the under skin, very severe disorder, and sometimes death, ensue. Some, who thought themselves

safe because they had no perceptible wound, and therefore exposed themselves to very virulent poison, have been infected by the poison's insinuating itself through the very slight rupture of the cuticle on the end of a finger, where a mere hangnail had been raised.

458. Other parts, that grow out of this cuticle, have the same structure, and are endowed with the same properties. The nails of our fingers are productions from this membrane, condensed and made firm. Yet they have the same power to protect, and the same insensibility. The hoofs of horses, the horns of cattle, are similar; they have the same protective and the same negative character.

459. The nail grows from the cuticle. It has its root (Fig. XXVII. *c*) in the inner layers of this membrane, and its under surface is closely attached to the true skin. It grows from the root forward.

Fig. XXVII. *Vertical Section of the Thumb and Nail.*

a, Nail.
b, Cuticle.
c, Root of the nail.
d, True skin.
f, Fatty matter under the skin.
g, Bone.

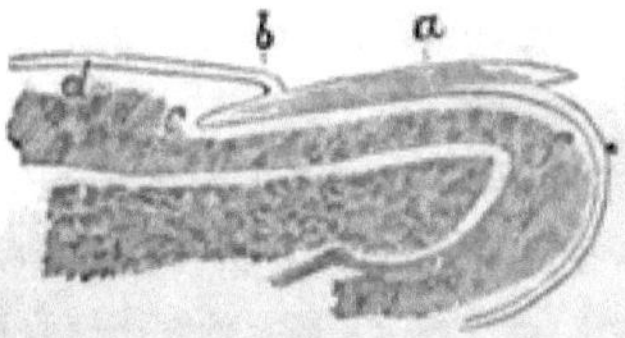

460. The hair (Fig. XXVIII. *b, c, b,*) is composed of a substance similar to that of the cuticle. It takes its origin in a pulpy bulb, (Fig. XXVIII. *d,*) which is situated below the true skin, (Fig. XXVIII. *f.*) It is fed by an artery, (Fig. XXVIII. *a,*) which supplies it with the material of growth. Within the skin, it is a tube containing a pulpy matter, (Fig. XXVIII. *c.*) In ill health, or in the later periods of life, this nutriment diminishes and the coloring matter ceases, and then the hair is white. Still later, the nutriment entirely fails, and then the hair falls out, and the aperture in the skin closes.

461. *The cuticle is continually casting off its outward layer in the form of little scales, so minute as to seem like dust;*

Fig. XXVIII. *Hair highly magnified.*

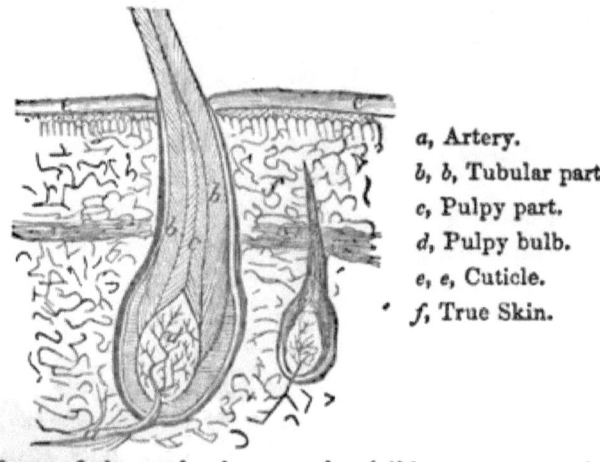

a, Artery.
b, b, Tubular part.
c, Pulpy part.
d, Pulpy bulb.
e, e, Cuticle.
f, True Skin.

the form of the scales is scarcely visible, except on the head. This scurfy dust is constantly gathering upon the surface, and needs to be frequently washed off with soap and water. And when, after a long neglect of this duty, we rub the skin vigorously in a warm bath, we feel this matter gather in little rolls under the hand. After some fevers, this skin comes off in little flakes, like scales of bran, but not in health, except on the head. The loss of these outer layers is continually supplied by the new growth of layers underneath from the inner skin. This process of change gives the scarf-skin a constant freshness of substance.

462. The cuticle is composed of several layers of thin scales. The outer and the oldest are transparent. The *pigment cells*, which are the *seat of color*, are situated in the innermost and the last-formed layer of the cuticle. The contents of these cells give the different shades to the various races of mankind, and to various individuals. This matter is white or flesh-colored in the European and North American, black in the African, yellow in the Mongolian, and copper-colored in the American Indian. It is this which is darkened or tanned by exposure to the sun, and bleached by pro-

tection. Carpenter says, "What has been termed the *rete mucosum* is simply the last-formed portion of the cuticle."

463. The various coloring of this inner layer of the cuticle gives to some animals their varied hues — to the serpent, the frog, and the lizard, and some fishes, which have a splendor of hue almost equal to polished metal. Goldfish and the dolphin owe their difference of color, and the brilliancy of their hues, to the color of this layer of skin.

464. *Underneath the cuticle is the true skin*, (Fig. XXVI. *b, b,*) the seat of all the active functions of the cutaneous membrane. This layer is a dense and thick membrane, and composed of firm and strong fibres, that are interwoven like the felt of a hat. It is almost filled with minute blood-vessels, so many that a large proportion of the blood of the whole system flows in them. If we cut the outer skin, no blood flows, because no blood is there; but if we cut through that and into the inner skin, we cannot fail to wound some of these vessels.

465. In health, when every thing goes on well in the animal body, the blood is properly distributed in all the organs, and each receives its due proportion; then it flows freely through the vessels of the skin, and the surface is florid and the cheek is rosy. But cold contracts the cutaneous vessels, and lessens their capacity for blood. The cutaneous circulation is sometimes influenced even by the state of the mind and the affections: the blood flows more abundantly in the capillaries of the face when the modest youth blushes, or when one is excited; and it is easily driven away, and the cheek turns pale, when one is oppressed with fear, or is overcome with anxiety.

466. This skin is furnished with a great quantity of nerves, for it is endowed with an exquisite degree of sensibility to pleasure and to pain; and it is also the seat of the sense of touch. In man, the nerves are more abundantly distributed to the skin than to the other organs. But those animals which are covered with hair, feathers, or scales, have

not this large supply of cutaneous nerves, nor this acute sensibility of the skin.

467. Underneath this skin there is a layer of fat, which varies in thickness in different parts of the body. It is very thick in the palm of the hand and the sole of the foot, and affords a cushion to meet the pressure that comes upon those or other parts that need this support; while in the forehead and on the back of the hand it is very thin, for there it is not needed.

CHAPTER III.

Functions of the Skin. Exhalations: Perspiration.— Sensible and insensible Perspiration.— Quantity.— Experiment at Phœnix Gas Works.

468. THE skin is the outlet for a good proportion of the waste of the body. Some goes off in the form of carbonic acid, some in the oil, but the greater part in form of perspiration. Sanctorius, a celebrated medical writer, carefully weighed himself and all his food, and drink, and excretions, daily, for thirty years; and, after all this observation, he concluded that, of every eight pounds which were taken into his system each day, five passed out through the skin.

469. Seguin, a philosopher, weighed, and then enclosed himself in a bag, which was glazed so as to prevent the perspiration from passing through it. He found that the largest quantity of perspiration that passed off in a day was four and a half pounds, and the smallest quantity was twenty-four and three quarter ounces; the medium was thirty-three ounces.

470. This is the insensible perspiration; for, although it amounts to about two pints a day, it is not usually perceptible; it passes off in such minute portions, and is so completely dissolved in the air, that we do not perceive it. Yet it can be perceived by holding the hand, apparently dry, near

a cold mirror, which will condense the invisible vapor and soon be covered with a slight dew; or if we put the hand into a large tumbler or glass pitcher previously wiped dry, and wind a towel about the wrist, so that nothing can pass out at the mouth of the glass, we shall then soon see the moisture gather upon the inner surface. This can be nothing more than the condensed exhalation from the hand.

471. This is called the *insensible* perspiration, in distinction from the *sensible* or visible perspiration, which flows in drops from the skin when we are excited or are unusually warm. The insensible perspiration never ceases to flow during health; and under all circumstances, if the skin is in a good condition, it is not interrupted. But the sensible perspiration flows only occasionally, and, though more abundant sometimes than the other, yet the whole amount is much less. In the cold-blooded animals, — the toad, serpent, &c., — the insensible is six times as great as the sensible perspiration. The difference in man is not so great as this, yet it is considerable.

472. The sensible perspiration — the sweat — is at times very great, and occasionally we saturate our clothing with it in a very short period. Some experiments were performed, and observations made, at the Phœnix Gas Works, in London, Nov. 18, 1836, to determine how large a quantity would be thus thrown out from the body under favoring circumstances.

473. "Eight of the workmen regularly employed at this establishment in drawing and charging retorts and in making up the fires, which labor they perform twice a day, commonly for the space of one hour, were accurately weighed in their clothes immediately before they began and after they had finished their work. On this occasion, they continued at their work exactly three quarters of an hour. In the interval between the first and second weighing, the men were allowed to partake of no solid or liquid, nor to part with either. The day was bright and clear, with much wind. The men

worked in the open air, the temperature of which was 60° Fahrenheit. The barometer was 29° 25′ to 29° 4′."

474. These eight men lost, during these three fourths of an hour, by perspiration from the skin, various quantities.

The first lost 2 lbs. 8 oz.	The fifth lost 3 lbs. 12 oz.
second, 2 " 9 "	sixth, 3 " 14 "
third, 2 " 10 "	seventh, 4 " 2 "
fourth, 3 " 6 "	eighth, 4 " 3 "

The average loss of all was 3 lbs. 6 oz.*

475. This constant perspiration, and the exhalations from the lungs, maintain the permanency of the weight of man; so that, although he eats and drinks from four to six pounds a day, his body at night weighs no more than on the day before; and, if one man eats and drinks more than another, he has more pulmonary and cutaneous excretions, and the superabundance is thus carried away.

476. This whole amount of cutaneous exhalations, sensible and insensible, will average about the same in a healthy individual, from day to day or from month to month. Yet there are many circumstances that cause it to vary. Climate and season influence it; it is more in summer than in winter; philosophers estimate it to be forty ounces in the south, and twenty ounces a day in the north of Europe. Active exercise — running, hard labor — will increase it, and make the sweat visible, so as to run abundantly in drops. Unusual quantities of clothing prevent the radiation of heat, and cause it to accumulate; the skin then becomes warmer, until an increase of perspiration occurs and relieves by its evaporation.

* Smith's Philosophy of Health, Vol. II. pp 391, 392.

CHAPTER IV.

Perspiration differs in various Temperatures. — More in dry than in moist Air. — Prepared in Glands for the Purpose. — Some Animals do not perspire. — Oily Excretions from the Skin. — Tight Clothing, Hats, &c., prevent Removal of these Excretions.

477. *This perspiration differs in various conditions of the atmosphere.* Heat increases the activity of the cutaneous blood-vessels, and the perspiratory action, and also the capacity of the air to receive vapor. (§ 349, p. 152.) Evaporation is therefore more rapid in warm than in cold weather. A moving atmosphere brings to the body a constant succession of layers of dry air, which absorb the moisture more rapidly. The perspiration is therefore more free in a windy than in a still day; and, if the wind is very dry and hot, this evaporation is still further increased. The sirocco, which comes over Sicily from the south, is so dry and hot as to produce in the skin a parching and painful dryness, and sometimes excites disease.

478. When the air is saturated or loaded with moisture, the evaporation is checked, and the perspiration is not carried off so freely; and sometimes this interruption causes a serious disturbance and burden to the system. The waste that is carried off through the skin gives important relief to the whole body, and, if this is interrupted, oppression follows, and the frame is languid. In some of the sultry dog-days, we are languid, because the atmosphere, being already filled with vapor, does not carry off the watery exhalations of the skin, and relieve the body. From this cause, internal diseases prevail more in low and damp situations of warm climates than in dry. On the banks of the southern and western rivers, this is most painfully manifested.

479. In this work of perspiration, the inner skin performs all the active duty. It not only throws this fluid out, but it originally forms it out of the elements which are found in the blood. This work of formation is done in little glands

which are placed (Fig. XXVI. *c,c,c,*) just beneath the skin; and the perspired fluid is carried from each one of them to the surface, through a minute tube which is attached to the gland, and leads outward (Fig. XXVI. *d,d,d,*) There are three thousand five hundred and twenty-eight of these little tubes on each square inch in the palm of the hand, and twenty-eight hundred on each square inch throughout the whole body, making seven millions of perspiratory tubes on a man of average size.

480. The power of relieving the body of its superfluous matter or moisture by perspiration is not common to all animals. Men and horses sweat, and thus find an outlet for these matters, and for the excess of heat. But dogs have no such means of relief; when they are heated by exercise, they loll their tongues, and the evaporation from their surface aids in the cooling process. Cattle, when heated in the summer, effect the same purpose in the same manner.

481. *There are other secretions of the skin beside the watery perspiration.* The skin is soft and oily, it is supple, and in health it is never dry and hard. To produce this condition, there are numerous little glands placed within the skin, whose business it is to gather out of the blood the elements of an oily matter, and with them compound this substance, and then throw it out upon the surface. If the preparation of this oil is checked, the skin is dry and hard; but, when this work is well performed, the skin is soft and supple, and pleasant to the touch. "It is this fluid which soils the linen, and which causes the water to collect in drops, when we come out of the bath." These oil-glands are more abundant in the face and in other parts exposed to the air, and in the arm-pits, &c., where one part of the skin comes in contact with another. But through all the skin there are enough to keep the surface soft, and in a natural condition.

482. This oily secretion is sometimes odorous, and in some parts unpleasant to the smell, and even in some men fetid. It is more so under the arms than elsewhere; but in some persons the whole surface throws out an offensive mat-

ter. This odor is not the same in all persons; it is said that each one has his own peculiar smell, by which the dog can scent his master at a great distance.

483. *These excretions are intended to be carried, not only out of, but away from, the body.* If suffered to remain, they are mixed with the dust in the air, and the particles of the scarf-skin that scale off; and they, together, form a thick, crusty matter, which fills the pores and interrupts the transmission of the natural fluids. Then the skin becomes comparatively stiff and hard, and loses its suppleness and agreeable feeling, and is also more liable to suffer from the effects of cold. The perspiration is usually carried away by mere evaporation. If, therefore, it have sufficient access of air, it will generally be removed.

484. *The air is an important agent in the action of the skin.* It gives it oxygen, and takes from it some carbonic acid. It removes the perspiration and some of its superfluous oil. It is necessary, then, that the air should reach the body. For this purpose, the clothing should be loose and porous. Tight clothing, water-proof dresses, oil-cloth, India rubber garments, glazed coats, and even leather clothing, prevent the access of air, and the transpiration of the perspired fluids; and, inasmuch as they thus interfere with the functions of the skin, they are unhealthful, and improper to be worn. India rubber shoes retain the perspiration, and the feet that wear them are often wet. It is a common complaint that glazed caps worn in summer, however light they may be, make the head ache. This is caused by the closeness of their texture, which prevents the free passage of the vapor. The cap fits so closely to the head, that no vapor can escape, and its impervious texture offers no avenue through which the perspiration can pass away.

485. For the same reason, hats made of felt are too close for health and comfort. Ventilated hats are made on true physiological principles, because they allow the cutaneous excretion free passage outward. If a tight felt hat is worn, it is better to be sufficiently large to afford room for much

of this vapor to escape. And even then we find great relief from frequently taking off the hat to air the head, or rather to air the hat itself, by letting the enclosed air, which is saturated with vapor, pass away, and fresh air take its place.

CHAPTER V.

Connection between the Skin and the internal Organs.— Stomach, Lungs, Muscles, &c.— Effects of Cold on different People various.

486. *There is a very intimate connection between the skin and the internal organs of the body.* The blood flows from one common centre through various channels to the minute vessels of the skin, and all the parts included within it. If the outer vessels are closed, and the circulation is interrupted there, the blood must flow in greater abundance into the inner vessels. On the contrary, if the inward flow is impeded, it must find passage outwardly. In either case, the balance of the circulation is disturbed, and the overburdened part is disordered. If we expose ourselves to sudden or long-continued cold, the surface becomes chilled, the cutaneous vessels contracted, the perspiration checked, and then some of the internal organs bear the burden which usually belongs to the skin. The check of the perspiration is not the cause of the disturbance; it is merely one of the consequences of the previous interruption without, and a sign of other troubles within.

487. We have seen the influence of tea, taken into the stomach, upon the cutaneous circulation and functions. (§ 441, p. 188.) Certain medicines, taken into the stomach, have the same effect. A man under the operation of an emetic often sweats profusely. Some kinds of food,— such as shell-fish,— when eaten, will cause the skin to break out with the nettle rash. In some cases of indigestion, the same effect is produced. In September, 1845, I saw a child

which had suddenly broken out with this rash, from eating indigestible food. Its skin from head to foot was covered with this scarlet eruption. But, as soon as the stomach was relieved of its disturbing cause, the rash departed, and the skin resumed its natural color.

488. *The lungs and the skin are intimately connected* by their mutual sympathies. They coöperate together in carrying off much of the waste of the body. They bear each other's burdens. When the circulation is checked in the skin, it may be thrown more upon the lungs; and an increase of the flow of blood in the skin relieves the lungs when they are oppressed. Every one is familiar with the character and operation of a cold, and with the common remedy of a sweating process. A man puts on a thinner dress, or goes into a colder atmosphere than he has been accustomed to. The cold of the air diminishes the capacity of the blood-vessels of the skin, interrupts the cutaneous circulation, and checks the perspiration. The balance of the circulation is disturbed, and the lungs are compelled to receive more blood than usually belongs to them. These organs are then oppressed, and the breathing becomes somewhat difficult; or the blood-vessels in the mucous or lining membrane of the air-cells and air-tubes of the lungs may become enlarged, and carry more blood, and throw out more mucus, which is coughed up. The sufferer then takes hot teas, or other stimulating remedies that excite the cutaneous arteries, and he covers himself under an unusual quantity of bed-clothing. The skin then is heated; its arteries are enlarged and more active, and carry more blood; the balance of circulation is restored, and the perspiration breaks out profusely; and then the lungs are relieved.

489. *A similar connection exists between the skin and digestive organs.* In summer, and in warm climates, an interruption of the cutaneous circulation more frequently disturbs the organs of nutrition, and excites them to excessive action, and produces a painful disturbance; and, on the contrary, the restoration of the external circulation and action

is one of the means of relieving the internal disorder. Exposure to cold is sometimes followed by disturbance in the organs of locomotion, and then we have rheumatism, pain, and sometimes swelling in the muscles and joints. In this, as well as in the other cases, the disorder is removed when the balance of circulation is reëstablished; for, when the natural perspiration and the other cutaneous functions are restored, the rheumatism diminishes, and the limbs and muscles become easy.

490. We thus see that the skin stands not alone, but is intimately connected with the internal apparatus of life, and does not suffer, without their sympathy, nor enjoy the full measure of health, without their participation in a greater or less degree.

491. Persons differ in the degree and distribution of their health and strength; all their organs and systems may not have the same power of action, or of resistance to disturbing causes. In one man the lungs, in another the organs of locomotion, and in a third the digestive apparatus, and in a fourth the nervous system, may be weaker than the other organs. It is the weaker internal organ that is in the most danger of suffering, when the balance of circulation is disturbed. Several men may be exposed to a storm together, and all may be drenched with rain and chilled. In all, the cutaneous circulation is disturbed, and the perspiration checked, the blood is thrown inward, and some internal derangement may follow. But this differs according to the previous state of the constitution. One of these men takes a cold in his lungs, the second is attacked with a pleurisy, the third with a disorder of the digestive organs, the fourth with rheumatism, the fifth has a fever, the sixth a headache, while the seventh has sufficient vigor of constitution to resist the internal disturbance, and to produce immediate reaction in the vessels of the skin, and restoration of all its healthy functions.

492. There are many conditions of the body that affect the insensible as well as the sensible perspiration. All

diseases that prevent the circulation in the skin interrupt the flow of this fluid. In some of the stages of fever, the skin is dry and parched. But, without disease, the perspiratory action is never stayed, so that, whenever we find the skin dry, we may be assured that all is not right in the body.

CHAPTER VI.

Skin is an Absorbent. Food and Drink sometimes taken into the Body through the Skin. — Medicines. — Contagion. — Poisons absorbed by the Skin. — Absorption more active in the Night than in the Day.

493. *The skin has other duties to perform, besides that of carrying off the waste of the body; it is an absorbent as well as an exhalent.* In certain conditions, it takes some matters into the body, while it throws others out. But this is not usually done in a period of health; it is rather when in a state of disease. Nevertheless, absorption is not always indicative of disorder. It may be used to prevent or relieve derangement of the system. It is the most active when the fluids of the system are reduced in quantity, and when nutrition is not well sustained.

494. Sailors, when destitute of fresh water, wear their clothes wet with sea-water. Then the skin absorbs and carries some of this fluid into the body; and thus their thirst is allayed, and sometimes entirely relieved. Dr. Currie relates a case of a patient, who, from disease of the throat, was unable to swallow any thing, and was therefore in danger of immediate death from starvation. His flesh was rapidly wasting away; he suffered extremely from thirst, and was nearly exhausted. While in this state of suffering, he was placed, night and morning, in a bath of milk and water. After this was begun, his body ceased to waste; and, while

this course was pursued, he maintained his weight, and the thirst ceased to be troublesome. In this case, the skin absorbed sufficient fluid and nourishment to maintain life.

495. One of the men who were subjected to the sweating experiment of Dr. Smith (§§ 473, 474, p. 199) — the one who lost two pounds and fifteen ounces — went into a hot bath at 95°, where he had remained exactly half an hour. He was reweighed on coming out of the bath; and then it was found that he had gained half a pound. This must have been by the absorption of water. In the case of Ann Moore, (§ 97, p. 50,) there was some matter constantly passing off through the lungs, and doubtless some perspiration. Yet, for years, she took nothing through the mouth but a little tea, and not enough of this to sustain life and to meet the wants of respiration, and yet she did not waste away. The body must have been sustained by matter which was absorbed from the atmosphere through the skin.

496. Other substances beside fluids may be thus absorbed. The odor of camphor or of garlic may be perceived in the breath, when a plaster of one of these substances is worn upon the skin. Medicines are sometimes thus introduced into the system; some liniments may be rubbed into the skin, and entirely absorbed. Antimony rubbed over the stomach is said to produce vomiting. Mercury, in the same way, may bring on salivation. Men at work in lead mines, or in an atmosphere of lead or lead paints, are often troubled with what is called the *lead colic*, from the absorption of particles of lead through the skin.

497. But what should set at rest all doubt of the absorbing power of the skin is the effect of contagion. The slightest quantity of matter from the pustule of the kine pox, when applied to the inner skin under the cuticle, excites disease in the whole system. So other contagious diseases — such as small-pox — are conveyed by the bare touch. Even the matter of the latter disease which may be rubbed from the skin, and lodged on the clothing or the bed, will be ab-

sorbed, and convey the disease to any one who should next sleep in that bed or wear that garment which had been thus infected. The poison of dogwood or ivy is absorbed by the skin of the susceptible, if they but touch the plant; and the disease, being excited within the skin, extends beyond the point of contact, and sometimes over the whole surface.

498. The poison of bad air is supposed to be thus absorbed. In marshy countries, where the exhalations from the earth infect the atmosphere with the seeds of fever or other disease, the people whose lungs breathe, and whose surface is in contact with this air, receive the poison by the absorbing power of their skin and their air-cells

499. *This absorbing power is more active at night:* then contagion of disease and infection of bad air act with more readiness and vigor, and men are more liable to be attacked by prevailing epidemics through the air, or by contagious diseases from contact with those already diseased, than in the daytime. It is more active when the body is badly nourished than when it is well fed. Hunger and thirst increase the absorbing power of the skin, and good nutriment diminishes it. So that one is more susceptible of disease before than after breakfast, (§113, p. 56;) and cautious physicians fortify themselves with nourishment in the morning, before they visit patients who are suffering from epidemic or contagious diseases.

500. Any poisonous or offensive matter in contact with the surface stimulates the cutaneous absorbents. The natural excretions of the skin, — the perspiration, the oil, and the dead cuticle, — being the offensive waste of the body, if not removed from it, excite this tendency to absorption; and when they are not washed away, or are confined too much by impervious clothing, they themselves are often taken back, to irritate and disturb the system.

CHAPTER VII.

Skin Seat of Touch. — Sensibility of Skin differs in different Parts, and in different Persons. — If the outer Skin is thick or foul, the Sensations of the inner Skin are dull. — Sense of Touch can be cultivated — Blind have acute Sense of Touch.

591. *The sense of touch is situated in the skin.* It is not in the cuticle, which is insensible, but it is in the true or inner skin, which is very sensitive and exceedingly alive to pain, and suffers from contact with any matter, however soft and bland. Strip off the outer skin and expose the layer beneath, and this, which before was comfortable when protected, will now, in its nakedness, ache with pain. Even the air is disagreeable to it. But this sensibility to pain is unequally distributed. The sensibility to contact, or the acuteness of the sense of touch, also differs in the various parts of the skin. Some parts are more plentifully supplied with nerves than others. The ends of the fingers, the lips, and the face of man, and the end of the elephant's trunk, have more nerves and more sensibility than the back or the chest; these and all other uncovered parts have more than the head, which is covered over with hair.

592. The cutaneous sensibility is as unequally distributed as are the nerves. It is the greatest at the tip of the fingers, and the least in the scalp. The sensibility of touch is more acute in the right than the left, but the sensibility in regard to heat is greater in the left than in the right hand; for, " if the two hands were immersed in warm water of the same temperature, that in which the left was plunged would feel the warmest." The sensibility differs very much in different individuals, so much " that that which amounts to absolute torture in one is a matter of almost indifference to the other." The sensibilities are more acute in the young than in the adult, and in the latter than in persons of advanced life. They are greater in the female than in the male; in the sanguine and nervous than in the phlegmatic

THE SKIN.

and bilious temperaments, and in those enfeebled by disease, than in the sound and robust." *

503. *The facility, with which cutaneous sensations are received, depends upon the condition of the outer skin.* When it is thick, as on the palm of the hand or the sole of the foot, sensation is somewhat interrupted. The seamstress finds it more difficult to feel and distinguish minute differences of objects with the fingers with which she uses the needle than with the others. The difficulty is, not that the sensibilities in these fingers, as in the palm or the sole, are more blunted than in the others, but that a thicker shield of cuticle stands between the nerves in the inner skin, and the object which is to be examined.

504. The sense of touch differs very widely, not only in various parts, but in various persons. Beside the natural and original differences of sensibility from organization, there is a very great difference owing to education; for this sense can be educated to a very high degree, so that one person may be able to perceive objects and characters which another, whose sense of touch is less cultivated, could not recognize.

505. It is a remarkable provision of a benevolent Providence that, when one sense is lost or impaired, the others become more acute, so as to compensate in a good degree for the defect. Thus the blind have or acquire a niceness of touch which the seeing never possess. Their method of reading is a singular proof of the extent to which the cultivation of the sense of touch may be carried. Their books, instead of being printed on soft paper, and with colored letters, are printed on stiff paper, and with raised letters. Their pages are perfectly white, but the surface is not smooth. Their letters stand out as if carved in wood. The blind move their fingers over these, and, by the sense of touch, they recognize the shape and kind of each letter almost as readily as others, who see, recognize letters that are printed with ink. It

---

* Wilson on the Skin.

is interesting to notice with what rapidity these sightless children can read. They must of course perceive one letter at a time, and, at the end of each word, determine what the several letters spell. Yet, with this additional mental process, they read nearly as fast as we do with the use of our eyes. This seems very easy when we see them do it; but if we shut our eyes, and then apply the fingers, not to a whole word, but to a single letter, — the letter *a*, for instance, — we shall find it is not so easy for the untrained to decipher the raised marks. If we further attempt to read a word or sentence, we shall be lost in the mazes of indistinguishable characters.

506. The blind are compelled thus to cultivate the sense of touch, to compensate for their deficiency of sight. But the power so to do is not confined to them. We all can do the same, if we apply the same diligence; and this we could do if we had as strong a motive as they have. The cloth-dresser learns to distinguish, by aid of the sense of touch in his fingers, the qualities of material, or minute differences of texture, which others cannot detect. The miller, in the same way, detects the various qualities of meal and flour, which escape the notice of others. In a great many of the arts of life, the sense of touch is thus educated to be used for minute and useful purposes.

507. This sensibility is blunted by several causes. Cold remarkably diminishes it. Our skin is numb when exposed to a very low temperature, so that men sometimes cut or bruise themselves, in winter, without feeling it; and the first intimation which they have of their injury is the sight of their flowing blood. This sensibility is also impaired by the natural excretions of the skin, by the mixture of the dead scarf-skin, oil, and perspiration, with the dust and dirt, if not removed from the surface. The blind man will wash his fingers before he attempts to read his raised letters; and the cook will pass through the same process when she leaves her ordinary work, and takes up her fine sewing.

CHAPTER VIII.

Animal Heat permanent. — Skin regulates it. — Excess of Heat carried off by Evaporation of the perspired Fluids. — Cold Sweats. — Sensations of Heat and Cold comparative.

508. *The skin is itself a bad conductor of heat;* that is, it does not allow heat to pass easily, either outwardly or inwardly, and therefore it is a good protector against high or low temperatures. The natural and usual temperature of the body is 98°; but the surrounding air is often at 100° in summer, and at 0 in winter; and, in some extreme climates, it is 30° warmer, or 150° colder, than our bodies. In the experiments of Sir Charles Blagden, (§ 402, p. 172,) it was more than 160° higher than the standard of 98°; and yet in neither case is the heat of the body materially changed. In the heated room, a thermometer placed in the mouth was hardly raised, and, beyond the arctic circles, it scarcely fell below our usual temperature.

509. There is, of course, a constant tendency to radiation of heat from the skin when the air is colder than our bodies, as from any other substance; and, on the other hand, there must be a tendency to receive heat from the air when that is warmer than the body. In the first case, in cold weather, more internal heat is produced, (§ 443, p. 189,) to supply loss from increased radiation. In warm seasons, the evaporation of the perspiration absorbs the excess of animal heat, and thus the equilibrium of the internal temperature is maintained.

510. The evaporation of the cutaneous fluids is the outlet of much of the surplus heat. (§ 441. p. 188.) Every one is familiar with the fact, that a wet skin is colder than a dry one, because the evaporation carries off more of the heat. The inhabitants of hot climates make use of this principle, and put water into porous jars, the surface of which is constantly wet with the moisture that oozes through; and

the rapid evaporation of this cools the water within. Even ice may be thus produced. We are therefore cooler when we sweat. Blagden found great relief, in his oven, from the profuse perspiration which was rapidly evaporated.

511. The cooling power of the air is influenced by other states besides its temperature. A dry atmosphere, by increasing evaporation, cools the body more rapidly than air saturated with vapor. Winds have the same effect. Even if the air is warmer than the body, if it is in motion and dry, it cools us; so that a lady's fan, at summer's noon, when the thermometer stands at 100°, two degrees warmer than the flesh, affords a pleasant coolness, by moving the air, and hastening the evaporation. So slight a motion of the air is thus perceptible by the increasing coolness, that men, when they cannot distinguish the direction of the wind by its force upon their bodies, or even by the movements of leaves of trees, often wet a finger, and, holding it up to the air, discern, by their sensations, which is the colder side. This determines the course in which the air is moving.

512. In order that the skin should passively permit the heat to pass off by radiation, or actively throw it off by perspiration, it must itself be in good health. It must be able to prepare within itself just as much fluid as will, by its evaporation, carry off the surplus heat, and no more; otherwise we may be too hot or too cold.

513 In some states of disease, men suffer from a constantly dry and parched skin. Their flesh burns within, and the accumulating heat finds no outlet, for the skin affords no relief. In other disorders, they are prostrated with a profuse and cold sweat. The skin pours out the perspiration like water, and this, by evaporating, creates a constant and painful demand for heat; and then the patient finds it impossible to keep warm.

514. Although it is through the sensibility of the skin that we perceive things to be hot or cold, yet this is by no means an exact measurement of the degree of heat; for the apparent and sensible temperature of any substance is

merely relative to the previous sensations. If, after we have been handling snow, we take a piece of iron heated to 50°, it feels to us warm. But if another, who had been holding his hands in warm water, at 98°, should take up the same iron at 50°, it would seem to him cold. If one should come from the outer air of a cold day in winter, where the thermometer is at 0, and enter a cellar where the temperature is at 50° or 60°, he would feel a pleasant sensation of heat, and call the cellar warm. But if, at the same time, another should descend from his parlor, heated to 70°, into the same cellar, he would complain of cold.

515. It is no uncommon circumstance for two travellers to meet midway on the side of a high mountain. One is coming from the top, where snow covers the ground and the air is wintry; the other is going up from the valley below, where summer reigns. The descending traveller, coming from the cold region, and finding the air warmer than that which he has just left, complains of the oppressive heat, and throws off his woollen clothes and puts on his summer garments; while the ascending traveller, coming from another atmosphere, much warmer than the present, complains of the cold, and changes his summer for his winter clothing. Both these men are exposed to the same temperature, but have very opposite sensations.

516. *Every thing which depresses the power and energies of life diminishes the production of internal heat,* (§ 434, p. 186,) and also lessens the protective power of the skin against the external cold. Under the influence of hunger and fatigue, and the consequences of exhausting disease, and when overborne by the depressing passions and emotions — grief, despondency, anxiety, and fear, — the skin has less power to defend us from the extremes of heat and cold, and we are then more uncomfortably hot in high temperature, and suffer more from the low. But the contrary happens when we are well-fed and fresh, when we are vigorous and cheerful, and when we are animated with hope or exhilarated with confidence.

CHAPTER IX.

Clothing needed to prevent excessive Radiation of Heat. — Parts usually clothed need more Protection than others. — Habit of Dressing affects the Necessity. — No positive Law for the Amount of Clothing.

517. THE skin is thus shown to perform three offices. It carries off much of the waste of the system, by means of perspiration, oil, and carbonic acid. It absorbs some matters from the atmosphere and other contiguous substances. It regulates the transfer of heat from within outward, and prevents its coming inward from without.

518. It is a natural question to ask, whether the skin can do this alone, or does it require our aid to enable it to perform these functions faithfully and successfully? We are so much the creatures of habit, we have been so accustomed, through many years, and even from generation to generation, to cover the body with clothing, that it is not easy to tell how great a degree of cold could be borne upon the naked surface. As it is, there are not many days, even in summer, when we should feel as comfortable as we now do, if those parts of the body which have always been clothed were left unprotected.

519. Certain it is that the heat is constantly prepared within the animal system; and it is equally evident that, when the body is warmed to its natural and usual degree of 98°, the excess beyond that must pass off. As much is then to be thrown out as is added, and this is done mostly through the skin. But it is not so certain that the skin could, unaided by clothing, regulate this transmission of heat so exactly that the internal temperature would not vary from its usual standard. Whatever the natural protective power of the outer surface might have done, if we and our fathers had from the beginning, lived in a state of nature, there can be no doubt that we and all civilized men, in temperate and

colder climates, now need the aid of clothing to protect ourselves from cold during most of the year.

520. There is a very great difference in the present protective power of different parts of the skin; and this varies, also, in different persons, according to their various habits. The air is seldom so cold as to compel us to cover the face, or even the upper part of the neck. These parts have always been exposed to the severities of winter, and they have borne them, and do now bear them, without suffering. But the chest and the back would hardly bear the open exposure to the weather of the warmest day of the summer, without suffering from chill.

521. The female costume usually exposes the neck and the upper part of the chest, and even sometimes a portion of the back and shoulders. But women do not complain of suffering materially from this exposure. The dress of men covers the entire chest, shoulders, and back, and most of the neck; and they seem to be none too warm. But if a man accustomed to dress thus should expose his skin as women do, or even if he were to leave off his cravat, after wearing it, in winter, he would immediately feel uncomfortably cold; and, if this exposure were continued for any length of time, he would so change the balance and direction of the circulation that the blood would be thrown inwardly upon the lungs or throat, and he would take cold, and perhaps severe disease would follow.

522. The North American Indian wears much less clothing than his civilized neighbors. While we cover ourselves from neck to feet, and leave no part of the surface exposed, the Indian is satisfied and comfortable with his blanket for his back and shoulders, his girdle for his loins, and his moccasons for his feet. His limbs and his breast are bare. In the costume of the Highlander, who lives in the northernmost parts of Scotland, the kilt, or the short petticoat, scarcely meets the stockings; and, as he wears no pantaloons, his flesh about the knees is bare and exposed to the cold of his severe climate; and yet he seems to be as com-

fortable as his southern neighbors, whose limbs are more carefully protected.

523. There is a great difference in the habits of clothing of individuals. One always wears thick clothing, and from the first approach of cold weather in the autumn till the warmth of spring, he never ventures abroad without a great coat; and if by chance he is compelled to go out without this protection, he is chilled, and perhaps disordered; while others dress much lighter, and find few, perhaps no days in winter so cold as to require any such extra covering.

524. Some men never wear gloves or mittens; others always wear the warmest they can obtain. Some wear flannels next to their bodies; others never wear any. Some always put on a tippet to cover the neck in any weather in winter, and suffer if they leave it off before the warm season returns; others wear only low cravats, or even none, and suffer no more. Men wear stout boots and thick stockings through the winter, while most women are kept apparently warm with worsted or cotton hose, and shoes as thin as men wear in the dryest and warmest days of summer.

525. Thus we see that there is no positive and fixed law for the quantity of protection which we should give to the external surface. There is a very great difference in mankind in this respect, without a corresponding difference of health and comfort. This is due, in a great measure, to difference of habit of clothing. Men cannot change this habit suddenly without suffering; yet, if they do this cautiously and gradually, they may nearly reverse their habits, and still retain their health.

526. Those who accustom themselves to wear but light clothing, and exercise actively, in the cold season, acquire and maintain the winter constitution. (§ 444, p. 189.) They have more radiation of heat outwardly, but they generate more heat inwardly to sustain it. But those who are always careful to cover themselves heavily, retain partially their summer constitution through the winter. Their radiation is then increased, but their internal fire does not burn more

vigorously. They are therefore tender in respect to cold, and cannot bear what others do without suffering. An undue anxiety to guard against exposure, manifested in excess of clothing, frequently disarms one of the natural protection against the effects of a low temperature. Those who are over-careful to dress warm, and never walk abroad in winter without the thickest outer garments for their bodies, over-shoes for their feet, and tippets for their necks, make themselves tender, and are more liable to be affected by changes of the weather, and to take cold, than those who clothe themselves more judiciously, and develop and depend more upon their own internal resources. The very common practice of schoolboys wearing woollen tippets about their necks has caused more sore throats than it has prevented.

527. Some differ very widely in their habits of dress in various periods of life. I know of men who once were accustomed to clothe themselves in the warmest woollens and furs, and who never went into the open air, in winter, without extra garments; and these were doubled in the severest weather. In this manner, they became so tender as to suffer if they infringed in the least upon their law of habit. But these same men, by slow degrees, have left off their extra dresses, and now find them to be seldom or never needed. They were before so delicate that they felt a chill, or a sore throat, or pain in the muscles, or joints, or the lungs, if they even entered the street, without a great coat, in cold weather. Now they walk boldly for hours without extra clothing, and suffer no bad or uncomfortable consequences. Precisely the reverse sometimes happens, and the hardy become delicate from an opposite change of habit.

CHAPTER X.

Those who have poor or insufficient Food, or Dyspepsia, or breathe bad Air, need more Clothing. — More Clothing needed in dry and windy Weather than in damp and still Air. — More needed in travelling than when quiet. — Every one should be clothed comfortably. — Hardening. — Old People and Children must be well clothed.

528. THE quantity of clothing depends, not only upon habit, but upon many other circumstances which are connected with the health, and which affect the generation of internal heat, and the healthy actions of the skin. If one is not supplied with sufficient fuel for the internal fire; if he is ill-fed, and has insufficient or poor food; if he is dyspeptic, and his stomach is unable to convert his food into the chyle for the blood; if nutrition goes on heavily, and the changes of particles are slow; or if he exercises but little, and the energies of his life are dormant, — he can bear less exposure unprotected, and he therefore needs more clothing.

529. Or if the lungs are supplied with insufficient air; if one sits in a crowded and unventilated lecture or school room for hours; if he ascends to the top of a mountain, where the rarefied atmosphere contains a smaller quantity of oxygen; or if the chest is encased with tight dress, so that it cannot expand and receive sufficient air; or if the lungs are diseased and their air-vessels partially closed; if in any way the blood receives less than its due amount of oxygen, — then there is less heat to be given out, and more protection is required.

530. If the fuel of good and nutritious food is not supplied for the internal fire, there must be greater external fire, or more protection; for then the body cannot sustain the loss of so much heat as would pass from flesh at 98° to a surrounding atmosphere at the ordinary temperature (65° to 70°) of comfortable rooms. Insufficiency of food thus creates a necessity for a greater expenditure for clothing and warming.

531. *It is manifest, then, that there can be no positive and universal law for the quantity of clothing.* This must be as diverse as are men's habits, health, and exposures. What is enough for one man may be too much for another; and what is only sufficient for comfort and for security from disorder at one time, or in one assemblage of circumstances, may be oppressive at another time, and in other circumstances.

532. Dr. Wilson says, "I have endeavored to establish as a law of health *the necessity of preserving an agreeable temperature of the body.*" "I should wish it to be understood, also, that the feelings, if the nervous system be sound, are a proper channel for arriving at a knowledge of the state of the warmth of the system." * It may be said, then, in general terms, that every one should wear sufficient clothing to make himself comfortable, and to secure his body from disturbance of health. More than sufficient clothing prevents the free radiation of heat, and causes its accumulation in the skin. The blood-vessels of the surface being stimulated to over-exertion, the perspiration is increased from the over-action of the blood-vessels, and carries off the surplus heat by evaporation. If a person wears less than this, the heat is carried off too rapidly by radiation, and he is chilled; the perspiration is checked, the cutaneous blood-vessels are contracted, the balance of the circulation is disturbed, and internal derangement follows.

533. In good health, a sudden and momentary chill from exposure to cold air, or a cold shower-bath, is not followed by these unpleasant consequences. On the contrary, reaction takes place in the cutaneous blood-vessels, and a glow of heat follows; and one is, perhaps, the warmer for this sudden transition. But continued cold is injurious both to the cutaneous circulation and to the internal health.

534. There is a great difference in the power of bearing cold, which comes from the habit of exposure. The driver

---

* On the Skin, p. 108

of a stage-coach sits on his elevated and unprotected seat in face of the severest winds of winter, for one or two hours, or even more, without apparent suffering, while his passengers, less hardy than himself, and perhaps much more heavily clothed, are shivering with cold. He has endured this exposure daily through the entire winter, and for successive years, and has become hardened; but they have been accustomed to the mild temperature of houses and shops, or, if they lived a while in open air, they kept themselves warm with active labor.

535. This coachman is a man of robust constitution; he eats heartily and digests easily, and is well nourished. He attained gradually to this power of endurance, and now he does not suffer. The pilot, the market-man, &c., who enjoy equally good original health, and have gone through a similar training, may bear the cold as well as he does. They too are hardened. But more feeble and less active men cannot thus expose themselves, without danger. There is a common but erroneous notion that any one can harden himself by exposure, and become able to endure severe cold without much outward protection. I have known some sedentary men, whose days were spent in warm rooms or shops, attempt to harden themselves by going abroad in the winter without outward garments; but they failed to accomplish their purpose. They did not begin with slight trials, and, proceeding gradually, go by slow degrees from small to greater and greater exposures; but they began with the greatest. They had not the robust health, the hearty appetite and vigorous digestion, nor the energy of muscular power, that belonged to laboring men, and consequently they did not generate an increase of internal heat to maintain the extraordinary radiation. Instead of returning from their cold walks or rides with a glow upon their cheeks, and the flush of ruddy health, they were pale and cold. Instead of a reaction afterwards, their cutaneous circulation continued languid, and they were not easily warmed. They became more susceptible of cold, rather than more

able to resist it, and in some instances their health failed, and they sank under the experiment.

536. *More clothing is necessary in infancy and in old age, when the generation of internal heat is more feeble than in the middle periods of active life.* It is a mistake to suppose that infants should be lightly clothed, or that the necessity of warm garments for them is the mere creation of habit. They can give out no more heat than is prepared within; and as this is less in them than in others, they cannot bear an equal loss without reducing their temperature below the natural standard. They must therefore be protected with more caution. The same law applies to the aged, and even more strictly, inasmuch as their sensations are so blunted that they cannot so easily tell when they are cold. And oftentimes they are suffering serious disturbance before they are aware of it.

537. It must not be inferred, from the preceding sections, that men who are engaged in sedentary employments, or who are otherwise than robust, cannot, by discreet exposure, acquire a power to endure the weather of cold seasons. Precisely the reverse is the fact. But this exposure must be just in proportion to the power of the body to bear it, and increased only as fast as the energies of the constitution increase. It should always be accompanied with so much clothing that the body shall not suffer while abroad, and the chill must not be so great that reaction will not take place immediately after returning to the house. With these precautions of suiting the exposure to the powers of the constitution, wearing clothing sufficient for comfort, or exercising actively enough to sustain the increased demands for heat, even the feeble can generally acquire a power to endure all the weather of this climate.

CHAPTER XI.

Clothing should be of loose Texture, and fit loosely to the Body. — Various Materials of Clothing. — Linen, Cotton, Silk Wool. — Flannel, next to the Skin.

538. The great object of clothing being to defend the body from cold, by preventing the radiation of heat, the materials should therefore be bad conductors of heat. This non-conducting principle is not so much in the material itself as in the air which is retained within its loose textures. "In every case it is the power which the coverings possess of detaining atmospheric air in their meshes which is the cause of this warmth."* Clothes of loose and open texture contain more air than those which are close and firm, and garments that are lined and wadded with very light material offer the same advantage of holding layers of air within the spaces of their texture. The loose and light kinds of wadding are the warmest, because they afford the largest space for air. The old-fashioned bed-quilts, which were made of double layers of old and worn woollen cloth, and a small layer of wool, very closely quilted, were much cooler coverings than the modern quilts of cotton cloth, with very light wadding of cotton or eider down. For the same reason, threadbare garments are colder than new, from which the nap is not worn off; and those which have a long and shaggy nap are much warmer than those which are well sheared and nicely dressed.

539. On the same principle, the garments should be made to fit loosely to the body, so as to leave a space for the air between them and the flesh. "Every one is practically aware that a loose dress is much warmer than one which fits closely; that a loose glove is warmer than a tight one; and that a loose boot or shoe is more comfortable in the winter than a tight one."* The loose sack is a warmer outer garment than the close-buttoned surtout. If there are

* Wilson on the Skin.

several layers of dresses, each one should be considerably looser than the next one within, so that a layer of air may be kept between them. In all these cases, the several strata of air between the different garments, and in the meshes of the loose textures of cloth, acting as non-conductors, prevent the passage of heat. From this cause, the attic chamber, which has nothing but the roof between it and the burning sun or freezing air, is much hotter in the summer and colder in the winter than the chamber below, which has the air of the attic between it and the solar rays or the outer atmosphere.

540. The various materials of our garments — linen, cotton, silk, and woollen — have different qualities, and are consequently suitable for different persons and seasons. The fibre of *linen* is round, pliable, smooth, and soft to the skin; it therefore makes a most agreeable garment. Yet it is a good conductor, and allows the heat to pass off rapidly, and therefore feels cold when it touches the skin. Moreover its fibre is porous, and absorbs and retains the water of perspiration. Water being a still better conductor than linen, those who wear this cloth are chilled after sweating, even in a hot day. For this reason, linen is more and more abandoned as an article for under-wear in hot climates.

541. *Cotton* is a worse conductor, and therefore warmer than linen. It is also soft, though less so than linen, and less pleasant to the touch, for its fibres are not rounded, but " are flat and have sharp edges," which irritate some delicate skins. But it does not absorb moisture, and for this reason it is the favorite and proper under-dress of all climates.

542. *Silk* is not so good a conductor, and is warmer than cotton. Its fibres are round and pliable, and it makes a pleasant garment for the skin. It attracts no moisture, and gives a sensation of freshness to the surface when it touches it. But, " on the slightest friction, it disturbs the electricity, and then becomes a source of irritation," and in very delicate and irritable constitutions it sometimes produces eruptions.

543. *Wool* is the worst conductor of heat, and is there-

fore the warmest for winter garments. It absorbs no moisture, and defends the wearer from the chills that frequently succeed perspiration in a hot but changeable climate. Its fibre is porous, and contains minute portions of air, and it makes cloth of loose texture. But its fibre is rough and scaly, and is very irritating to delicate skins. It also disturbs the electricity even more than silk. For these reasons, many cannot bear any woollen garment next to their bodies. However fine and delicate the fabric, it always irritates them.

544. It is desirable to guard the warmth of the skin, not only from the permanent influence of the atmosphere, but against any sudden changes which would produce a chill. If our clothing is filled with water, the heat is carried off very rapidly, as water is a good conductor. Wool is therefore a more appropriate material to be worn next to the skin than linen; and if the garment is made loose, and of fine texture, such as thin flannel, it is a great safeguard against the effects of changes in hot climates and hot seasons; and the feeble and delicate would be safe to wear it at all times.

CHAPTER XII.

Advantage of Flannel in hot Climates. — Cutaneous Excretions received on the Clothing. — Foul Clothing offensive to the Sense of Touch — Clothing and Beds should be aired.

545. Dr. Andrew Combe quotes the example of a British ship of war, which, after sailing for two years among the icebergs on the coast of Labrador, was immediately ordered to the West India station. On this change of location, every man was provided with flannel shirts and drawers, which they wore while in the hot climate. "The ship proceeded to the station with one hundred and fifty men, visited almost every island in the West Indies and many of the ports in the Gulf of Mexico, and, notwithstanding the sudden transition from extreme climates, returned to England without the loss

of a single man, or having any sick on board." The same commander had, at another time, the charge of the gun-brig Recruit, which lay about nine weeks at Vera Cruz, and used the same precautions in the clothing of his crew, and thus preserved the health of his men, while the other ships of war, which were anchored in the same harbor, and exposed to the same influence of climate and labor, lost two fifths of their men.*

546. It should be stated, that this wearing of flannel was not the only precaution taken by this provident officer for the health of his crew. Every kind of pains was taken to secure a dry and pure atmosphere in the seamen's sleeping apartments; and every means of cleanliness used, so that they should neither breathe foul air nor be exposed to foul exhalations from the walls and floors of their rooms.

547. A commander of a merchantman, who had sailed much to St. Petersburg, in Russia, and to the West Indies, East Indies, and Brazil, from Boston, informed me that he provided flannels as carefully for his southern as for his northern voyages, and he found them as effectual a safeguard against the diseases of the warm climates as against the chills, colds, catarrhs, and rheumatisms of the north.

548. Those who practise the cold water system in the treatment of disease, and who seem to bear exposure to cold water and cold air with remarkable ease, discard flannels as injurious. But their experiment has not been sufficiently tried to establish a universal law. It is therefore safe, at least for the old and the delicate, to adhere to their custom of wearing flannels next to the skin.

549. The cutaneous excretions are first received upon the clothing, and then a part of them are carried away by the atmosphere, and a part of them are retained upon the garments. It is easy to perceive this, by seeing the dark and dingy color of the white cotton or linen which has been worn next to the skin, and so closely covered by the outer clothing, that no dust nor dirt could come to it from abroad.

* Combe's Physiology, Chap. III.

This coloring matter upon these under-dresses could only come from the skin. If further proof were needed, notice the foul odor, on Saturday, of the inner garments of some laborious men, who do not bathe, and who change their shirts but once a week.

550. *Clothing that is soiled by being worn next to the flesh is offensive to the touch* as well as the sight and the smell. Shakspeare makes the merry wives of Windsor, when they wished to throw the greatest indignity on Sir John Falstaff, put him into a basket of foul linen, which was covered with the cutaneous excretions of the body. We feel a sensation of comfort when we put on clean linen, and of dissatisfaction when we put on that which is otherwise. And without the aid of the eye or nostrils, the sensitive skin can determine whether a garment is pure or foul, when we put it on. And, however dark it may be, we can tell by the feeling whether our sheets are fresh and clean, or soiled and worn. No children are more particular to put off foul clothing and put on clean, than the blind at the Institution at South Boston.

551. To prevent this accumulation of the cutaneous excretions, which, being retained, become foul and offensive, the garments which come in contact with the body should be frequently changed and washed. None of the clothing of the day should ever be worn in the night, nor ought the clothing of the night to be worn in the day. Morning and evening there should be a complete change of every article of dress; and each garment, when taken off, should be separately spread, in order that the air may come in contact with all their surface. By this airing, much of the foul excretions is carried away from them. The clothing which is taken off at any time to be reworn should not be hung up in a close closet, nor packed in drawers or trunks, until it shall have been thoroughly aired by a similar exposure.

552. That thrifty housewifery which requires the beds to be made up in the morning as soon as vacated by the lodgers, is prejudicial to health. The beds and bedding need airing more than the day clothing. This last is ex-

posed to some changes of air most of the time while it is worn, and a portion, at least, of the excretions is dissipated. But as the body, while sleeping, continues in one place, and with no change of air through the night, the bed-clothing loses none of the animal excretions, and there they remain in the morning. The bed, therefore, should be opened, its several parts separated, and the mattress, the feather-bed, and the under-bed, should be laid apart one from another, and the sheets, blankets, and all the other bedding hung on chairs or other things which will allow the air to reach both their surfaces. And thus should the chamber be left, and the bed be aired for some hours each day, with a window open, however cold the air.

553. It is an uncomfortable as well as an unhealthful custom to use the single cabin of canal boats for day as well as for night room. There all the work of life is carried on. There the passengers sit and eat during the day, and sleep during the night. To prepare for lodging, the beds are fixed to the walls by means of hooks and ropes, every evening; and, in the morning, in order to make room for the breakfast table, these beds are all taken down and packed in as small a compass as possible; without opportunity of airing, or any means of purification, they are closely compressed through the day, until they are needed again at night. In addition to the unavoidable excretions of the present night, these beds retain the accumulated excretions of several nights, and perhaps of successive passengers during a whole trip.

554. Some dwellings of the poor in cities present the same seeming necessity of piling the beds and the night clothes into one close heap, to allow room for the day operations of the family. Press-beds and sofa-beds in sitting-rooms, which are shut up immediately after being left in the morning, are liable to the same objection. They have no opportunity of being aired, and the foul excretions of the night are retained during the day, to irritate the skin of the lodger when he returns.

CHAPTER XIII.

Dead Particles of Cuticle are lodged on the Skin. — *We bathe the Hands and Face, but the Body is not generally bathed.* — *Those who bathe daily have soft Skin.* — *Bathing a religious Rite in ancient Times and in Oriental Countries.* — *Much practised in Russia and Finland.*

555. THE cuticle is constantly casting off its outer layers in scurf or minute scales. Some animals cast their skins entire, and others cast their shells once a year; but man is incessantly casting his skin. The outer scales, which are the dead particles of the cuticle, lie loosely on the surface, and can be scraped off with a knife at any time; they have then the appearance of branny powder. Some animals cast their hair or shed their coats, and birds moult their feathers annually. But the hair and nails of man grow from their roots, and thrust out their outer extremities, which, if not trimmed, would be continually breaking and dropping off.

556. If any one who had for a long time deprived himself of the needful luxury of a warm bath should remain for several minutes in one, he would be surprised to see how large a quantity of this accumulated matter of the dead skin he could rub off with his hands or a flesh-brush. The removal of this gives to the skin a very agreeable sensation of comfort. This is more perceptible after taking a warm than a cold bath.

557. We bathe the face and hands daily, and oftener, and know how comfortable the skin upon those parts feels after this operation. But if this duty is neglected, the skin is irritable and irritated; it seems stiff and loaded, and we feel disposed to scratch and rub it to remove the disagreeable burden. But the other parts, which are not so frequently washed, are not so easily offended. They bear the burden of accumulated excretions and dust with less complaint. But, if they were cleansed as faithfully as the hands and the face, they would be equally sensitive, and feel as keenly

the comfort of a bath and the discomfort of neglect. This sensibility of the skin of the hands and face is a mere matter of cultivation, and might as well be cultivated in the skin of the other parts which are covered with clothing. But those parts which are not exposed to sight are with most people rarely, and with some never, bathed; and the great majority of mankind leave so much of their surface unwashed and untouched with water, from summer, through the entire cold season, until summer again returns.

558. The consequence of this negligence of ablution is, that the skin becomes overloaded with the gathered excretions of months and years; it loses its exquisite sensibility; it is less able to throw off the waste of the body; the cutaneous circulation is not so well sustained; the skin is less supple and elastic, and less able to maintain the equilibrium of heat; and the whole body is comparatively dull and inactive. In those who are accustomed to take their daily entire bath, the whole skin is soft and elastic; the cutaneous waste is carried freely away. They are consequently enabled to bear the heat and cold, over their whole frame, with much more ease than others do who wash their hands and face alone; and they enjoy a more acute sensibility of skin, a general lightness and buoyancy through their frame.

559. *To maintain the most perfect health of the skin and of the internal organs, the whole surface should be daily cleansed of all its excretions,* — the oil, the scales of the cuticle, and the salts of the perspiration, — and also of the other matters which lodge on the body and become mixed with these. No part should be neglected. Water will remove the salts, and soap the oily excretions. Nothing can be substituted for soap. Some have attempted to use wash-powder, and others sometimes use alcohol or spirits; but none of these combine with the oily matter, or dissolve the others, and cleanse the skin. "Soap," says Dr. Wilson, "renders the cutaneous product of the skin freely miscible with water, and hence it is an invaluable agent in purifying the skin.]

may affirm that it is an indispensable aid; for in no other way can the cutaneous substance, and the dirt which adheres to it, be thoroughly removed from the surface."* No other matter applied to the surface will give it the healthy glow, the comfortable sensation, and the natural and lively look, and beautiful hue, that are left by soap and water. Various kinds of powders are sometimes used upon the face, with the mistaken notion of improving its beauty. These mix with the oily excretions, and form a pasty compound. They increase the burden upon the skin, and impair its vitality, deaden its liveliness of expression, and sully the brightness of its color.

560. Some nations have practised bathing as a religious rite. It was a good custom of the ancient Israelites, the Egyptians, and the inhabitants of the East Indies, to bathe as a part of duty, as typical of moral purification. The Greeks and Romans considered bathing so essential, that their public bathing establishments were large and magnificent, and their private baths were as splendid as the means of the owners would permit.

561. The moderns have not improved upon the ancients in the care of their skins; nor have the civilized nations of Central and Western Europe, and America, improved upon the less cultivated Hindoos, Persians, and Turks. The Russians and the Finlanders indulge themselves very much in this matter, and baths are attached to houses of all classes in Finland, Lapland, Sweden, and Norway.† But the English and the Americans do not generally use the bath.

562. There are various kinds of baths, — the cold and the warm, the shower and the vapor bath, — all of which have their appropriate uses. The Russians are very fond of their peculiar vapor bath. This is one great hall, warmed by stoves; large red-hot stones are placed on the stone floor, and water poured upon them; the room is then filled with vapor heated up to 120° or 130° Fahrenheit. Dr. Grenville says he found that the Finlanders, in some instances, sat half an hour in vapor baths heated by hot stones to 169°

---

* On the Skin.   † Bell on Baths. p. 33.

The bathers, covered with the steam, sit on benches until they break out with a profuse sweat. Then they are washed with soap suds, and next buckets full of warm water, and lastly of cold water, are poured upon the head. Sometimes the Russians will run from this steam bath and plunge into a bank of snow, and feel no injury; on the contrary, a comfortable glow of heat comes from the vigorous circulation, which the cold of the snow stimulates.*

CHAPTER XIV.

Cold Bathing. — Sponge Bath. — The most Laborious need Daily Bath. — Some cannot bear Cold Bath.

563. If the health is good, and the body is full of animal heat, the cold bath answers the purpose of health in summer; and if used with energy and perseverance, it is also sufficient for winter. It is an excellent habit of some to take a cold bath every morning, both winter and summer, not omitting it even in the coldest weather. Those who do this find it not only very endurable, but they usually enjoy a glow throughout all the surface afterwards. So far from suffering from cold, the reaction of the cutaneous circulation produces an increase of heat, and they are made the warmer by this ablution.

564. It is desirable that every one should be able to take, daily, a plunge bath. But this is impossible for all. The convenience of a large bathing tub cannot be provided in every house. Yet a good substitute is within the reach of all. The sponge bath is very easily taken, and requires but a very limited and simple apparatus. Provide a large wash-bowl and a piece of extra carpet, which should receive the drops that fall to the floor, — or, what is much better, a large tin basin in shape of a hat, with a shallow crown and very broad brim, — and then a soft towel for wiping, and a crash towel for friction. These are all that are needed, and

* Bell, Chap. II.

with these, this very grateful and invigorating ablution can always be performed on getting out of the bed.

565. The cold bath is most conveniently taken as soon as one gets out of his bed. It is best to take it when one is warm, when there is sufficiency of heat to bear the shock and to produce the reaction. This bath should not be so long continued in winter as materially to reduce the heat and energies of the circulation in the skin; and immediately after it the surface should be dried and rubbed until the reaction commences. The exercise necessary for this rubbing, which the bather should do for himself, and the friction on the surface, excite the circulation, and produce a very pleasant glow of warmth upon the whole frame.

566. None need this bath more than the most industrious laborers, who have the greatest demand for their strength, and therefore need to take the greatest pains to develop it. Yet it is generally urged by them, as a reason for the neglect of this duty, that their avocations allow them no time for this, and, however well it may do for the wealthy and the men of leisure, it cannot be performed by the poor and laborious, who are always in haste in the morning to go to their work. This is certainly a mistake of calculation. The mechanic considers no time lost that he devotes to putting his machine in good order; and the wagoner thinks it an advantageous disposition of his time to rub and curry his horses faithfully. Both of these believe that they will be enabled to accomplish so much the more for this preparatory care. So it is with the laborer's body. In order that it should be able to accomplish the most work, it must be put in the best working order; and this is done, in part, by cleansing the skin of all impurities, unloading it of its burdens, and so preparing it for its functions that its work will not only go on well during the day, but contribute its portion to the general health and the muscular power.

567. This cold bathing is a general rule, but not a universal one for mankind; for some cannot take it with safety. If the body is in full health, and the circulation vigorous, —

if, after the bath, there is reaction and a glow of heat throughout the surface, — then the cold bath is both safe and useful; but if the body is feeble, or the flesh cold, and not easily warmed after the bath, or even if the heat does not naturally and spontaneously return, then this bath is injurious, and a tepid bath should be substituted in its stead. Any one can tell, by his own experiment, what temperature he can best enjoy, and how great a degree of cold will be followed by the comfortable sensation of warmth in his skin.

568. Even for those who take the daily cold bath, the warm bath is occasionally necessary. This is a more effectual cleanser of the surface. It softens and allows the removal of the dead scurf of the skin more readily than the others; and for those who do not practise the daily ablution, the occasional use of the warm bath, for mere cleanliness as well as for health's sake, should not be omitted.

CHAPTER XV.

Effect of Cold Bathing. — Protects against Cold. — Feeble and consumptive Persons should bathe. — Time for Bathing. — Conditions of Bathing.

569 The effect of the cold bath is not only to invigorate the body, and give a tone and activity to both the skin and to the internal organs, but it fortifies the skin, so that it is better able to endure the exposures to the cold abroad. I formerly clothed myself very carefully, seldom went abroad without an overcoat in the winter, and often wore a cloak over this. Seven years ago, I began the practice of cold bathing, and have followed it without intermission since, breaking the ice in the coldest weather, and bathing in my chamber, where was no fire. I now wear lighter clothing in winter, a great coat much less than formerly, and never an extra cloak; and, with so much less protection, I suffer less from cold than when I was clothed more, but did not bathe.

570. Dr. Andrew Combe confirms this effect of the daily cold bath by his own personal experience, and by the observation of others. "Instead of being dangerous, it is, when well managed, so much the reverse, that the author of these pages has used it much, and successfully, for the express purpose of diminishing this liability, both in himself and in others, in whom the chest is delicate. In his own instance, in particular, he is conscious of having derived much advantage from its regular employment, especially in the colder months of the year, during which he has uniformly found himself most effectually strengthened against the impression of cold, by repeating the bath at shorter intervals than usual. Few of those who have steadiness to keep up the action of the skin by the above means, and to avoid strong exciting causes, will ever suffer from colds, sore throats, or similar complaints." *

571. For the weakly, — for those who are liable to pulmonary complaints, who may have any hereditary disposition to consumption, or who are subject to rheumatism, — the cold bath, or, if this cannot be borne, the warm or tepid bath, is one of the means of protection, and should never be omitted by people of such tendencies. It should be begun in summer, and practised, without intermission, through the autumn and winter; and the gradual increase of strength and the power of endurance will keep pace with the gradual approach of the cold season.

572. It has before been stated (§ 565, p. 234) that one should take his bath in the morning, on rising from the bed. This is a matter of convenience; for then the labor of undressing is spared, and the means may be ready. This, however, is not to be universally practised. To some, — the feeble and the debilitated, and to others of peculiar temperament, — the bath upon an empty stomach, when the system wants nourishment from the fast of ten or twelve hours, would be injurious. On the other hand, the bath is not the safest and best, when taken upon a full stomach, imme-

* Physiology, Chap. III.

diately after meals. For then the cold bath might drive the blood too much within, and oppress the stomach, which is already excited with the work of digestion; and the warm bath, by relaxing the cutaneous vessels, might draw too much of the blood outward, when it is needed for the work within. In either case, the balance of the circulation is disturbed, and the digestion is interrupted.

573. The best time for the bath is in the forenoon, afternoon, or evening, when the system is well nourished, and the stomach is not full, nor the duodenum empty; nevertheless, the vigorous and robust may take it in the morning, before eating, with impunity, and even with advantage.

574. There is a very common notion that it is injurious to go into the water when the body is warm. We have seen (§ 562, p. 233) that the Russians go from the hot vapor bath, in a profuse perspiration, into the snow. It must not be supposed that the case of one of these, heated by the vapor bath, is strictly analogous to that of one who is profusely perspiring with running or other exercise; and therefore the practice which is safe for the Russians in one case, may be unsafe for one who is differently heated. Yet the contrary is not true. The rule which is often enjoined upon boys and men, that, when they go to the river-side or sea-shore to bathe, they should first sit on the bank, in the cool air, until their temperature is reduced somewhat toward that of the water, lest they be injured by the sudden change, is not a good one. One should not go into a cold bath when he is already cold — when he has lost so much heat that he can spare no more, for any further reduction would be injurious.

575. The practice of two young men, who, several years since, bathed in the Connecticut River daily, during the summer and autumn, even through the month of November, was contrary to this; and it certainly was successful in their case, and doubtless may be in others similarly situated and with similar constitutions. They lived rather more than half a mile from the river, and on an elevation from which there was a descending slope of about a hundred feet to the

water. As soon as they could see daylight, during the colder months of their bathing, they ran down this slope to the water's side, undressed, and plunged in immediately; and, after remaining a few minutes, they came out of the water, dried, rubbed, and dressed themselves, and then ran back to their home, as they said, "in a delightful glow."

576. The same principle directs us to warm ourselves well before going abroad in winter. There is no ground for fear of taking cold by going from a warm room to the cold air, if the body is properly clothed, or if the exercise abroad is sufficiently vigorous. The practice of some to cool the body partially before going out in the winter, so that the changes shall be neither great nor sudden, is altogether needless and unphilosophical. Let one sit for a time in a cool room, and reduce his temperature as low as he can bear it, without much discomfort, and then go abroad into the colder air, and he will begin to suffer much quicker, and be much more liable to take cold, than another who has been sitting in a well-warmed room, and goes out into the cool air, with a comfortable heat in his body.

CHAPTER XVI.

Nervous Sensibility increased by Bathing. — Sense of Touch made more acute. — Nervous System affected through the Skin. — We must aid the Skin in the Performance of its Functions.

577. *Another effect of bathing is to heighten the nervous sensibility.* The whole human surface, amounting to fifteen square feet, being bespread with the terminations of the innumerable nerves, exposes a wider extent of the nervous system to the influence of external substances than any other organ; and through this the body receives stronger and severer impressions than through any other avenue. An injury to the eye, the nose, or the ear, is generally limited in its consequences to the injured organ: this may be de-

stroyed, while the rest of the body remains sound; but a burn upon the skin — certainly one that covers the entire surface, though not deep, and even slight — is fatal.

578. If this great extent of skin is covered with the gathered excretions of days, and months, and years, — if its pores and its excretory apertures become filled, and the surface agglutinated with the compound of perspiration, and oil, and dust, — the sensibility of the nervous extremities must be blunted, and the power of receiving impressions materially diminished. But when the skin is cleansed and unburdened of its load of impurities, these nervous points are more free to receive impressions, and are more easily acted upon by external objects; the skin has then a more lively sensibility to pleasure and pain, to heat and cold, and a keener sense of touch. The blind would not attempt to read his raised letters, nor the draper to discriminate the qualities of cloth, with soiled fingers; nor would the accomplished performer play on his violin with unwashed hands.

579. Laura Bridgman, whose senses of sight and hearing are lost, has a most delicate sense of touch. The increased sensibility of her skin compensates, in good measure, for her other privations, and enables her to maintain a fastidious neatness of person and dress. Nothing of the kind can exceed the purity of her skin, or the acuteness and liveliness of her cutaneous sensations. She has, therefore, the nicest power of discerning and comparing minute objects. She can, with unusual correctness, discriminate the various textures of cloths, and distinguish the different degrees of fineness of dresses. She enjoys the delicacy of workmanship upon wood and metals, and discovers the cleanness and uncleanness of her clothing. Few detect more readily any blemish upon her garments, or seem to be more averse to wear unwashed linen, or more desirous to enjoy the change of the worn for the fresh dresses from the laundry.

580. Some produce the greatest effect on the nervous system through this avenue of the skin. Esquirol, one of the ablest writers upon insanity, and the physician in a very large

Lunatic Hospital in Paris, calms the excitement of the furious maniac by pouring cold water over the whole surface for a considerable period, and, by thus cooling the skin, depresses the external nervous system, and, through this, tranquillizes the agitated brain. This method is practised with wonderful success in his hands, and is elsewhere found to have a similar effect upon many excited patients.

581. Such is the structure of, and such are the offices performed by, the skin. We have seen that the former is complicated, and the latter are numerous. It is manifest that this whole organ is intimately connected with the operations of all the other organs, and that the freedom of action, the health, and the very life of the inner man, depend very materially upon the healthy condition of the outer man. We have seen, also, that this organ, while it is ready to perform its own part well, cannot do it alone, but stands in need of our direction and aid to help it in its work, and our faithful watchfulness to guard it from suffering and evil.

582. We have therefore a responsibility to sustain in regard to the skin. We must bathe it, and purify it from all foulness, and cleanse it from all the excretions that adhere to it; we must give it tone and vigor, and power of resistance to external injury, and make it more capable of receiving impressions and conveying sensations; and, lastly, we must clothe it, and otherwise defend it from excessive radiation of heat, and over-active evaporation, which would cool the body below its healthy and natural temperature; and yet we must not so overclothe it as to lessen its power of self-protection.

PART VI.

BONES, MUSCLES, EXERCISE AND REST.

CHAPTER I.

Bones, Composition of. — Flexible in Childhood. — Brittle in Old Age. — Strongest in Middle Life. — Supplied with Blood-Vessels and Nerves. — Subject to Growth and Decay. — Grow strong by Use. — Should be used cautiously in Childhood. — Rickets.

583. The bones are hard, stiff, and very strong. They are externally solid, but are somewhat hollow within. They are composed of such materials, that, without being heavy, they are very firm, and formed in such a manner, that, without being large or clumsy, they are very strong. The composition of the bones is twofold — the earthy and the animal. The earthy part of the bones is lime, or rather a phosphate of lime. This gives them solidity and firmness. The animal part is composed of gelatine, which is a substance similar to glue. This gives the bones their strength and life. Either of these alone would make imperfect and weak bones.

584. When these two elements — the lime and the gelatine — are united in due proportions, the bones are very strong, and will bear very heavy shocks; but, if either is deficient, the body is not supported. When the lime is deficient, the bones will bend; and, when the gelatine is deficient, they will break. We see these different conditions of the bones in the different periods of life. In early infancy, the gelatine predominates, and the bones are soft and yielding. They are then easily bent, but not easily broken.

585. As the child grows, the lime is added, and the bones become stronger, until the full maturity of life, when the composition is the most perfect and the frame has the greatest power of resistance, and will bear the hardest blow and support the greatest burden without suffering. This state continues until the approach of old age. Then the gelatine diminishes and the lime preponderates; the bones consequently become brittle, and are more easily broken.

586. *The bones are supplied with blood-vessels and blood;* they are subject to growth and decay, to deposition of new matter and absorption of the old particles, as the other textures are. The change is shown in the experiment of feeding sheep with some coloring matter. (§ 248, p. 114.) The bones of these animals were red while they ate madder, and became white when they returned to the usual food of hay and grain. In the first case, the red particles of the food were deposited in the bones unchanged. In the second, these red particles had been absorbed, and others, of the natural color, had been deposited in their stead. The bones are supplied with nerves, and are therefore susceptible of pain under some circumstances. In some states of disease, the patient complains of pain in his bones. When the bone is sawed in amputation, it does not seem to suffer; but, if it becomes inflamed, the pain is very severe.

587. When a bone is broken and the parts divided, the textures of the severed ends repair the breach. They, in the first place, throw out at the broken extremities a quantity of adhesive matter. This unites the parts with a soft bond, which would prevent their being drawn asunder, but would not prevent their bending at this place. After this flexible union is formed, the blood-vessels throw into it earthy matter, which combines with the jelly, and forms a new bone, of composition similar to that of the original bone. At first, this new structure is not quite so firm as the old; but nature provides for this by increasing the quantity of new deposit, and making a bulbous projection all around; thus

the bone is at the place of junction larger than either portion of the shaft, above or below. When, in the lapse of time, this new bone becomes condensed, and as strong as the old bone, then the external deposit is absorbed, and the whole shaft is nearly restored to its original form.

588. The process of absorption and deposition of the particles of bone is shown in their change of shape; as when a tumor or enlargement of the arteries presses upon the ribs within, and causes some parts of them to spread; or as when the close dresses press upon the same bones without, they contract to meet the necessity of the case. (§ 340, p. 150.)

589. *The bones grow larger and stronger by use, like the other systems.* Exercise of the parts quickens the circulation and increases their nutriment. Disease and inaction weakens them. If any one in good health should lie upon his bed for a long period, — months or years, — at the end of this time he would not find it easy even to stand. The bones would not easily support his weight. The exercise of the bones favors the deposition of the earthy particles; and for this reason the bones of the laborer are dense and hard. They have the due proportion of the animal matter and of the earthy matter in their composition, and have, consequently, great strength. But those who are unaccustomed to labor, or even exercise, have not the full proportion of lime in their bones, nor the strength that belongs to the working man.

590. The bones in early life, being more gelatinous and earthy, are consequently weak; and the child, although able to exercise, is incapable of hard labor. If he is put to hard work, the deposition of the earthy matter is hastened, and the bones become consolidated before they attain their full size; and the boy, not being allowed sufficient time for growth, becomes a stunted man. The bones of the child require more care for their shape and their growth than those of the man; and if not supported, or if made to bear too great weight, they are liable to become distorted. In sitting, the child should either find rest for the entire thigh bone, from the hip to the knee, upon his chair or bench, or the lower part should

be supported from the knees and the legs by the feet resting upon the floor. Too many of the school-rooms are furnished with seats built upon one uniform model and of the same height. If these are high enough for the older and larger children, they are too high for the younger and smaller. When, therefore, these sit, the lower leg and foot hang from the lower part of the thigh, which projects beyond the seat, and may cause it to suffer. (Fig. XXX.) The weak bones

Fig. XXIX.   Fig. XXX.

of this age do not well bear long continuance of any posture; the attitudes, therefore, should be very frequently varied.

591. This process of consolidating the bones from infancy to old age is gradual, and is one of the evidences of good health. But in some feeble persons the lime is not deposited in the usual proportion, and the gelatine prevails through life, as in childhood; this is the disease familiarly known as the *rickets*. The bones are then weak, and liable to be bent. The heads of such bones are generally enlarged and misshapen, and the shafts frequently crooked. The spine is curved, and sometimes the skull is enlarged. This disease happens mostly among those who are badly nourished, who have poor and insufficient food, who live in damp and dark rooms or hovels, and breathe foul air.

592. The bones are not solid. Their inner parts are loose

and porous, but the external layers are arranged in cells, like a honey-comb. This is the most observable in the heads of the long bones, as in that of the thigh, (Fig. XXXI.) This arrangement gives the bones the greatest strength with the least weight.

FIG. XXXI. *Head of the Thigh Bone sawed open.*

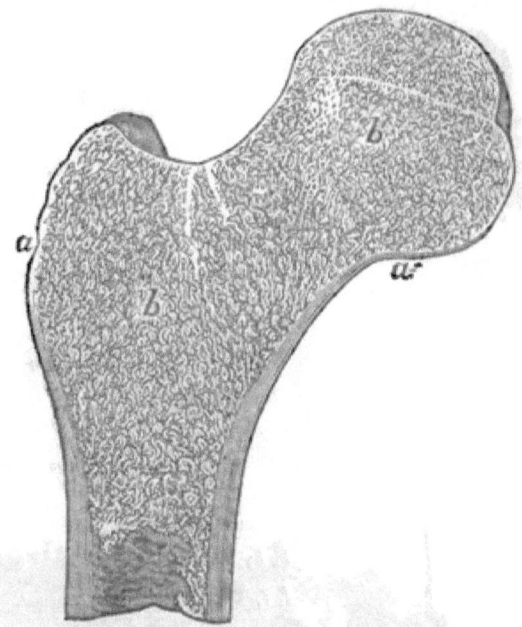

a, a, Outer layer of the bone.
b, b, Inner, or honey-comb structure.

CHAPTER II.

Skeleton. — Number of Bones. — Head. — Chest. — Spine. — Vertebræ. — Cartilages. — Pelvis.

593. *The skeleton* (Fig. XXXII.) *is composed of two hundred and forty-six bones, including the teeth and the parts of the head.* Some of these bones are thin and flat, as the shoulder-blade, (Fig. XXXVII.,) and the parts of the head.

Some are long and flat, as the ribs, (Fig. XV.) The bones of the arms and legs are long and roundish. The vertebræ, (Fig. XXXIV,) or the parts of the spine, and the bones of the wrist (Fig. XXXVIII.) and ankle, (Fig. XL.,) are more compact. The bones of the pelvis (Fig. XXXII.) are irregular

FIG. XXXII. *Skeleton.*

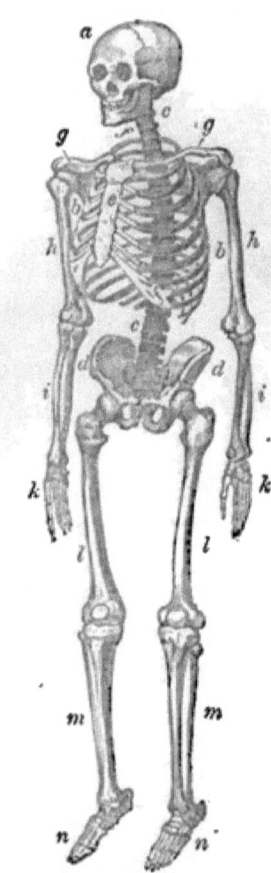

a, Head, or *cranium*.
b, b, Chest.
c, c, Back-bone, or *spine*.
d, d, Pelvis.
e, Breast-bone, or *sternum*.
f, f, Ribs.
g, g, Collar-bone, or *clavicle*.
h, h, Upper arm-bone, or *humerus*.
i, i, Bones of the fore arm.
k, k, Bones of the hands and fingers.
l, l, Thigh-bone, or *femur*.
m, m, Bones of the leg.
n, n, Bones of the foot.

in shape. The bones have various projections, which serve for the attachment of the muscles, and apertures for the blood-vessels and nerves to pass through; and all are made to fit their places in the structure, and fulfil their purposes in the animal economy.

594. *The skeleton is divided into the head, the trunk, and the upper and lower extremities.* The *trunk* includes the spine, or the back-bone, the ribs, the breast-bone, and the pelvis. The *head* is composed of the eight bones of the skull, the fourteen bones of the face, and the six little bones of the ears. The eight bones of the skull are arranged to form a

Fig. XXXIII. *Bones of the Head.*

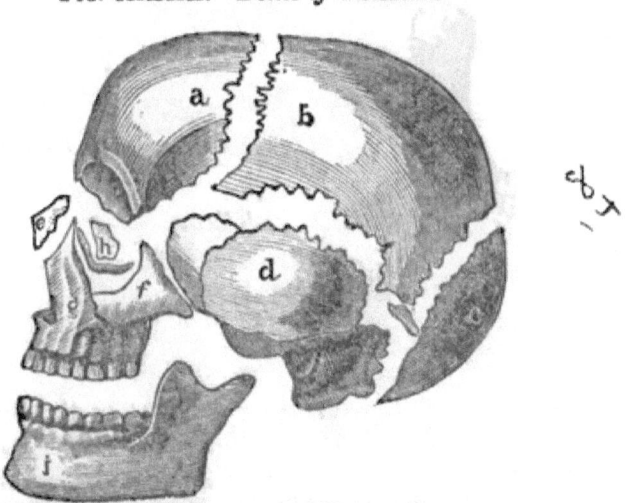

a, Frontal bone.
b, Parietal or side bone.
c, Occipital bone.
d, Temporal bone.
e, Nasal bone in the nose.
f, Malar, or cheek bone.
g, Upper jaw.
h, Thin bone below the eye.
i, Lower jaw.

hollow box, to contain the brain. They are united and held together by a sort of dove-tailed joint, (Fig. XXXIII.) This arrangement and union of these bones give great strength to the skull, and enable it to bear heavy blows without breaking. The brain has, therefore, a secure resting-place.

595. *The chest* is composed of twelve of the vertebræ, or bones of the spine, the twenty-four ribs, and the breast-bone, (Fig. XV.) The *spine* or *back-bone* is composed of twenty-four distinct bones, called vertebræ. Each *vertebra* consists of a body, a ring, and various processes or projections,

(Fig. XXXIV.) These are placed one upon another, from the pelvis, at the bottom of the back, to the head, on the top of the neck. These bones vary in thickness, from about an inch in the loins, to about a quarter of an inch in the neck.

Fig. XXXIV. *Vertebra of the Neck.*

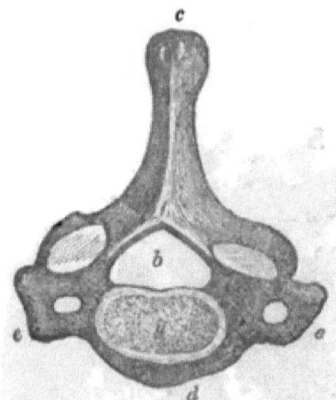

a, Body of the bone, upper surface.
b, Ring.
c, Process of bone extending backward.
d, Front surface of the bone.
e, e, Processes extending to the right and left.

596. There are seven of these bones in the neck, called *cervical vertebræ.* These are thin, and have long projecting or *spinous processes* extending directly backward, (Fig. XXXIV. *c.*) There are twelve bones in the back, against the chest. These are called *dorsal vertebræ.* They are con-

Fig. XXXV. *Vertebra of the Back.* (*Side View.*)

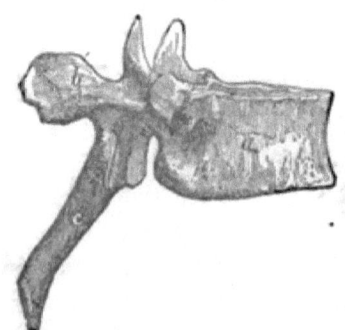

a, Body of the bone.
b, c, Processes of bone extending backward.

nected with the ribs. Their long spinous processes extend downward, (Fig. XXXV. *c.*) The other five are called *lum-*

bar vertebræ. These are in the hollow of the back, between the pelvis and the chest. They are very thick, and have short, club-shaped spinous processes, that extend directly backward. These processes can be felt through the skin.

597. Between these vertebræ are layers of very tough and elastic cartilage, which adhere very firmly to the bones, and hold them together. These layers are very thick in the loins, and thin in the neck. They are capable of compression and expansion, like India rubber. They may be compressed on one side and stretched on the other at the same time. When we bow, these cartilages are pressed and flattened on the front side, and stretched and thickened on the opposite side. In this way, we bend the back in any direction, and the spinal column is made exceedingly flexible with this succession of joints. This cartilage is very strong, and capable of sustaining great weights. The head, the arms, the chest, and most of the abdomen rest upon, and are supported by, the spine. All this weight resting upon the back-bone from morning till night, while the body is in an erect position, brings so much pressure upon these intervertebral cartilages, especially those at the lower part of the spine, that they become somewhat flattened in the course of the day, and thereby the length of the spine is diminished, and a man is from a half to a whole inch shorter at night than he is in the morning. But as soon as the body is placed in a horizontal position, and the pressure taken from the cartilages, they begin again to expand, and the column recovers its length, and the man regains his ordinary height by morning. Nevertheless, the continued pressure of the weight of the upper part of the body overcomes, in some degree, the elasticity of the cartilages in the course of a long series of years, and a man is consequently somewhat shorter in old age than in his youth.

598. *The bones of the spine are arranged so as to form a column with a double curve,* somewhat like the Italic *f*. At its lower end it is curved outward; as it ascends it is curved inward at the loins, and forms the hollow of the back; again, it is bent outward at the upper part of the back to enlarge

Fig. XXXVI.
Spine, or Back-bone.

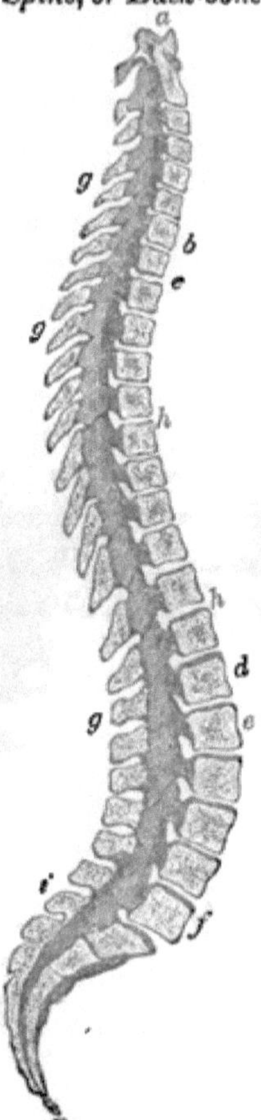

the chest and give room for the lungs; and, finally, at the neck it is erect. Notwithstanding these curves, the top of the spine, the resting-place of the head, is vertically over the sacrum, on which the last bone of the spine rests. This arrangement of the bones of the spine gives this column great strength and flexibility.

599. *The pelvis forms the base of the trunk,* (Fig. XXXII, *d.*) It is composed of three bones — the two hip bones and the sacrum, which is apparently a continuation of the spine. These bones are spread out to form a sort of basin, on which the abdomen rests. The spine stands on the sacrum, and the thigh bones are attached to the hip bones.

a, Resting-place of the head.
a, b, Seven cervical vertebræ.
c, d, Twelve dorsal vertebræ.
e, f, Five lumbar vertebræ.
g, g, g, Spinous processes.
h, h, Intervertebral cartilages.
i, Sacrum, a part of the pelvis.

CHAPTER III.

Upper Extremity. — Arm. — Wrist. — Hand. — Lower Extremity. — Leg. — Foot. — Arch of the Foot. — Shape of the Foot. — Natural. — Deformed.

600. THE *upper extremity* includes the collar-bone or *clavicle,* the shoulder-blade, upper arm, fore-arm, wrist, hand,

and fingers. The shoulder-blade or *scapula* (Fig. XXXVII.) is a broad, thin bone, of triangular shape. It lies flat on the back of the chest, imbedded in the flesh, and held in its

FIG. XXXVII. *Shoulder-Blade, or Scapula.*

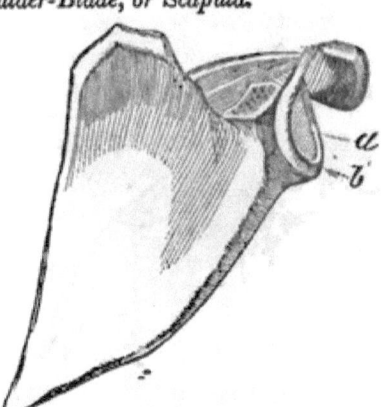

a, Socket for the head of the bone of the upper arm. These form the shoulder joint.

b, Border of the shallow socket.

place by the muscles. It has no direct attachment to the trunk, but at its upper and outer corner it is connected with the collar-bone. At this upper and outer corner it has a shallow socket for the head of the bone of the arm.

The collar-bone or *clavicle* (Fig. XXXII. *g, g*) extends

FIG. XXXVIII. *Bones of the Wrist.*

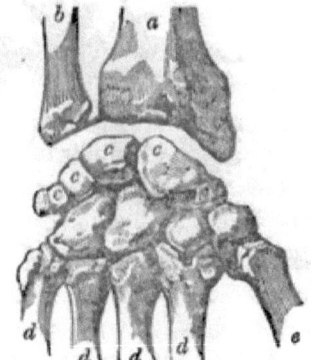

a, Radius.
b, Ulna.
c, c, c, c, Bones of the wrist.
d, d, d, d, Bones of the hand.
e, Thumb.

from the upper end of the breast-bone to the upper and outer corner of the shoulder-blade. It keeps the shoulder in its place.

The upper arm has a single bone, *the humerus*, (Fig. XXXII. *h*.) The fore-arm has two bones, *the radius* and *the ulna*, (Fig. XXXII. *i*.) The wrist has eight bones, (Fig. XXXVIII. *c, c, c, c,*) which are held so firmly together by ligaments that they are rarely displaced. The hand is composed of four bones, to which the fingers are attached, and the bone to which the thumb is fixed, (Fig. XXXIX.) These

FIG. XXXIX. *Bones of the Hand.*
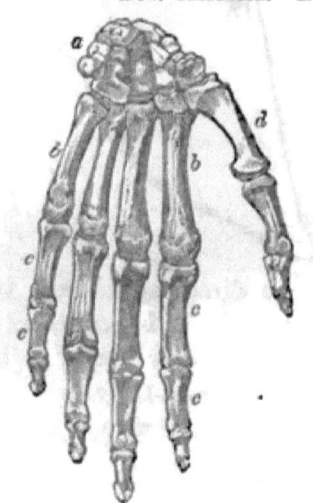

a, Wrist.
b, b, Hand.
c, c, c, c, Fingers.
d, Thumb.

are held more loosely together by ligaments, and enclosed in one sheath of skin. The *fingers* are each composed of three bones, (Fig. XXXIX. *c, c*.) The hand is beautifully and skilfully arranged and adapted to an almost infinite variety of purposes. Its wonderful structure and multiplied uses are suited to the exigencies of the mind which directs it, and gives to man a superiority over all other animals.

601. *The lower extremity* (Fig. XXXII.) is composed of the thigh-bone, which is a single shaft, the knee-pan, the two bones of the leg, the bones of the ankle, foot, and toes. The foot (Fig. XL.) is composed of twelve bones. Seven of these bones are of irregular shape, and are arranged to form the ankle and the arch, (Fig. XL. *a, b, c, d, e, f, g*.) The other five bones are long. They are joined to the instep

behind, and support the toes in front. The great toe has two bones, and each of the other toes has three bones, corresponding to the bones in the thumb and fingers.

Fig. XL. *Bones of the Foot.*

a, b, c, d, e, f, g, Bones of the ankle joint.

i, i, Bones of the anterior part of the foot.

k, Great toe.

l, l, Other toes.

602. The arch of the foot extends from the heel to the ball, (Fig. XLI.) The bones are exactly adapted to each

Fig. XLI. *Arch of the Foot. (Side View.)*

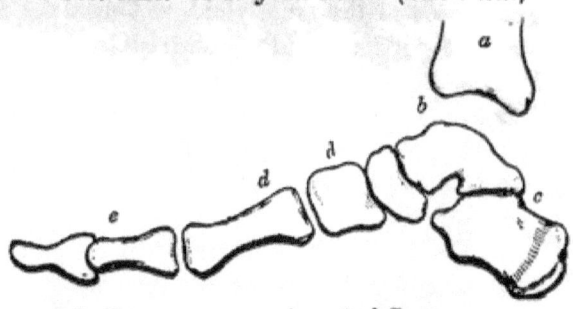

a, Bone of the leg.
b, Upper bone of the ankle.
c, Os calcis, or heel bone.

d, d, Instep.
e, Great toe.

other, and are held together by ligaments, very firmly, but

not immovably. The arch, therefore, is not unyielding like the skull, but it is somewhat loose, and allows a little spring to the foot when the body rests upon it. This arrangement gives to the foot both strength and elasticity. It admits great ease of motion, and saves the jar that would otherwise happen, when we step upon the ground. The foot rests, not upon its whole under surface, but upon the heel and the ball near the toes, which are the ends of the arch. The hollow of the foot bears none of the weight. The bones of the leg (Fig. XLI. *a*) rest upon the top of the arch, *b*.

603. When we step the hollow of the foot upon a round stick, so that the ends of the arch do not support the weight of the body, as when walking on the round steps of a ladder, or when a boy walks on stilts, we feel an unpleasant jar, and the want of that elasticity and ease in the step which we feel when we walk naturally on a flat surface. When we walk, we first place the heel upon the ground; this receives a part of the shock; next, the ball comes to the ground, and the force or weight comes upon the arch. Thus the shock is so divided that it is hardly felt.

604. When we jump down from any high place, we throw the toes downward, so that the first force of the blow is received upon the ball of the foot. The ankle then bends, and the second force is received upon the heel; and again these two, being the ends of the arch, yield, and thus the force is divided into three portions, and is received in part upon the ball, the heel, and the arch; and thus no violent jar is communicated to the general frame above. A sailor, falling from the mast in a rolling ship, struck the hollow of his foot upon the railing, and received a very severe shock and much injury. But another, falling the same distance, struck with the ball of his foot upon the level surface of the deck, and received no great shock and suffered no material damage.

605. The natural shape of the foot is somewhat broad in front, with the toes spread, or, at least, lying loosely, and the inner side of the great toe in a line with the heel and the ball. The greatest length of the foot is along this line,

BONES, MUSCLES, EXERCISE, AND REST.   255

(Fig. XLII.) This gives to all the bones freedom of motion, and to the whole foot its greatest elasticity. But the shoes that are usually worn are narrowed in front of the ball of the foot, and the toes are rounded, and even pointed sometimes. This shape carries the toes inward from both sides,

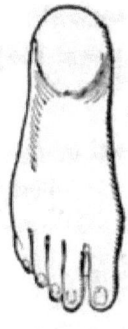

Fig. XLII.
Natural shaped Foot.

Fig. XLIII.
Compressed Foot.

and causes some of them to override others, (Fig. XLIII.) This diminishes the freedom of motion, the elasticity of action, and the usefulness of the feet, and creates a limping and awkward gait.

CHAPTER IV.

Joints. — Hinge. — Elbow. — Knee. — Ball and Socket. — Shoulder. — Hip. — Cartilages. — Self-oiling Apparatus. — Ligaments. — Capsules. — Sprains. — Dislocations.

606. *The joints unite the bones together*, and yet allow them to play upon each other. They are so strong that the bones cannot be separated without great violence, yet they do not interfere with their motions upon each other. There are various kinds of these joints, suited to the wants of the several parts in which they are placed.

The hinge joint allows motion in only one direction, forward and backward, as in the elbow, (Fig. XLIV.,) the knee, (Fig. XLVII.,) and the connection of the lower jaw with the head.

Fig. XLIV. *Elbow.*

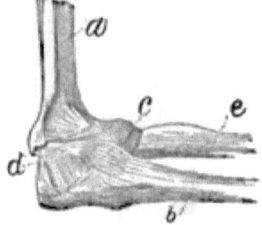

a, Bone of the upper-arm.
b, e, Bones of the fore-arm.
c, Inner angle of the joint.
d, Ligaments of the joint.

The ball and socket joint is composed of a ball on the end of one bone, and a cup or socket in the other, in which the ball plays, as in the hip joint, (Fig. XLV. *c, d,*) or as the

Fig. XLV. *Hip Joint.*

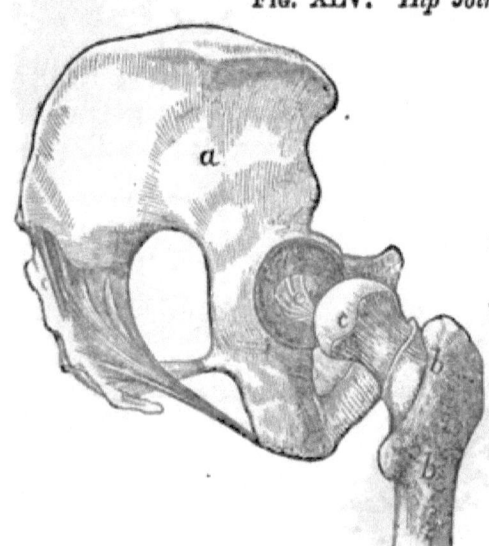

a, Hip bone.
b, b, Thigh bone.
c, Head of the thigh bone.
d, Socket in the hip bone.
e, Ligament attached to the bottom or centre of the socket and the head of the thigh bone.

shoulder, or the attachment of the thumb to the hand. This joint allows motion in every direction, forward and backward, upward and downward, and in a circular manner.

607. The thigh bone has almost a globular head at its upper extremity, (Fig. XLV. *c,*) which is fitted into a very deep

socket in the side bone of the pelvis. The head of the bone of the upper arm is less than half a ball, and is fitted into a very shallow socket in the upper corner of the shoulder blade, (Fig. XXXVII. *a*.)

608. *The head is connected with the upper vertebra by a hinge joint*, which allows it to bend forward and backward. The upper vertebra has a ring on its posterior side, in which a tooth or pin from the second vertebra is inserted. This allows the first vertebra to turn upon the second, as a gate turns upon a hook-and-eye hinge. By this joint we turn the head from side to side, and by the other we nod and lift the head.

The cartilages between the several bones of the spine allow the column to bend in every direction.

609. The head of every bone is covered with a very dense, but somewhat elastic cartilage, which is sufficiently soft to break the force of pressure or jars upon the bones, but not soft enough to be loose and interfere with the movements of one bone upon another. These cartilaginous facings of the joints are not very thick — not more than a sixteenth or eighteenth of an inch. They are covered with an exceedingly smooth lining, that presents the most polished surface imaginable.

610. It is one of the admirable provisions in regard to the joints, that they never wear out. Though they are in such frequent use, and exposed to so much pressure and motion from infancy to extreme old age, even eighty or ninety years, yet they never wear out. The tough, cartilaginous coverings of the ends of the bones, and the delicate and glairy facings of the joints, are as thick and as smooth at the end as at the beginning, or at any period of life. These substances, if worn at all, are perpetually renewed. They pass through the same changes, they are subject to the same death of particles from exhaustion, and the same renewal of living particles, as the other organs and textures.

611. *Not only do these joints and their parts wear well, but they have a perpetual self-oiling apparatus*, that keeps

their faces always moist and slippery. This is the living joint, which prepares and pours out this oily fluid, as the skin pours out perspiration; and if the joints are properly used, and in good health, this fluid is of the due proportion and consistency. But in some cases of disease or injury, it becomes abundant, and fills the sac of the joint, as a bladder, with water. This happens most frequently in the knee, from a blow or a strain. Then this lining membrane takes an unnatural and increased action, and throws out much more fluid than is needed, and so fills the sac of the joint that the knee is swelled and lame.

612. *The ligaments and the capsules hold the bones together at the joints.* The capsule encloses the whole joint: it surrounds the end or the seat of junction of each bone, and is

FIG. XLVI. *Knee, Capsule, and Ligaments.*

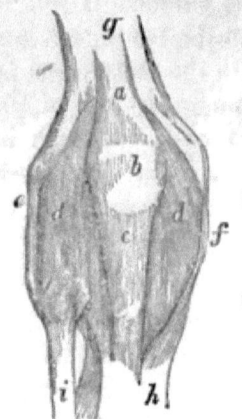

a, Tendon of the great muscle of the thigh, attached to the knee pan.

b, Knee pan.

c, Tendon connecting the knee pan with the bones of the leg.

d, d, Capsule covering all the joint.

e, f, Lateral ligaments extending from the thigh to the bones of the leg.

g, Thigh bone.

h, i, Bones of the leg.

attached to both. In the knee, (Fig. XLVI. *d, d,*) the capsule surrounds, and is attached to, the lower end of the thigh bone: it passes over the space between the bones, and in like manner it surrounds, and is attached to, the upper ends of the bones of the leg. The *synovial membrane* within this capsule prepares and throws into the joint the synovial fluid that moistens and oils it.

There are other ligaments within and without the joints. Some of those within the knee are seen in Fig. XLVII. The wrist (Fig. XLVIII.) and the ankle are supplied with a great variety of ligamentous bands, that hold their small bones in

FIG. XLVII. *Knee and Internal Ligaments.*

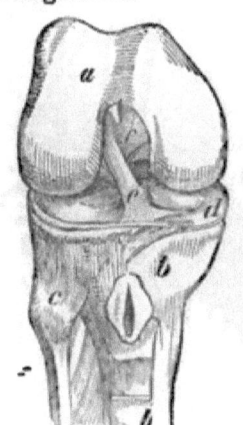

a, Thigh-bone.

b, c, Bones of the leg.

d, e, f, Ligaments connecting these bones together.

their several places, and yet allow them due freedom of motion. The hip joint has a round ligament, which is attached to the bottom of the socket and to the middle of the head of the thigh bone, (Fig. XLV. *e.*)

613. These capsules and ligaments are strong and inelastic. They are so distributed and arranged as to allow all the requisite movements, and yet to retain the bones in their respective situations. When they are exposed to great or sudden violence, they are sometimes strained, and some of their minute fibres are broken. *Then the joint is said to be sprained,* and requires a long healing process for restoration. This happens most frequently to the foot and ankle from any misstep or fall, causing a greater strain upon the ligaments than they can safely bear.

614. *A bone is dislocated or thrown out of joint* when it is exposed to still greater strains or violence. The shoulder is more liable to this injury than any other joint. The cavity is so shallow, that when the arm is stretched forward or backward, and any pressure or blow comes upon it in front or behind, it has little power of resistance, and the head is thrown out of its socket. Persons falling forward, as when thrown

from a horse or a carriage, are apt to throw their arms out to save themselves; the weight of the body comes upon the

FIG. XLVIII. *Ligaments of the Wrist and Hand.*

a, a, Bones of the fore-arm.
b, b, b, Bones of the hand.
c, c, c, Ligaments of the wrist.
d, d, Ligaments of the hand.

hand, and, of course, upon the arm and shoulder; the force of the blow presses the bone backward, and sometimes thrusts the head over the edge of the socket.

CHAPTER V.

Muscles. — Motive Power. — Number. — Arrangement. — Action. — Description and Use. — On Front of the Body. — On Back. — On Side.

615. THE bones are merely the framework. They are the rigid parts upon which the action is made, but they have no active power. *All the motive power is in the muscles.* These perform all the motions in the animal body.

The muscles form the most abundant part of the body. They constitute the great bulk of the limbs, the back, and neck. They cover the face and chest, and form the principal portion of the walls of the abdomen. In lower animals, they are the lean meat which we eat upon our tables. They consist of stringy fibres, that usually lie parallel with each other, and are fastened, by a strong whitish-looking substance, into bundles. Each bundle, thus fastened together, forms one muscle.

616. These muscular fibres have a power of drawing up or contracting, like the earth worm; and when they thus contract, they draw their ends toward each other, and draw together or move toward each other whatever parts or bones may be fastened to these ends.

617. *All the parts of the body that move are furnished with some of these muscles,* or bundles of lean flesh. There are five hundred and twenty-seven muscles in the human body. Five hundred and fourteen of these are in pairs, being the same on the two sides, and thirteen others are single muscles, as the heart, &c. These are arranged in layers, and in some parts of the body, as on the back and the large limbs, there are several layers between the skin and the bones. Some of the outer layers are shown in Figures XLIX., L., and LI. The inner layers are shown in Figures LII., LIV., LV., and LVII.

618. The *cutaneous muscle,* (Fig. XLIX. *a,*) is attached to the skin and flesh of the cheek above, and of the neck below. It aids in drawing the mouth downward.

The *deltoid,* XLIX. *b,* or triangular muscle, has its upper side, or ends of the fibres, attached to the collar-bone and shoulder-blade. These fibres meet in a point, which is attached to the humerus or bone of the upper arm. Its front fibres draw the arm upward and forward; its back fibres draw the arm upward and backward; and all, together, lift the arm.

The *biceps,* or two-headed muscle of the arm, XLIX. *c,* and LI. *b,* is attached above to the shoulder-blade and the upper part of the humerus, and sends a cord to the lower arm near the elbow. It bends the elbow.

Fig. XLIX. *External Muscles of the front Part of the Body.*

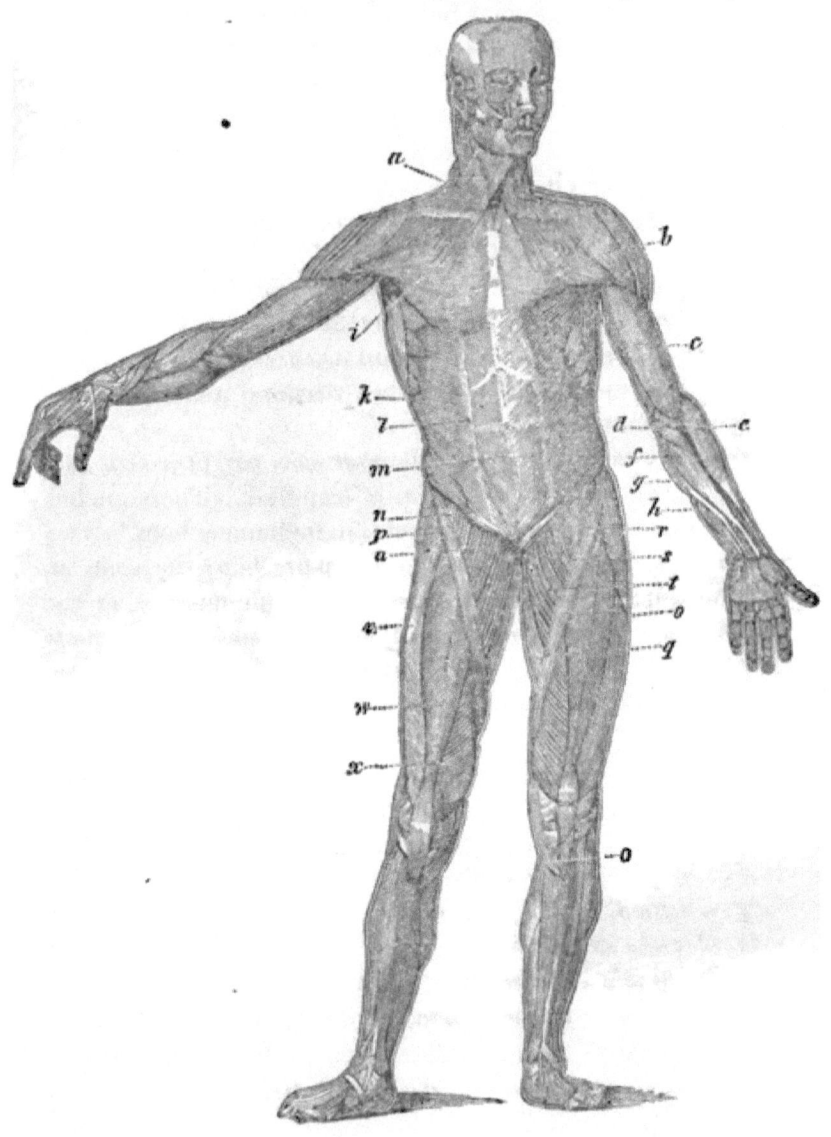

The *pronator*, XLIX. *d*, extends from the inner side of the arm to the outer bone, and rolls the arm inward, and turns the hand downward.

The *supinator*, XLIX. *d*, LI *d*, extends from the outer side of the humerus and the outer bone of the fore-arm to the inner bone; it rolls the arm outward, and turns the hand upward.

The *flexors* of the wrist, XLIX. *f*, *h*, and the long *palmar* muscle, XLIX. *g*, extend from the lower end of the upper arm, and send tendons to the bones of the wrist and hand. They bend the wrist.

The *pectoral muscle*, XLIX. *i*, is attached to the breast bone and to some of the ribs in front, and extends back to the upper and inner part of the humerus. It moves the arm forward and upward obliquely.

The *external oblique*, XLIX. *k*, LI. *i*, arises from the edge of eight of the lower ribs, and passing downward and forward it covers the abdomen in front, and is attached below to the pelvis. It supports the abdomen, and, by its contractions, it presses upon its contents and forces them and the diaphragm upward in expiration, (§ 286.)

The *semilunar line*, XLIX. *l*, and the *linea alba*, XLIX. *m*, or white band, are layers of gristly matter extending from the breast bone to the pelvis. They give support to the abdominal muscles.

The *sartorius*, or tailor's muscle, XLIX. *o*, LI. *m*, extends from the hip bone over and in front of the thigh, and is joined to the inside of the bone of the leg, below the knee. It aids in bending the thigh: it rolls the hip joint, and lifts one leg over the other, as tailors sit.

The *gracilis*, XLIX. *q*, assists the tailor's muscle.

The muscle, XLIX. *p*, aids in turning the thigh outward.

The *psoas* and *iliac* muscles, XLIX. *r*, *u*, are attached to the back bone, and, passing over the pelvis, are joined to the thigh bone. They bend the thigh on the trunk.

The *pectineus* and *triceps*, or three-headed muscles, XLIX. *s*, *t*, extend from the front part of the pelvis to the inner side of the thigh bone. They assist in bending the thigh, and in rolling it outward.

The *rectus*, XLIX. *w*, LI. *p*, is attached to the front part

264   PHYSIOLOGY AND HEALTH.

Fig. L. *External Muscles of the back Part of the Body.*

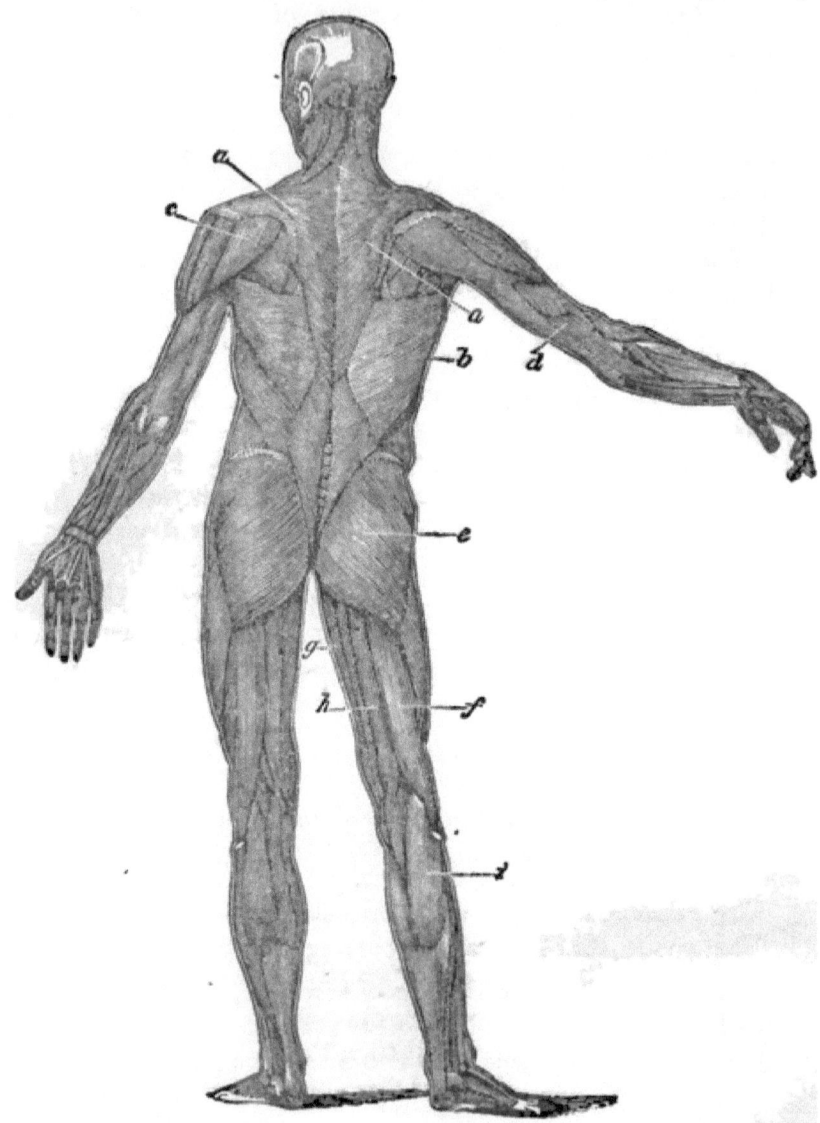

of the pelvis, and extends along the whole length of the thigh bone, and is fixed by a cord to the knee pan. It straightens the knee, and lifts the body upon the leg when sitting down.

The *vastus internus* and *externus*, XLIX. *v, x,* LI. *n, o,* are two muscles that are attached to almost the whole length of the thigh bone, and send tendinous bands to the upper end of the bones of the leg. They assist in straightening the leg.

619. The *trapezius*, (Fig. L. *a, a,* LI. *f,*) is attached above to the head, and the bones of the neck and back: it extends outward and part downward and part upward, and its opposite end is attached to the collar-bone and shoulder-blade. Its upper fibres lift the shoulder; its lower fibres draw it downward and backward; and its middle fibres, and the whole acting together, draw the shoulder backward.

The *latissimus dorsi*, or the broadest muscle of the back, L. *b,* LI. *g,* extends from the middle and the lower half of the back bone, and the back part of the pelvis, and extends forward and upward to the upper arm near the shoulder. It draws the arm backward and downward.

The *deltoid,* L. *c,* XLIX. *b,* lifts the arm.

The *triceps,* or three-headed extensor of the fore-arm, L. *d,* is attached above to the shoulder-blade and humerus, and below to the extreme point of the elbow. It straightens that joint.

The *great gluteus,* L. *e,* LI. *k,* extends from the back part of the pelvis to the upper part of the thigh bone. It straightens the hip joint, and raises the body up on the lower limbs from the sitting position.

The *biceps,* or two-headed flexor of the leg, L. *f,* is attached above to the pelvis and the thigh bone, and sends a tendon from its lower end through the outside of the ham to the outer bone of the leg. It bends the knee, and its tendon forms the outer ham-string.

The *semitendinous* and the *semimembranous* muscles, L. *g, h,* are attached above to the lower part of the pelvis, and, extending down the back and inner part of the thigh, send cords through the middle of the ham to the inner bone of the leg. They aid in bending the knee, and their tendons form the inner ham-string.

The *gastrocnemius,* or great muscle of the calf, L. *i,* is at-

266 PHYSIOLOGY AND HEALTH.

Fig. LI. *External Muscles on the side of the Body.*

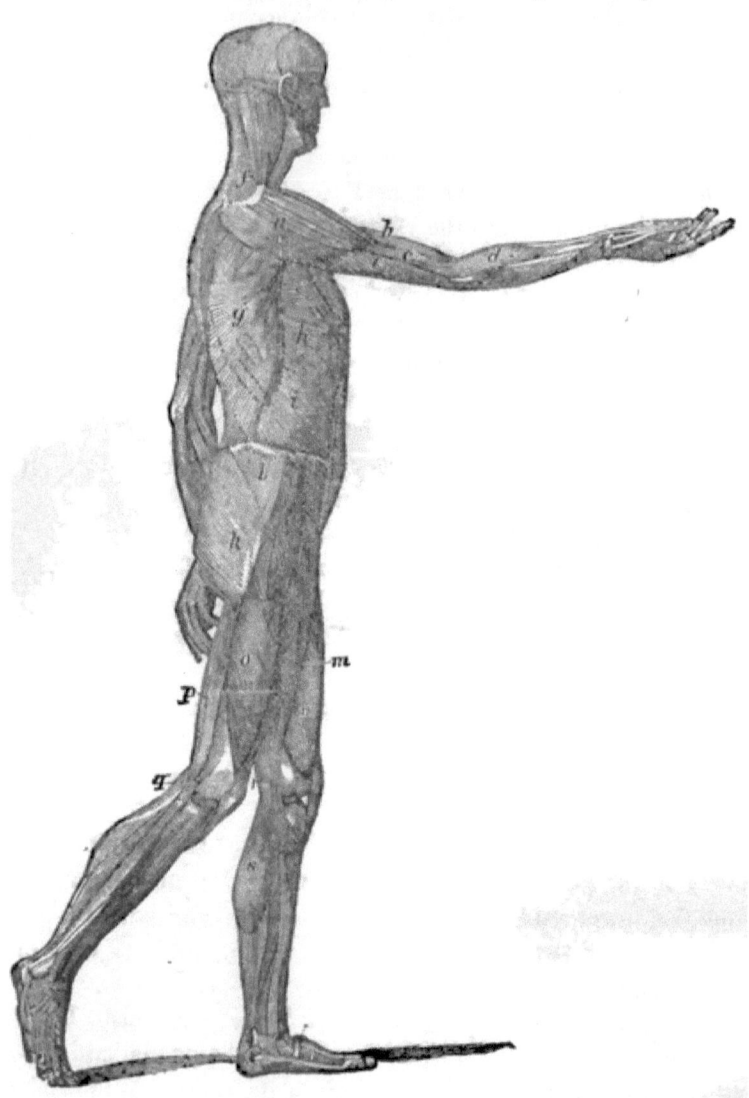

tached above to the back of the thigh bone, and, from its lower end, sends a tendon to the ankle. It straightens the ankle and lifts the body on the foot.

620. LI. *a. Deltoid,* lifts the arm.

b. Biceps, or two-headed flexor of the arm.

c. Brachial muscle, extends from the middle of the humerus to the front side of the lower arm. It aids in bending the elbow.

d. Supinator, turns the hand upward.

e. Triceps, or three-headed extensor of the arm.

f. Trapezius, draws the shoulder upward, backward, and downward.

g. Latissimus dorsi, or broadest muscle of the back, draws the arm backward and downward.

h. Serratus, or saw-edged muscle. One side has nine fleshy teeth, which are attached to the upper nine ribs in front: the opposite edge of the muscle is attached to the front and outer edge of the shoulder-blade. It draws the shoulder-blade forward, and when this bone is fixed by the other muscles, it aids in lifting the ribs.

i. External oblique of the abdomen, aids in expiration.

k. Gluteus maximus, or great gluteal muscle, straightens the hip.

l. Gluteus medius, or middle gluteal muscle, extends from the back part of the pelvis to the back and outer part of the thigh bone. It draws the thigh outward and backward, and rolls it outward.

m. Sartorius, or tailor's muscle, crosses the leg.

n. Vastus internus, straightens the leg.

o. Vastus externus, straightens the leg.

p. Rectus, straightens the leg.

q. Tendon or cord forming the outer ham-string.

r. Tendon forming the inner ham-string.

s. Gastrocnemius, forms the calf of the leg, and straightens the ankle.

CHAPTER VI

Muscles. — Shape. — Attachment. — Situation. — Swell in Action. — Arrangement. — Antagonism. — Coöperation.

621. The muscles are of various shapes, and their fibres are arranged in various directions, to suit the wants of the places where they are to operate, and the convenience of distribution. On the limbs they are long and roundish; on the trunk they are mostly flat. Sometimes the fibres are arranged in the shape of a fan, the broad end being fixed to an immovable bone; and, at the other end, all the fibres are gathered into one point, and are fixed to a movable bone, as in the *deltoid*, (Fig. XLIX. *b*,) and the *trapezius*, (Fig. L. *a*,) (§§ 618, 619.) When the fibres of one side of this fan-shaped muscle act, they draw the bone in that direction; and when the fibres of the opposite side act, they draw the bone in the opposite direction; and when they all act together, they draw it in a line with the middle fibres.

622. The diaphragm (§ 273, p. 124) is a muscle, and acts by the same contractile power as the muscles of the arms and legs. But, unlike them, it is attached to no joint, and moves no bone. It is fixed to the bottom of the ribs, and is arched up into the chest. (Fig. XVII.) Its fibres, like the spokes of a wheel, extend from the circumference to the centre; and, when they contract, they draw the centre and the whole arch downward towards the line of attachment of the outer edge. This descent of the arch leaves room in the chest for the expansion of the lungs and the inspiration of air. The heart (§ 206, p. 94) is a hollow, muscular bag; when its fibres contract, they lessen the cavity within and expel the contents of blood. The muscular coat of the œsophagus, (§ 23, p. 17,) the stomach, (§ 28, p. 20,) and of the alimentary canal, winds about them, and is attached to no fixed point. It presses upon the contents of these organs, and aids the digestive operation in the stomach, and carries the contents onward through the channel.

623. These muscles, distributed and arranged in nature's most skilful and benevolent manner, are the source of all our power of motion. By these we walk, we lift, we strike, we eat and swallow, we breathe and cough, we speak, wink, nod the head, bend the back, and do all our work. A palsied limb, whose muscular power is lost, is as motionless as the limb of the dead.

624. The muscles are mostly attached to different bones at their two ends, and by their contractions move these bones on or toward each other, as at the joints. But some muscles are intended to move only the flesh, and are, therefore, attached only to the flesh, at least at one end, as in the face.

FIG. LII. *Muscles of the Face.*

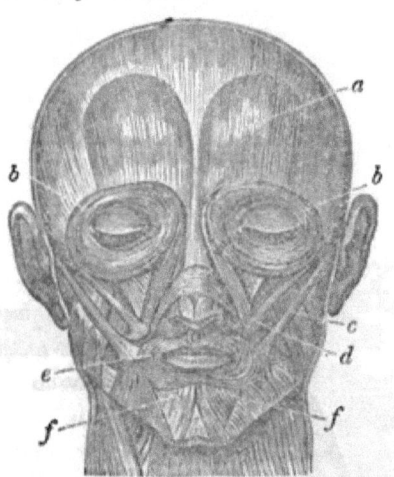

a, Frontal muscle, raises the eyebrows, and wrinkles the forehead.

b, b, Circular muscles, close the eye.

c, d, Muscles that raise the corners of the mouth.

e, Circular muscle, closes the mouth.

f, f, Muscles that draw down the lower lip.

The muscles (Fig. LII. *a, a*) that wrinkle the forehead and draw up the eyebrows, are attached to the bone above, and to the skin and flesh below. The muscles that raise and draw down the corners of the mouth and the lips (Fig. LII. *c, d, f*) are attached to the bones of the two jaws and the cheek by one end, and to the lips by the other. The muscles of the eye ball (Fig. LXVIII.) are fixed by their inner ends to the bone within the socket of the eye, and by their outer ends to the ball. By their various contractions they roll the eye.

Some are circular, as those that close the mouth and eye, (Fig. LII. *b, c.*) The heart has no attachments; it is suspended in the chest, and acts only on its contents. The muscular coat of the œsophagus, stomach, and alimentary canal surrounds these organs, and, by its contractions, it forces their contents onward.

Fig. LIII.

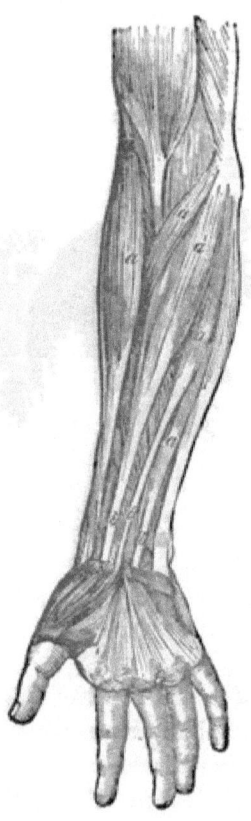

a, a, a. The fleshy parts of the muscles.

b, b, b. Tendons of these muscles passing over the wrist to the hand and fingers.

625. The muscles are not placed directly between the bones which are to be moved, — this would make the joints clumsy, — but at some little distance, and attached directly at one end to one bone, and then generally fastened to a tendon or cord which moves over the joint to the other bone. The muscle that bends the elbow (Fig. LIV. *d*) is placed entirely on the front part of the upper arm. The upper end of this is fixed to the upper bone, but no part of the muscle touches the lower arm. A cord passes from the lower end, over the elbow-joint, to the lower arm. So, also, the muscles that move the hand, the knee, the foot, &c. Sometimes the muscle is placed at a considerable distance from the bone to be moved. The muscles of the fingers are placed, not on the hand, nor even on the wrist, but on the fore-arm, (Fig. LIII. *a, a, a;*) and the long cords or tendons (*b, b, b*) can be easily felt as they pass along the lower part of the fore-arm, and over the wrist, and over the hand to the fingers. By a similar arrangement, the toes are moved by muscles situated on the leg. The cords pass through grooves or pulleys at the

ankles, and there change their direction and pass to the toes.

626. When the muscles draw up, they increase in size, and swell out at their middle; and when they cease to act, they are again drawn out in length, and their size is reduced. If we place the right hand on the front of the left upper arm when we bend the left elbow, we shall feel the biceps muscle (Fig. LIV. *c, d*) swelling out and becoming hard. If, at the same time, we place the fingers in the angle of the elbow, we shall feel the cord drawn tight and moving. If we put the hand on the same place when we straighten the elbow, we shall feel the muscle growing soft, the swelling going down on the upper arm, and the cord moving backward and becoming loose in the elbow.

Fig. LIV. *Flexor Muscle of the Elbow.*

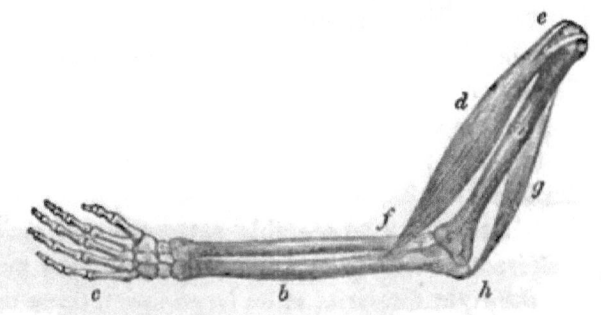

a, Humerus, or bone of upper-arm.
   b, Bone of fore-arm.
   c, Hand.
   d, Biceps muscle.
   e, Its upper attachment.

f, Its cord attached to the lower arm.
   g, Muscle that straightens the elbow.
   h, Its attachment to the elbow.

627. The muscle on the back of the upper arm that straightens the elbow is broader and larger, so that its swelling and reduction are not so perceptible as in case of the bending muscle. Nevertheless, its action and relaxation may be perceived, when the lower arm moves upon the

upper. The same action may not only be very distinctly felt, but even seen, on the temple, in the muscle that closes the jaw. The upper end of this muscle is fixed upon the bone of the temple, and passes down behind the projecting arch of the cheek-bone to the upper side of the lower jaw. When it contracts and draws up the lower jaw, it swells out on the temple so prominently that its increase and diminution are seen every time we move the jaw, either in mastication of food or talking.

628. *The muscles have only a power of contraction.* They have no active power of forcible expansion. They can draw the bones together, but they cannot push them apart. The muscles that bend the wrist or the ankle cannot straighten it out again. But nature has made beautiful provision for this, by affixing to every joint two or more sets of muscles for the various kinds and directions of motion required. Upon the hinge-joints there are two sets, one to bend, the other to straighten them. Thus the muscle on the front of the upper arm bends the elbow, and the muscle on the back of the upper arm, straightens it. The same arrangement is found in the muscles that move the wrist, the fingers, the knee, the ankle, and the toes.

629. It is interesting to see this antagonism of muscles, and their alternate working in the movements of the fingers. If we clasp the right fore-arm, at its largest part, three inches below the elbow, with the fingers and thumb of the left hand, and then drum rapidly with the fingers of the right hand, we shall feel the swelling and decline of the muscles on the opposite sides of the arm, alternating with each other, and precisely corresponding with the motions of the fingers. While the fingers are bending, the inside muscles swell and the outside muscles decline; and, while the fingers are straightening, the outside muscles swell and the inside muscles decline. The same alternation may be felt in the movements of any other joint.

630. The hinge-joints want two sets of muscles only — one to bend, the other to straighten But some other joints

are not limited to this single line of motion. The shoulder not only bends forward and backward, but upward and downward; and it can move in any direction, even round like the spoke of a wheel. It would seem necessary, then, that there should be many sets of muscles to produce this great variety of motions. The ball and socket joints are supplied with

Fig. LV. *Internal Muscles of the Back, Shoulder, and Hip.*

a, Muscle that lifts the shoulder-blade.
b, Raises the arm.
c, Rolls the arm outward.
d, *e*, Rhomboid muscles, draw the shoulder-blade upward.

g, *Serratus*, attached to the back-bone and the ribs, draws the ribs down.
h, Straightens the hip joint, and rolls the thigh outward.

many muscles, (Fig. LV.,) which act individually, successively, or in combination, and produce every variety of motion; by these we move the arm, the thigh, the wrist, and the thumb in any direction which we may desire.

CHAPTER VII.

Bones are Levers; Muscles are moving Powers. — Muscles act at Disadvantage. — Power sacrificed to Convenience of Action. — Especially in the Fingers and Toes. — Muscles coöperate to produce one Motion. — Rapidity and Precision of Muscular Action. — Illustrated by Piano-Forte Player. — Violinist. — Writing. — Carpenter.

631. THOSE muscles which are situated at a distance from the object to be moved work at great disadvantage. If that muscle which bends the elbow were placed in the bend of this joint, and its lower end attached to the lower end of the fore-arm (Fig. LVI. *b*) instead of the natural position at the elbow, it would act with more power than it now

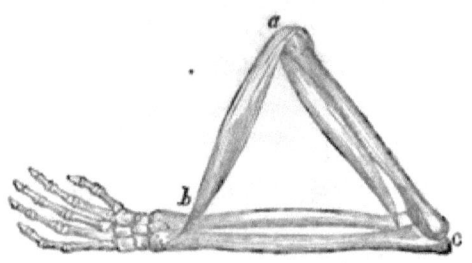

FIG. LVI. *Upper Arm Biceps misplaced.*

a, b, Biceps attached to wrist.

does. But muscles so situated would interfere very much with the freedom of the movements, and with the application of the limbs to a great variety of purposes, and be very awkward.

632. In motion, the bones are levers, and the muscles are the moving powers. The bones of the fore-arm are fixed very near its end, to the extremity of the bone of the upper arm. The upper bone is the fulcrum or point of support; the muscles that move this lower arm are attached to it within less than half an inch of the fulcrum, while the whole bone or lever is ten to twelve inches in

length. Consequently, the muscles act with a disadvantage of twenty or twenty-four to one.

633. *Nature makes sacrifice of power to convenience of motion, to grace of structure, and to beauty of limb.* The muscles are therefore made very powerful. When we lift a pound in the hand, and bring the fore-arm to a horizontal position, the muscle bending the arm exerts a force which would lift twenty or twenty-four pounds, if it were attached to the point where the weight rests. When a strong man lifts fifty or a hundred pounds with his hands, holding the fore-arm in a horizontal position, the muscle that raises the arm exerts a force equal to twenty or twenty-four times that which would be necessary to lift the same weight if it were fixed directly to the weight.

634. The muscles that straighten the ankle, and raise the body upon the toes, act with the same disadvantage upon the short end of the lever. The ankle is the fulcrum, the whole foot the lever, the heel the short arm, and the foot in front of the ankle is the long arm. It is plain that the power required to move the foot by the heel must be as much greater than that which could move it by the toes as the anterior part of the foot is longer than the heel, which is about twelve to one.

635. *Muscular contraction is effected with rapidity —* in some cases, almost instantaneously. This is seen when the musician executes rapid pieces of music on his flute; for every note that is made includes two motions of the muscles on the fore-arm — one to lift, and the other to carry down the fingers. It is an established fact, that some persons can pronounce distinctly fifteen hundred letters in a minute. The pronunciation of every one of these letters must require a distinct and double action of the muscles connected with the voice and enunciation, for each action includes both the contraction and the relaxation. Here are, therefore, three thousand actions in one minute. Insects exceed all other animals in the rapidity of their muscular motions. It is calculated that, with their wings, they strike

the air, not many hundred, but even many thousand, times a second.

636. These muscles work in concert, and produce, by their coöperation, just the motions that are required. It is interesting to see how exact is this harmony, and with what precision is the power measured from, and exertion made by, each one in creating any motion. We lift the hand to the chin, then to the mouth, then to the nose, to the eye, and to the forehead. In each of these motions, there is a combination of several or many muscular actions; each muscle and each fibre pulls just so much as, and no more than, is necessary to carry the hand exactly to the place appointed. A very slight change of force of one of the muscles in the shoulder carries the hand from the mouth to the nose, and from the nose to the eye, or lets it fall upon the breast.

637. Performers on the piano strike the varied notes with equal precision; — first *a*, then *b*, then *c* sharp, then an octave, or a fifth, or a third. In all these changes, the hand moves half an inch, quarter of an inch, four inches and one half, six inches and a quarter, — whatever distance is required, it moves just so much and no more, — and hits the exact note. This is done by the muscles of the shoulder, elbow, and the wrist, — all in harmonious coöperation. However small an extent of contraction of the muscle at the shoulder, at the short end of the lever, is needed to carry the hand a quarter of an inch to the right or to the left, just that extent, that hair-breadth of contraction, is made, — no more, no less, — and the unerring finger hits the note.

638. The complete government of the muscular action is still more remarkably manifested in the skilful violinist. He determines his notes by touching the tip of his finger on the strings. If he touch them a little higher or a little lower, he varies the note. The most accomplished players acquire such exact control and discipline of the muscles of their fingers, that they can produce at will the minutest

shades of difference of sound. To produce the precision of motion in the ends of the fingers, the muscles that move them must contract with still greater precision, without varying a fraction from the required amount of motion.

639. Another illustration will present itself to any one when he is writing. He moves his pen through the forms of his letters by means of the muscles of the thumb and fingers, and of the arm and shoulder. These combine their actions, and carry the pen now up, now down, in straight or curved lines, forward and backward. They make the exact contractions that are needed. The slightest variations of these contractions make an *e* instead of an *i*, a *d* instead of an *a;* yet these mistakes are not made, but the letters are formed in the exact shape that is required.

640 This control over muscular action enables the painter, the draughtsman, and the engraver, to produce the very pictures which they desire, and the mechanic to use his tools, not only with exactness, but with safety to himself. When a carpenter is cutting with a broad-axe, upon a small block, which he holds with his left hand, he strikes boldly and accurately. But a very slight difference of the contraction of the muscles of the shoulder that bring the upper arm downward and forward would carry the axe upon the fingers, and not upon his block. When driving nails, he holds the nail between the thumb and finger of the left hand, while he lifts the hammer with the right. He hits the nail upon the head; and yet how exceedingly small a variation of the muscular contraction would carry the hammer upon the thumb or the finger, and not upon the nail!

CHAPTER VIII.

Strength of Muscles not always dependent on Size. — Strength differs in various Animals, in Man and Insects, and in various Men. — Exercise of Muscles increases their Size and Power. — Muscles in Limbs that are exercised stronger than those that are not used.

641. *Muscles are, in general, strong in proportion to their size; but this is not a universal law.* Birds are very strong, but they have not very large muscles. These would add to their weight, and be very inconvenient for flight. On the contrary, fishes have large muscles, yet are not very strong. Living in the water as they do, which is nearly as heavy as themselves, great bulk is no impediment. They move about as well with a large as with a small mass of flesh.

642. The difference of muscular power in different classes may be shown by comparing man with some insects. A man must be more than usually active to be able to jump his own height. A man of ordinary strength can hardly lift more than twice his own weight; one of the strongest men on record could lift eight hundred pounds. But insects have astonishing strength of muscle. A flea will leap sixty times its own length, and one of the beetles can support uninjured, and even elevate a weight equal to five hundred times that of its own body. If a man were strong and active in the same proportion, he could jump three hundred and forty feet, or more than twenty rods, and lift about three and a half tons' weight.

643. Among animals of the same kind, especially among men, it may be safely considered that the muscular power corresponds with the size of the muscles. Though all men are endowed with the same muscles, and these are arranged in the same manner in all, yet it is manifest that all men are not equally strong, and that their strength is not distributed in the same proportion over the various parts of the body.

One man is very powerful, another is very weak; one is strong in the arms and weak in the legs, while another is strong in the legs and weak in the arms.

644. *This great and almost infinitely varied muscular power is given to men and animals for action.* It was no intention of the benevolent Creator that this should remain dormant. Some of the involuntary muscles work incessantly; day and night, asleep or awake, from birth to extreme old age, the heart beats, and the chest moves, and the digestive organs carry on their operations and find no rest; nor yet are the muscles which perform these labors exhausted or even weary. It is the design of our being that we use the muscles that are put under our control, and, by their action, both procure our subsistence and sustain our health. These two ends are attainable by the same means, and are made to correspond with each other.

645. *Exercise of the muscles increases the waste of their dead particles,* (§ 252, p. 115,) *and consequently the demand for nutrition.* The arteries then act more vigorously, carry more blood, and deposit more new particles. The active muscle, being better nourished, is composed of fresher atoms, and has more strength. When this exercise is judiciously taken, neither too little nor too much, the nutrition exceeds the absorption, more new atoms are brought than old carried away; then the muscle grows larger, and more dense and powerful. We find proof and examples of this too frequently to escape notice. The muscles of the inactive, those who neither labor nor exercise, are small, soft, and flabby, and hang loosely on the bones; consequently, they are weak. But the muscles of the active and laborious are large and firm; they are full and very strong, and endowed with great energy of action.

646. The robust and healthy laborer exercises his limbs and his trunk, and consequently has strong muscles. He can take his scythe and mow from morning till night, and follow this from Monday till Saturday; he can lay stone wall or cut wood with the same perseverance; and he does

all this without any great fatigue. But let a student, whose arms are unaccustomed to labor, and whose muscles are therefore soft, small, and feeble, undertake to do either, and he will, in a few minutes, be exhausted.

647. The student is not necessarily weak in body. If he is faithful to the laws of health, and takes daily exercise to a suitable degree, his muscular system is sufficiently developed, and he is strong enough for the support of health, though he may be much weaker than the farmer. He can walk several miles. He can do the work of gardening, rake hay, or perform the lighter labor of farming for an hour or two, without much fatigue. The muscles of the inactive and indulged girl, who has been taught that exercise — however useful to the laborer, and proper for men and boys — is not requisite for her delicate frame, are soft, like those of a babe, and not much stronger. If she attempts to walk a mile, she will fail of accomplishing her purpose, or suffer from fatigue, perhaps from exhaustion.

648. *Exercise develops the form and the strength of the muscles which are called into operation.* In order that the whole should be developed and strengthened, this exercise should be so varied as to use every one of them. The farmer is at one time cutting wood, and thereby using the muscles of the arms and shoulders; at another, laying stone wall, with the muscles of his arms, shoulders, and back; at another time, he is ploughing, and using the muscles of his arms, back, and legs; sometimes pulling, sometimes pushing, lifting, striking, treading; and in all these ways calling every muscle into action. His muscular energies are, consequently, universally developed, and he is strong in all his frame beyond other men. There are other occupations that have nearly the same effect, but none that give the variety of exercise and universality of muscular power that comes from the cultivation of the earth.

649. All employments that call for the use of only a part of the muscles, but not the whole, develop the size and the strength of those that are used more than the others. So

we find some men are very strong in some of their limbs, and weak in others; and these strong and weak parts differ according to the habits and employments of the people. A sailor uses his arms more than his legs. He pulls ropes, and lifts the anchor, but has little opportunity of walking. Consequently, the muscles of his arms and shoulders are large, strong, and hard, while his legs are smaller and weaker than those of other men. The blacksmith uses the same muscles, and has the same muscular development of his arms, and comparative weakness of his legs. The pedestrian and the dancer have large and strong muscles of the lower limbs, while their arms are comparatively small and feeble.

A similar difference of muscular development is manifested in the porter. His employment calls for the exercise of the muscles of the back more than that of other men. He carries his burdens on his head or his shoulders. It is necessary, therefore, for him to keep his spine erect by the constant and vigorous action of the muscles of that region. They, therefore, are used more, and grow larger and stronger, than the spinal muscles of men who are otherwise occupied.

This law of special muscular development from special use affects the lower animals as well as men. Hence we find in them a growth and an increase of power in the muscles that are used, and a comparative smallness and weakness in those that are not used. The wild birds use their wings mostly for locomotion. They sustain themselves very long in the air, and fly great distances. They have, therefore, very large and strong muscles on their breasts and wings. And, as they walk but little, the muscles of their legs are comparatively small and weak.

But the domestic fowls seldom fly; they use their legs mostly for locomotion. They have, therefore, much smaller muscles on their breasts and wings, and much larger on their legs. This difference is, perhaps, the most remarkably seen in the wild and domestic turkeys and geese.

CHAPTER IX.

Some Muscles strong and others weak in the same Person. — American Women walk little. — Muscles become weak by Disuse. — Whole System made stronger and more lively by muscular Exercise. This aids Digestion, Respiration, Circulation, and increases animal Heat.

650. The muscles being thus unequally used, and their strength unequally developed, in the same person, he may then be very strong in one part, and weak in another. The sailor or the blacksmith would be wearied with a walk of a few miles, while the pedestrian might not be able to carry the porter's burden, and the porter would soon be exhausted with swinging the scythe or the sledge-hammer.

651. Many women, however industrious at home, are not generally accustomed to much exercise abroad. Their muscles of locomotion, being little used, are neither large nor strong. They cannot move with a vigorous gait nor with an elastic step. They walk with so little ease and energy, and are so soon fatigued, that they find very little inducement to go any considerable distance on foot. Many who can spin and weave, wash, make butter and cheese, and perform all sorts of even the hardest household labors, without any great weariness, are overcome by a short walk.

An example of this was found in an unusually industrious and healthy wife of a farmer, in a country town of Massachusetts. Few women were more faithful and energetic in the management and labor of the house. Nothing was too hard for her strength, no household work was too great for her energy; all the domestic avocations, in their due course, were her ordinary exercise. But she never walked abroad. She lived about a mile from the village and the church, and went to these very frequently, on business or pleasure, and for worship. But she had always a horse and chaise at command, and always rode. In her later years she said that, for near forty years, she had not walked from her house to the village, and she thought she had not strength to do it.

652. The strength and size which the muscles gain by exercise are to be preserved by the same means. If suffered to remain inactive, they lose their fulness and power, they shrivel, and become soft and feeble. Whatever may be the cause of the disuse of the limbs, these consequences of waste and weakness must follow. An active and strong man receives an injury, perhaps a cut on the foot. This wound, without producing any general disease, may lay him up, and keep him still for some months. When his wound is healed, he finds his legs are weak, and unable to do their former labor.

653. Dr. Reid cut the great nerve that went to one of the legs of a rabbit. The limb was immediately palsied, and could not move. In seven weeks he killed the animal, and compared the muscles of the palsied leg with those of the opposite and sound one. They were paler, softer, and smaller, and weighed only about half as much as those of the other limb. The bones, also, of the inactive leg were diminished in size. So the muscles of the paralytic man, who does not and cannot walk, become shrivelled and weak. Mr. J., in consequence of an injury, has not been able to bend the foot on the leg for more than sixteen years. The muscles which formerly lifted his foot, not having been used for so long a time, are now shrivelled, and much smaller than in other men of his size.

654. Exercise not only invigorates its own apparatus of motion, but it contributes to the strengthening of all the other systems, and aids them in the performance of their functions. The man of active habits of body has a better appetite and digestion and is better nourished, he breathes more freely, he has a freer circulation of the blood and a clearer brain, than the inactive and the sluggish. The laborer, the farmer, the active mechanic, and sailor, seldom complain of want of appetite or of indigestion. They work

hard and eat heartily. Their food gives them a comfortable sensation, and they are well nourished. But the student, the clerk, the watchmaker, the engraver, all men engaged in sedentary employments, and men and women of no occupation, often complain of failing appetite and weak digestion. If they eat heartily, they feel now and then distressed after so doing, and give painful evidence that if a man do not work, he cannot eat satisfactorily. The food the inactive man eats gives him neither the nourishment, nor the elastic energy, nor the pleasurable sensations, that it does to the man of more active habits.

635. So closely is use of the muscular system connected with appetite and digestion, that exercise is usually one of the first means advised for their restoration when they are impaired; and thus we see dyspeptic students leaving college, and dyspeptic sedentary men giving up their business, and betaking themselves to travelling, to farming, or some other active employment, as the best method of regaining their lost health. But if the amount of exercise which the invalid takes as a means of recovery had been distributed through his previous days, and mingled with the hours of study and sedentary occupation, very probably it would have saved him from his present suffering and indigestion.

636. When we run, or walk, or labor in any way, the heart beats more rapidly than when we are at rest. The blood is carried, not only more frequently, but in larger quantities, through the muscular system, and through the whole frame. The alternate swelling and decline of the muscles, in their contraction and relaxation, press upon the veins, and force the blood out of them; and, as the valves in the veins do not permit this blood to go backward, it must go onward toward the heart. This is especially seen in the process of bleeding from the arm, when the patient holds a cane or ball in his hand, upon which he presses and relaxes his fingers in rapid succession. This action swells the muscles on the arm, the swelling presses upon the veins, and forces the blood onward and outward through the aperture.

All exercise of the muscles directly aids the circulation in the veins, and indirectly hastens it in the heart and arteries.

657. *When we walk, we breathe more rapidly than when sitting still.* (§ 342, p. 150.) If we run, or labor with great violence, we breathe very rapidly. During the process of exercise, more fresh blood is carried in the arteries to the organs, especially the muscles, and the more impure blood is brought away in the veins. The changes of living and dead particles are more rapid. (§ 252, p. 115.) There is then a greater quantity of old and dead particles — more carbon and hydrogen — to be carried out of the system, and consequently a greater demand for oxygen to convert these into carbonic acid gas and water. For this purpose the chest expands, and we breathe more frequently, and give the lungs the increased quantity of air that is needed. Hence we see that exercise aids respiration, and is most advantageously taken in the fresh air abroad.

658. When a boy is cold in the winter, he runs to get himself warm; when a passenger is insufficiently warm in his vehicle, he gets out and walks or runs by the side of his horse to warm himself. When a farmer sits in his house in a cold day, he has a large fire, sufficient to heat the room to near 70°, in order to keep himself comfortable. If he ride at the same season, he wears a great coat, and wraps a buffalo robe about him. But, when he takes his axe and cuts his wood in the open air, he wants neither fire nor great coat, perhaps not even his close coat. His exercise keeps him warm, and, if he labors violently in loading his sled or his wagon with heavy logs, he becomes uncomfortably hot, and may get into a free perspiration. The muscular action increases the amount of wasted atoms to be consumed in the body, and the more rapid respiration and circulation increase the fuel and oxygen for the internal fire, which burns more and more in proportion to the activity and violence of the labor. (§ 422, p. 181.)

CHAPTER X.

Muscular Action strengthens the whole System, and aids the Brain and Mind. — Neglect of Exercise debilitates the whole Powers — Various Persons need different Quantities of Exercise. — Too violent Exercise exhausts.

659. It will be easy now to understand how a muscle that is used grows and becomes hard and strong. The active contraction promotes the flow of blood and increases nutrition by the deposit of new particles of flesh. After the labor has ceased, there is less waste of particles by the absorption of the old. But the rapid motion of the heart does not cease with the exercise; the increased flow of blood is thus continued, and with it the increased deposit of new particles in the muscles; and thus it gains in size and strength by labor.

660. Thus we see that muscular action promotes digestion, respiration, circulation, and nutrition, and it assists the preparation and maintenance of animal heat. Finally, as the health and functions of the nervous system are connected with the condition of the other systems, we may safely add, that muscular exercise aids the brain also in its work, and that no mind can be the clearest and the most vigorous for study and reflection, unless the body is accustomed to action.

661. A sagacious physician, whose domestic economy was worthy of all imitation, when any one of his daughters complained of a headache, was accustomed to inquire, first, whether she had taken her usual exercise abroad; and, if this had been insufficient for the purpose, he frequently advised, not medicine, but another walk. Some judicious schoolmasters, when they find their boys and girls heavy and indisposed to study, send them out to play awhile. After exercise out of doors, they return to their studies with new alacrity. If older students, when they find it difficult to fix their minds upon the subjects before them, — when the reasoning powers are clouded and the imagination is dull, — would

leave their books and their studies, and walk, or otherwise use their muscles abroad, they might recover that energy of brain and mental clearness which they had struggled without success to gain while they were in their rooms.

662. A few years ago an unusually robust youth left his home in the country, and went to college. He was of a quiet, contemplative disposition, very fond of his books, and faithful to his plans of life. He did not enter into the active sports of the other students; and the walks in the neighborhood of the college did not interest him. Yet he was then in excellent health, had a good appetite, and ate heartily. Moreover, he was an industrious student, and a good scholar. While the first year wore away, he walked little, studied assiduously, and, at the end, his health was still good. The second year found him and left him about the same, except that he walked rather less. He was advised to pursue a different course, and the necessity of muscular action was urged upon him as a law of health. He acknowledged the law in general terms, but claimed to be an exception to its requirements. He had lived two years with very little exercise, and yet he was not only well, but was able to study as much as his fellows; he therefore supposed that he was exempted from the almost universal necessity of action abroad. His third and fourth years passed away, in much the same manner, except that he walked less and less, for the reason that it was more and more irksome to him; but his health was not very perceptibly deteriorated. After he left college, he taught school one year, and then commenced the study of his profession, and pursued this about two years, with the same habit of physical inaction and mental industry. But in this period his appetite began to fail, and he suffered from indigestion. His powers of mind languished, and his spirits grew dull. He lost his power of application and habitual cheerfulness. At the end of this period he broke down, and was unable to pursue his professional studies any longer; he then gave them up, and went to his home a confirmed dyspeptic. There he re-

mained several years an invalid, incapable of any business, or of engaging in any more study. That firm and inflexible constitution, which had held out seven years against the violation of the physical law, was equally inflexible with regard to recovery, and required more than seven years to be so far restored as to allow him again to engage in any pursuit.

663. The same quantity of exercise is not necessary for the health of all men and women, nor are all able to endure the same amount of labor. It must be measured out according to the constitution, the strength, and the habits of various persons. What is necessary for one may exhaust another. The quantity of action should be determined, not by any previously established theory, but by its results in each case — by its effects upon each individual.

664. If the exercise is too violent or too long continued, the body is rather exhausted than invigorated; the process of waste is carried on beyond the power of nutrition, and then the muscles grow thin and lank, rather than full and strong, and the individual suffers from languor, and is ill fitted for any other labor. But if, on the other hand, after the exertion, we are only a little fatigued, but not languid, — if we are ready then for any other occupation, for reading, writing, or conversation, — we may be assured that the increased waste is counterbalanced by the increased nutrition, and the labor has strengthened rather than weakened the body. The exercise should never go beyond a slight fatigue, never to exhaustion, nor produce that uneasy restlessness which unfits one for any other immediate duty, and which rather wears upon than adds to the general health.

665. Although the muscles have a power of contraction, they have not a permanence of this power. The muscle needs alternate relaxation with its labor. One can strike with more force than he can pull, and lift a much heavier weight than he can continue to hold up, for even a few minutes. It is one of the severest and most painful punishments to compel a boy in school to hold out the arm in a horizontal position, even without a weight in his hand, for any length of time.

CHAPTER XI.

Feeble Persons weakened by any Excess of Exercise, but strengthened by very moderate Exertions. — Exercise must be adapted exactly to the Strength. — In this Manner Strength may be increased daily. — It is an Error for dyspeptic Students or Invalids of the City to attempt to be Farmers or Sailors. — Gymnasium not adapted to Powers of those who use it.

666. We are told by some that they cannot walk or move in any way abroad, that it always gives them the headache or pain in the limbs, and that they return from their excursions sick and languid. This is, indeed, a truth; but it is very easily explained. These uncomfortable consequences flow, not from the mere exercise, but its excess. The walk which exhausted them may have been short compared with those which others take with ease, and return from with buoyant alacrity; but it was too much for their feeble and unpractised limbs. But, although a mile exhausts, half a mile will probably cause merely fatigue, and ultimately strengthen; and, if this be too much, a still shorter one will answer the desired purpose.

667. If the exercise be judiciously begun, with just the quantity that is sufficient, and no more, it will leave the person in slight fatigue; but in a little while, he will feel more fresh and vigorous, and capable of making a still greater exertion. Adopting this method, the feeble must begin according to the degree of his strength, however small; and with proper management and perseverance, he may go on adding a little exertion day by day, and accomplish more and more.

668. However small may be the person's strength, that must be the measure of the exertion. However low the power, that must be the starting-point. Any other measurement, any other point of beginning, would be fatal to the hopes of gaining strength by the effort.

A young man, in Waltham, Massachusetts, was very

feeble, but not sick. He was advised by his physician to set out upon a journey on foot, but was cautioned not to walk at any time until exhausted. He began his journey in the morning, and, with short exertions and frequent rest, he walked three miles on the first day, and was fatigued. The next morning, to his surprise, he felt more vigor and courage to go on, and started again. He walked on that day, in the same manner, and accomplished four miles before night. He thus gained strength and energy, day by day, adding little to little, and finally walked to Niagara Falls — more than five hundred miles. After viewing these to his satisfaction, he returned, in a much shorter time than he went. But he did not return by a direct course. He visited the interesting places in the neighborhood of his homeward route, and at the end of his sixth week, he reached home, having walked more than a thousand miles in forty-two days. On the last day he had walked forty miles, and was so little fatigued with the day's journey, that in the evening he felt sufficient energy to visit his young friends in the neighborhood.

669 There is a common notion, that, as great action gives strength to the strong, it will do the same for the weak; therefore the debilitated student and the languid child of the city, who have become so feeble as to be unable to carry on their studies or attend to their sedentary business in the counting-room, are advised to leave their occupations, and go into the country, and work with the farmers. They commence their labors with zealous courage; but they soon give up. They find that they are exhausted by the work, which the practised laborer accomplishes without apparent exertion or fatigue. Instead of being invigorated, they are weakened, and they abandon too frequently all hope of recruiting their wasted powers by muscular exercise. But if, instead of attempting to mow, plough, or dig, for several hours in succession, they undertake the lightest work, and do this for a few minutes or an hour or two, with frequent, and perhaps long intervals of rest, they gain power as the pedestrian just now described.

670. For the same purpose, and with the same mistake of means and ends, young men in ill health are sometimes sent to sea as sailors, and engage to do the work of the common and practised seamen. They err in their estimate of the effect of hard labor on the weakened frame, and are obliged to give up their purpose or alter their plans. But a much better and more successful method is, to enlist as weak sailors, without wages, and without responsibility. This allows them to work only so much as their strength gains upon them; and, if this be judiciously expended, they will add to it daily, and accomplish all they desire, and return, after some months, in more vigorous health.

671. It was supposed, several years ago, that the gymnasium would furnish opportunities and inducements to exercise for all such as were not required, by their business or their condition in life, to labor. In these establishments means were provided for using all the limbs and muscles. There were ropes to climb, parallel bars to walk upon with the hands, and wooden horses to mount upon or leap over. There were means for climbing, swinging upon the arms, leaping, vaulting, and for performing some of the feats of the rope-dancer, and some of the labors of the sailor. These exercises were active, and even laborious. Those who engaged in them made, or endeavored to make, the exertions which only strong men could make. But they were soon fatigued, and left the gymnasium; or, if they persevered, were nearly exhausted. The error was in not adapting the mode to, and measuring the amount of exertion by, the strength of those who needed it. The students of Cambridge, in 1826, (§ 165, p. 78,) complained that they were fatigued, and sometimes overcome, rather than invigorated, at the gymnasium, and were unfit for study for some hours afterwards. The final result of this attempt to introduce this system of exercises into our colleges, schools, and cities, was a general failure. But, if they had been arranged and measured so as to correspond with the little strength of sedentary men, they might have still been in general use, and productive of great advantage to health.

CHAPTER XII.

Kinds of Exercise. — Walking. — Sports of Childhood allowed to and beneficial to Boys. — Girls exercise more quietly, and with less Advantage. — English and American Women. — Exercise of the Arms and Chest. — Carpenters' Tools. — Time for Exercise. — Morning and Evening not the best.

672. *There are as many varieties of exercise as there are muscles in the human body.* It is not easy to determine which of these is the best, nor is it of consequence that we should settle the question in advance. There is no one kind that is better than all others, or can be substituted for all the rest. One kind uses one set of muscles, another uses another set. Walking employs the muscles of locomotion; cutting and sawing wood exercise the muscles of the arms and shoulders. Riding on horseback employs the muscles of the lower limbs, back, and arms, and agitates the whole frame.

673. *Walking is the most readily accomplished*, and is within the reach of every one. There is every where a road or a field to walk in; and if this exercise be taken with due energy, as boys and young men usually take it, moving with alacrity, swinging the arms and calling into requisition the contractile power of most of the muscles of the body, it will ordinarily be sufficient for the maintenance of health. Walking is the most advantageous when it is bold and easy. The body should be carried erect, the chest allowed the greatest freedom of expansion. The arms should hang and swing freely from the shoulders. A stooping posture interferes with the action of the lungs; and a confinement of the hands, the folding the arms on the chest, or carrying them in a muff, limits the muscular exertion, makes the movements unnatural, and causes an ungraceful gait.

674. The sports of boyhood, the games of the street and the playground, which not only require much muscular exertion but, are attended with exhilaration, answer all the

purposes of health. By custom and the general opinion of society, boys are thus happily indulged. They are allowed and encouraged to run, jump, and leap, and even to shout. They are consequently in good health, and have great vigor of body and activity of motion.

675. But the custom of society and the notions of propriety demand a different manner from the girls. They are not permitted to walk with that energy and vigor that their brothers are. There is a great fear of romping. They are required to be staid and quiet, and to confine themselves to walking. They are prohibited from the noisy plays, the bold activity of motion, and that free exercise of the lungs, which strengthen and delight the boys. And while the boys run and pursue any object of interest through the roads and fields, over rocks and hills, the girls are required to limit their movements to walking on the smooth and level paths which require comparatively little exertion.

676. A walk of three miles is not frequent for American women; and, when a lady of a country town of New England walked sixteen miles at once, in the year 1842, it was considered so extraordinary as to be made the subject of a newspaper article. But in England, a walk of some miles is an every-day occurrence for women; and thus they have means of locomotion ever at their command. When an American clergyman was visiting a family in England, it was proposed by the young ladies to visit a friend who lived at the distance of five or six miles. He cheerfully consented, but was surprised that no carriage came to the door. They walked, and spent a part of a day with their friend. On returning, the ladies proposed to the clergyman that they take another way homeward, which would make a walk of three miles farther, and call on another friend. He consented, and they went this long, and, to him, wearisome way home. The females neither regarded the walk, nor even seemed to think they had done any thing extraordinary, or out of their usual habit; but the American gentleman was unusually fatigued.

677. Some other employments give more exercise to the arms and upper part of the frame than walking. Garden-

ing, raking, hoeing, and digging, call into play the muscles of the arms, shoulders, and back. Working with the carpenter's and cabinet-maker's tools has the same effect, and when either of these can be combined with walking, the best effects upon the health are obtained.

678. Dancing, when practised at proper hours, and in sufficiently ventilated rooms, is an excellent exercise. It brings many muscles into action, and it is usually attended with cheerful exhilaration, that quickens the flow of blood and increases respiration. But the mere practising of attitudes, or the walking quietly through the figures, gives no exercise; and the late hours, crowded rooms, and night suppers, too often connected with this amusement, render it of very doubtful utility, if not certainly injurious.

679. To the hardy and laborious, it may seem a matter of indifference whether we take exercise at one or another hour of the day; and, for those who work from morning till night, all hours are alike in this respect. Still, for the invalid, and for those who only exercise for a short period, and for the maintenance of health, all hours are not equally advantageous. It is common to recommend the morning as the time to walk. The freshness of the morning air has been the song of the poet, the theme of the moralist, the faith of the philosopher. All have conspired in its praise, and in urging upon the feeble and the sedentary the beauty and advantage of early action abroad.

680. The morning may be the time for exercise of some, but it is not the best time for all. After the long fasting of the night, the body requires nourishment before it labors. (§ 113, p. 56.) It is apt to faint if it works before breakfast. Beside, the dews and dampness of the night, and the exhalations which have arisen from the earth, are upon the morning air, and must enter the lungs of those who are then abroad, and prevent their receiving the refreshing invigoration which a walk at another hour would give them. The same objection applies to evening and the night, and the sedentary should not then take their excursions in the open air.

681. The state of the digestive organs should be regarded when we exercise. We should not work just before eating, especially if we have long fasted and are hungry, for then the system is comparatively weak, and needs nourishment, and is therefore easily exhausted by exertion. And moreover, muscular action would expend the nervous energies that should be reserved to sustain the stomach in digesting the coming meal. (§ 164, p. 77.) Neither should we exercise immediately after eating, for the work of digestion requires all the energies of the system, until the food becomes thoroughly mixed with the gastric juice.

CHAPTER XIII.

Place for Exercise. — Should not be in House, but in open Air. — Exercise should be frequent and regular. — All need it, especially the Sedentary. — Consequences of Neglect.

682. We need an abundant supply of oxygen to sustain the increased demand for nutrition and discharge of waste which is caused by muscular action. Exercise abroad in the open air gives more health and vigor to both body and mind than exercise in the house. Some have prepared gymnastic apparatus in their garrets, or in their cellars, in order that they and their families may exercise without the trouble and exposure of going out of doors. Some gentlemen in cities saw and split wood in their cellars; but they fail of obtaining the full measure of good that action in open air would give them. Even those mechanics whose employments give them sufficiency of muscular exercise, especially those who work in close shops, would do well to add a walk abroad to their in-door labors; for they would gain in vigor of body, and freshness of spirit, and effective power, more than sufficient to compensate for the loss of time devoted to their renovation.

683. Whatever may be the weather or the season, the demand of the system for exercise abroad is the same; for

we have the same wants, and need the invigorating effects of muscular action both winter and summer — in fair weather and in foul. Nor is there any sufficing objection to it, for very few days of the winter are so cold that we cannot keep ourselves comfortable by rapid walking, or other exertion, and, indeed, the colder the weather the more dense is the air, and the greater quantity of oxygen is received into the lungs to sustain the internal fire. (§ 444, p. 189.) Very few days are so stormy as to prevent this exercise abroad, and on such days it may be taken under cover of the house.

684. This law for the health of the frame, and the necessity of exercise abroad, is one and the same for both male and female, for the rich and the poor. All can have a fuller development of strength, and health, and life, by taking it abroad; and all must suffer the same depreciation of life if they neglect it. There are none so favored in life as not to need it, none so high as not to be benefited by it, and very few so feeble as not to be able, in some degree or other, to obtain it.

685. Exercise should be frequent and regular. The system wants this means of invigoration as regularly as it wants new supplies of food for nutrition. Every day, therefore, should have its own, and no day should have more. It is not enough for health that we live inactively for several days, then devote one day to action of the muscles. But many do so; they have, in all, a sufficiency of exercise, but they take it irregularly. A clergyman, of very studious habits, devoted Mondays to walking, or riding on horseback, and the rest of the week to mental labor. While writing his sermons, he often for three days scarcely left his room. He became dyspeptic. Some teachers labor incessantly in their vocation for weeks successively. They teach six hours daily in school, and read and study the other waking hours out of it, with the intention of devoting their vacation to excursions and labor, and then, they think, they shall get exercise enough for another term of confinement.

686 The industrious seamstress, earning her scanty pit

tance by incessant toil; the shoemaker, working the whole day upon his bench; the mother, watching over her sick child; the faithful minister, writing for his people; and the judge, trying the issues of life and death — suffer as surely from indoor confinement, and want of daily exercise abroad, as the indolent, who have no occupation and no call for action. These must fall short of that full measure of power of body and of mind to do their present and pressing work, which a proper attention to the wants of the body would have given them. They may think they have no time for recreation abroad, and that an hour a day, spent in mere walking, is so much waste of opportunity of usefulness or of profit. But it is not so; the time required for the repair of the vital machine is not lost, for the body will not work the most easily, and with its fullest energy and most successful effect, if it is not in the best order. None need this daily recreation more than those who are compelled to produce every day the greatest result from mental or physical in-door labor, and who want, for that purpose, the fullest vigor, both of their muscular and nervous systems, and the most complete control of their powers.

687. *The evil consequences of neglect of exercise are not sudden nor immediately perceptible.* They are gradual and accumulative. They steal slowly upon, and secretly bind the strong man, and then take away his health. Dyspepsia, defective nutrition, muscular weakness, nervous irritability, and mental dulness, so manifest and oppressive as to compel the sufferer to change his pursuits or his habits, and betake himself to some means of relief, are remote results. But the immediate effects, however small and unnoticed, are none the less sure to come, and diminish the activity and force of life in proportion to the neglect. If this is continued, and violation of this law is frequent, weakness necessarily follows, until marked and acknowledged disorder is established.

688. It must be now considered as established that a certain quantity of muscular exercise is necessary for the maintenance of health, and for the best performance of the func-

tions of digestion, respiration, nutrition, and of the brain. The amount of this exercise may vary according to the constitution, and habit, and powers of the individual. It is best when so varied as to bring into play all the muscles of the body. It should be taken out of doors, and in the free air. This is necessary for all men and all women, of whatever occupation, and especially for those of no occupation.

CHAPTER XIV.

Amount of Exercise may be greater than Health requires. — Body grows strong with judicious Labor. — Limit to Man's Increase of Strength. — Fulness of Strength may be maintained to Old Age, with proper Care. — Man has a limited Power of Endurance. — No more Strength must be expended in the Day than is restored in the Night. — Men worn out by excessive Toil. — Length of Life differs with Amount of Labor.

689. THAT amount of exercise which is necessary for health is not the limit of muscular power. If it were, we should be able to accomplish but a small part of the work which we now do, and the labors of the farm and workshop would not be effected. We have a power of muscular contraction and of motion, which may be applied to the ordinary purposes of life. With this we cultivate the soil and carry on the operations of the mechanic arts, we navigate ships, and perform all our labor for pleasure or for profit.

690. It becomes a question to every man who works for profit, or who exercises more than is needed for the bare maintenance of health, how much can he work? Have the moving animal frame, the muscles and the bones, an indefinite power of endurance and action? It needs no physiological explanation to show that there is a limit to this power of labor. Then there comes another question — Where is that limit? How long can a man labor? How much may he labor each day, and not wear upon his permanent health, nor interfere with his continuance of life? To a certain

extent, the body grows strong with labor, and every exertion adds new particles and power to the muscular fibre. The man who walked to Niagara Falls (§ 668, p. 281) was stronger the second day than he was the first, and stronger the third than the second. He began with a power to walk three miles, and ended in six weeks with a power to walk forty miles. Whether he could have added still more to this power, or how much stronger he could have grown, is not known, for the experiment was not tried any further. In the same manner, any one unaccustomed to labor, if he has a good constitution and health, and if he proceeds gradually, and cautiously increases from small beginnings, can, in time, become sufficiently strong to do the ordinary agricultural labor.

691. In this process of invigoration there may be fatigue, but there must be no languor nor exhaustion. But if either of these happens, — if the working man finds himself exhausted after his toil, if he is uneasy, and restless, and unable to sleep, and awakes the next morning unrefreshed and unprepared for new exertion, — he may be sure that he has overworked his frame, and reduced rather than increased his strength.

But by faithful and prudent use of the power already gained, by never over-working on any one day, by always stopping short of exhaustion, additions are made to the strength day by day, for a certain period and to a definite extent. The laborer may increase his exertion as long as he feels this increase of power. But there must be, and is, a limit to this. No man grows infinitely strong, and sooner or later he must reach the end of his growth of power; then he possesses his fullest measure of strength.

692. Having arrived at this fulness of strength, he can maintain it if he uses it discreetly and temperately, and if he exercises daily, but never expends in any one day more than its due portion of power. By this self-management, a man can keep himself in the highest working order, and he will be able to accomplish the most labor, not in any one day, nor in a single year merely, but in the whole course of life, and

protract that life and its full working power to the natural period. For this end, he must attend to the first sensation of weariness, and never permit it to increase to exhaustion; and, whenever he begins to feel this, he may understand that the waste of life has reached the measure of the nutrition; and, if he then stops and rests, this last will go on, and he will in due time be refreshed and ready again for labor. In this way, he will always work at the full flow of his strength, and be able to perform the greatest amount of labor.

693. There is a common and mistaken notion, that man has an indefinite power of endurance, and may work until fatigue or exhaustion compels him to stop; and that whatever strength is not used in the hour and in the day, and is carried to the bed at night, is so much lost. Therefore some weak men labor as long as their strength holds out, and some strong men labor as long as the day will permit them. In consequence of this daily fatigue and exhaustion, they are never in full vigor, they are always reduced somewhat below their natural standard, and commence each day with a lower energy, and work upon a lower tide of power, than they otherwise might have done.

694. By this excess of labor, a man expends more strength in the day than he recovers in the night, and rises unrefreshed in the morning. He is wasting his constitution, and, if he perseveres, he reduces himself to a lower standard, and then he is compelled to limit his exertions and perform lighter labors. In this reduced and weakened condition, he may, by proper management of his diminished strength, lead a life of considerable action, and perhaps regain his original vigor. Or he may, by still overworking, reduce himself below the power of labor, and, becoming decrepit, suffer the pains and debilities of old age long before his time.

695. Those overworked and exhausted men, completely broken down and unable to labor, are not very common, yet they may be found. A farmer, within my observation, began his life with small means, but with great energy and large hope. He seemed to think there could be no end to his

power and labor. He rose with the sun in summer, and let it go down upon his toil. He allowed himself a few hours for sleep, and a few minutes for food, and no time for digestion. But in a few years he was worn out, and was then and afterwards, in health and strength, an old man. For the rest of his years he was an invalid, and unable to undertake even the lightest labor of his farm. Such instances of complete waste of power by excess of labor may be rare, yet the lesser degree of exhaustion is very prevalent. We find many who, after the middle period of life, feel obliged to favor themselves, and do lighter work, and with more frequent intervals of rest than others, because they have overworked and expended their power; and there are not many laborious men who do not thus begin to slacken in their labors soon after, perhaps before, they have passed their fiftieth year.

696. This constant labor, not only wastes the strength, but breaks up the constitution, and finally disarms it of its power of resistance to the causes of death. Hence the life of the overwrought laborer is, in general, shorter than that of the moderate worker. According to the registration of deaths in Massachusetts, for the 20 years and 8 months ending with 1863, it seems that the average length of life is the shortest with those classes of men whose days are spent in the severest toils. The ages and professions of men are recorded in these registers; and from these we learn that 19,252 farmers died at the average age of 64 years and 37 days; 14,733 common day laborers, at the average age of 45 years and $11\frac{1}{4}$ months; and 5070 sailors, at the average age of 45 years and 8 months.

697. The farmer is not obliged to make so continued and unremittingly severe exertions as the day laborer. He has a capital in store, and, in most cases, can have help at command, and suspend his own labor, or take the lighter tasks. But for the laborer, each day's bread must be obtained by that day's toil. To him there is allowed no rest nor choice of work; and to him is usually assigned the heaviest and the hardest. The sailor is presumed to be, in all cases, a man of full strength. He must assume his share

of the ship's labor, and perform a stout man's duty. He must take his turn, and watch night as well as day. He can never expect to have an entire night's rest.

Making all due allowance for the difference of habits of many in these several classes, — admitting that the poor laborer is often an invalid, and is worse clothed and fed than the farmer, and that more accidents happen to both the laborer and the sailor than to the farmer, and even other circumstances that interfere with the full enjoyment of health, — still much of this great difference of longevity — a measure of years on earth given to the farmer about forty per cent. greater than that given to the laborer and the sailor — must be charged to that excess of toil which wears out the man prematurely.

CHAPTER XV.

Languor always succeeds great Efforts. — Serious Injury may follow. — Growing Children need various and light Exercises. — Youths cannot endure the full Labor of mature Men.

698. THE period of vigor and of a man's productive power is shortened, the sum of the whole life's exertion is materially diminished, and its last stages rendered comparatively useless, by attempting to work beyond the power of permanent endurance. This is the result of long-continued over-exertion. But even if it is not long continued, if it is merely temporary, the same result of languor, and lessened ability for labor, follows for a proportionably short period. An extraordinary labor of one day is followed by extraordinary languor the next; and whether this greater exertion, this excess of waste of the powers over the nutrition, be more or less, it must be followed by a proportionate feebleness.

699. Excessive temporary exertions are followed, not only

by a corresponding languor, but sometimes by serious injury to the constitution. We are sometimes told by men that, since they strained themselves by lifting, or running, or by violent working at a fire, or some other occasion, they have not enjoyed their former health, nor have been able to work with their former energy. These evil consequences more frequently follow the excessive efforts of boys and young men. Dr. Hope says, that violent corporeal efforts, of every description, accelerate the circulation, and cause an unnatural pressure of blood upon the heart. "In growing youths, excessive rowing is one of the most efficient causes of this disturbance of the heart. Violent gymnastics produce the same effect."*

700. The natural exercises of children, their sports and games, which they enter upon and go through with boisterous zeal, give them light, and varied, and sufficient muscular action. None of the muscles are tasked too much, none called to labor for any length of time. Children and youth are not made for hard labor; whatever work is required of them should be light, varied, and short. Great exertion during the forming and growing period would very soon exhaust their strength, and, if continued, it would prevent the full development and growth of the muscular system, and even wear down the constitution.

701. A man does not reach his full measure of strength and power of endurance until he has passed his 25th year. And if, under that age, he is exposed to the hard-labor and privations which older men seem to endure with impunity, and in situations where can be no relaxation nor favor shown, the health and strength of the youth fail; and, if there is danger from the labor and exposure, the youth is the first to die. Young soldiers sink under the labors and privations of the camp sooner than mature men. Napoleon complained that boys were sent to supply his army, rather than men who could endure the toils and the sufferings of the campaign.

---

* Diseases of the Heart, Part III. Chap. I.

The development of strength is progressive through the several periods of early life, and each period has its appropriate means and opportunities. The plays of childhood prepare the muscles for the light employments of youth; these increase the muscular energies, and prepare them for the full power of labor in manhood. Each is necessary, and none can be omitted, nor can they be interchanged. Some children are exclusively devoted to study, and have no inclination to active play. They prefer their books, while other children are playing abroad. These develop their nervous systems, but their muscular powers are dormant. They become good scholars, but are weak in body, and ultimately their mental energies sometimes falter.

CHAPTER XVI.

Labor requires healthy Organs of Digestion and Nutrition. — Good Food, sound Lungs, and fresh Air. — Healthy Skin, sound Condition, and Coöperation of the Brain and Nervous System. — Exercise most beneficial when the Brain is lively and Spirits cheerful.

702. Muscular action presupposes waste, (§ 252, p. 115,) which is supplied immediately from the blood, and remotely from the food. The active have, comparatively, new flesh, and the inactive old flesh. (§ 257, p. 117.) In order to supply this waste, and make new muscular atoms, the organs that effect the changes in the food, and the channels through which it passes in its progress from its condition as food in the mouth to its new condition as flesh in the tissues, should be in good health. A working person must, therefore, have a sound stomach and good digestion. The organs of circulation and nutrition must also be in good order; for otherwise the food cannot be converted into blood, nor the blood into flesh, to meet the changes that the action of the muscles demands.

703. Next in importance to a sound condition of the organs of digestion and nutrition is the supply of good food. Food alone is the source of all our bodily strength; and it

gives this power only in proportion as it can be converted into new atoms of flesh. The laboring man wants rich and nutritious food to sustain him in his exertions. He cannot work with his fullest energy with poor and innutritious aliment. Just in proportion as his diet is low or lacks in nutriment, must his strength and his power of labor fail. The English trainers, who develop the greatest muscular force, eat, or give their men the best of bread and the best of meats, — beefsteak and mutton. On the same principle, the judicious but economical farmer feeds his cattle, and gives the oxen that work better hay than those that lie still.

704. *The laborer not only wants nutritious food, but that which can be converted into flesh with the least cost of power.* Digestible food requires but little exertion of the stomach to convert it into chyle; but heavy bread, tough meats, matters that are badly cooked, and all other sorts of food that are hard to be digested, are not converted into chyle without much labor. If more strength, or more nervous power, is expended in digestion, of course less can be expended upon the muscles of motion. This is well understood when one has eaten a very heavy dinner, which absorbs all the energies of the system for its digestion. Then he can no more work with his hands than he can when he is using his feet with all his force. The same is true, though in a lesser degree, when any food that is of difficult digestion is taken. While this is going on, the man may work, but his power of labor is reduced in proportion to the difficulty of the digestive operation. A person, therefore, can accomplish the most when he has eaten light bread and the best pieces of meat; and the laboring man, whose life is in his power of labor, can afford to eat no other.

705. The waste caused by exercise must find free outlet through the skin and the lungs. Both these organs must, then, be in a healthy condition in the laborer. His surface must, by frequent bathing, be kept free from every thing which would clog its pores, or obstruct its operations. The increased waste through the lungs demands a greater supply

of oxygen, and, consequently, a more frequent respiration during exercise. But if the lungs are unsound, or inflamed, or in any way impaired, or if their motions are impeded by any external incasement around the chest, they cannot receive air sufficient to carry off the excess of wasted particles which are thrown into them. For this cause, a man laboring under asthma or consumption cannot run, or pump at a fire-engine, or mow grass, or perform any labor that requires great and continuous effort. An abundant supply of pure air is equally necessary to carry off the excess of waste. Men who work in close shops, mines, and the holds of ships, have less power than those who work in the fields.

706. The heart, being the engine that propels the blood, is required to work with greater force and rapidity when exercise demands a greater supply of nutriment in the tissues. If this organ is unsound, it cannot make this extraordinary exertion; and those who have diseased hearts cannot perform very active labor. Mr. H., a very industrious farmer, once complained to me "that he could not cut his wood, nor mow his grass, as he had done, for the exertion immediately caused great distress about the region of the heart, and then he could not move." He was suffering from a disease of the heart, of which he afterwards died.

707. The brain is connected with all the muscles through the medium of nerves, and directs and sustains their motions. The state and health of the brain and mind affect, very materially, the value of exercise and the power of labor. We will, or determine, to move the finger; at the same instant the volition is sent from the brain to the muscle that moves the finger; this then contracts, and the finger moves. This is the case with all voluntary motion; yet the volition of the mind and the action of the brain are not always observed. Some motions are performed so much from habit, that we are unconscious of the volition. When we walk, we contract the muscles, those which bend the joint of the hip, the knee, and ankle, of one side; and, at the same moment, we contract the muscles which straighten these joints on the other side. At

the next moment we reverse this action, and contract the straightening muscles of the first side and the bending muscles of the opposite side; by these motions on opposite sides, the limbs are lifted and carried one before the other, alternately, and the body moves onward.

708. While we walk easily over a smooth and familiar path, we may direct the energies of the brain partially, but not exclusively, to other matters, and talk or think of agreeable and interesting subjects. But if the walk becomes difficult, — if the mind is required to pick the way through wet or stony places, or perform any severe labor, — all the energy of the brain is required to direct and sustain the muscles of motion, and then conversation and thought are suspended. But if, on the contrary, the brain is all absorbed with other subjects, — if we are engaged in deep thought or oppressive anxiety, — the muscular contraction is not so easily effected and controlled, the strength of the muscles is not so well sustained, the exertion wears more upon the body, and consequently exercise under such circumstances is not so invigorating, and labor not so effectual and profitable.

709. Although the brain and mind must not be absorbed or oppressed with care or thought when we exercise or labor, yet they should not be dormant. They should be lively and engaged. The exercise should have an object. Walks through pleasant scenes and among interesting objects, excursions in pursuit of flowers, minerals, or other natural objects, do more for the health than those which have no object.

CHAPTER XVII.

Labor should have an adequate Object. — Hope and Confidence give, and Doubt and Fear diminish, Strength. — Cheerfulness and Melancholy have similarly opposite Effects. — Effect of Passion, Alcohol, on Strength.

710. *Labor for profit, as well as exercise for health, should have an adequate object.* The mind must be satis-

fied, or else the brain will not coöperate with its full energy, and give the muscles full power. When the farmer fears that the cultivation of his fields will produce no crop, when the laborer believes the wages are inadequate to his services, and the mechanic thinks his wares will return him no profit, they cannot make their greatest exertions, or, if they do, it is at a cost of the permanent power. But the laborer who is well paid generally feels a motive that stimulates the brain, and strengthens the muscles, and enables him to work vigorously and successfully.

711. *Hope and confidence give almost unmeasured strength;* but despair weakens, almost paralyzes. When a man falls overboard at sea, he swims for his life as long as he has hope of rescue or can move his limbs, until, fatigued with his labor, and in despair of obtaining relief, he seems unable to swim any farther, and suspends all exertion, and gives himself up to death. But if, at this moment, a boat comes in sight, or land appears to him, a new hope is excited, new strength is given, and he swims again with a power which was impossible a few minutes before.

712. The same effect of confidence that strengthens the muscular system, and of doubt that weakens and sometimes paralyzes exertion, may be seen in any of the labors of common life. If the student walks grudgingly, with doubt as to the efficacy of the exercise, and fear that he is misappropriating his time; if the over-cautious girl walks with fear lest the exercise flush her cheek too much, or the perspiration soil her garments, or the exercise derange her dress; if she moves with timid anxiety lest she assume ungraceful attitudes, or in any way transcend the becoming delicacy of a lady, — the brain will not coöperate earnestly with the work, nor send full stimulus to the muscles; the limbs then labor languidly, and the exercise fails to invigorate the system.

713. *Cheerfulness and melancholy have the same opposing effects on muscular power as hope and despair.* Whatever depresses the spirits, depresses the energies of the brain, and consequently the energies of the locomotive ap-

paratus. The slow and measured step of the funeral procession, and the light, elastic step in the merry dance, are both equally indicative of the energies of the muscles. There is as certain a difference between the power of muscular contraction in the dancer and in the mourner, as there is between the buoyant spirit of the one and the oppressed spirit of the other. In one, the heart is joyous, the brain is active, and the motion quick and easy; while in the other, the heart is sad, the brain is heavy, and the movements are slow and wearisome.

714. *All alcoholic liquors stimulate the muscular energies temporarily.* A man under the influence of these can make greater exertions, and for a short time accomplish more work. But this increased power, like the effect of passion, is of very short duration. While stimulated by excitement, a man may strike heavier and quicker blows; but then the unnatural labors soon weary and exhaust him. All his excessive exertions make a draft upon his permanent constitutional power, and leave him weaker than they find him; and he who habituates himself to depend upon the stimulus of spirit to give him strength for labor, like all others that overwork, wears himself out early, and brings on the infirmities of age before the natural time.

CHAPTER XVIII.

Attitudes. — Spine supported by Muscles equally on both Sides. — Spine very strong. — Porters. — Pedlers carry Burdens on Head, and Spine erect. — Centre of Gravity over Line of Support. — Head so carried. — This Attitude easiest and most graceful.

715. *The attitudes assumed in exercise or labor are of great importance.* The structure of the spine, and the arrangement of the muscles which support it, give this column its greatest strength and flexibility when it is held erect. The spine curves from back to front, and from front to back; yet these curves are so balanced that the upper end of the column, the

resting-place for the head in its natural position, is vertically over the base at the pelvis.

716. The spine, as well as the other bones, is held in its erect position by a double series of muscles, (Fig. LVII.,)

Fig. LVII. *Internal Muscles of the Back.*

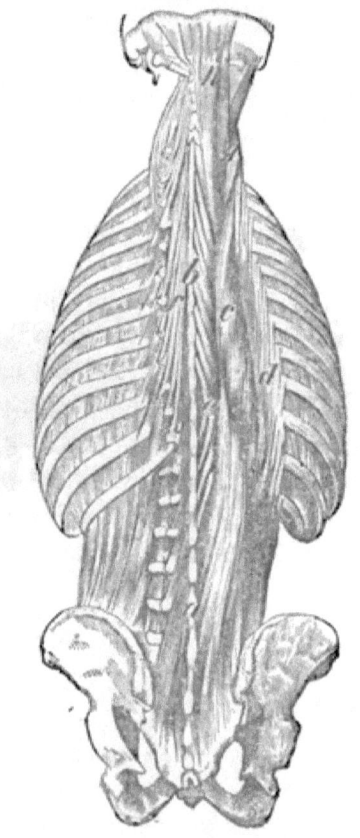

a, b, Spinous processes of the back-bone.

c, Longissimus dorsi, longest muscle of the back.

d, Muscle extending from the pelvis to the ribs.

e, Muscles extending from the vertebræ of the loins to those of the back.

f, g, Muscles extending from the vertebræ of the back to those of the neck.

h, Muscle extending from the back and neck to the head.

which are placed on both sides of the back-bone, where they form a cushion of flesh, and are easily felt. The lowermost of these are fixed by one end to the hips, and by the other to some of the bones of the back; others are attached to one and then another of these little bones; and others still are attached to the ribs and to the back-bone. All these serve to bend the back to one or the other side, and forward or backward.

The uppermost bend the head. They serve also to keep the head and the back in their erect position.

717. The muscles on the opposite sides of the spine are naturally of the same size and length, and equally strong. They give to each side of the back the same support, so long as they are accustomed to the same amount of exertion. But if we allow the back to bend to one side, the muscles within the curve will be shortened, and those on the outside will be lengthened. The muscular action is then increased on the convex, and diminished on the concave side; for the muscles on the outside of the arch are obliged to exert a constant and much greater force to prevent the further curving of the spine than was necessary merely to keep it balanced in the erect position; while very little action is required of those which are on the opposite side and within the curve.

718. *The structure of the back-bone gives it great strength as well as flexibility.* Being composed of alternate bones and cartilage, and held together by strong ligaments and supported by many muscles, it is capable of bearing great burdens. All the upper parts of the body, more than half of its weight, and all the burdens that we carry on the head, the shoulders, the back, and the arms, rest upon it. Porters who are long trained to their business will carry upon their shoulders, or upon their heads, a burden of some hundreds of pounds. The Turkish porters in Smyrna, Asia Minor, carry enormous loads on their backs. A friend who has been there writes to me, " The porters in Smyrna have a pack on their backs, about twenty inches wide, flat on the outside, so that the load lies on it steady without fastening. I saw one take a box of Havana sugar on his back, to carry from the boat up to the warehouse, a short distance from the water. The sugar weighed four hundred and fifty pounds, or more. Captain N. of the navy said to him, ' You had better add a bag of coffee,' which weighed one hundred and thirty pounds. The porter said, ' Put it on,' which he did. He, the porter, then turned

to Captain N., and said, 'If you will give me a dollar, I will carry you on the top of these.'"

719. I once saw, in the streets of Louisville, Kentucky, a colored woman carrying a tub of water upon her head, and a pail of water in each hand. There it was not unfrequent to meet a woman with a large pail of water on her head, which she carried with apparent ease and without spilling. We often meet the Italian pedlers carrying a large tray covered with images, or flower-pots, or toys, upon the head. They carry this with as much apparent security as others would carry them in their arms. Those who bear upon their heads heavy burdens which require strength, or pails of water that must not be spilled, or fragile merchandise that must not be broken, carry their heads very erect and their back-bones very straight. They hold the upper extremity of the spine directly over the lower end.

720. This perfectly erect position of the spine affords the easiest method of carrying, not only burdens, but the head and the trunk, in the ordinary walks of life. Any one can try the experiment of holding a pole erect in his hand by one end. If it be vertically erect, he will exert no more strength than barely to lift the weight. But if it be inclined to either side, it will require considerable exertion to prevent its falling. So, if the head be bowed forward, if the chest be bent downward, then the weight is not immediately above the point of support; it does not rest upon this foundation, but it must be held up by the exertion of the muscles, which is a very wearisome labor.

721. In order to carry the head and body with the greatest ease, we must be governed by the same law as in carrying any burdens. We must bring the weight, the centre of its gravity, perpendicularly above the point of support. But, if we have any weight added to one side of the body, we must change the direction of the spine, so that it shall bring the centre of weight in the proper line. This we do instinctively. When a boy carries on his breast an armful of wood, he leans backward to bring the weight over the base of the

spine. The fat man with large abdomen does the same, and for the same reason. Otherwise they would necessarily exert great force of the muscles of the back, to prevent their falling forwards.

722. If a porter stands perfectly erect when he carries a trunk on his back, (Fig. LVIII.,) the line of gravitation falls behind his natural line of support, and tends to throw him backwards. To prevent this, and bring the burden over the point of support, he leans forward, (Fig. LIX.)

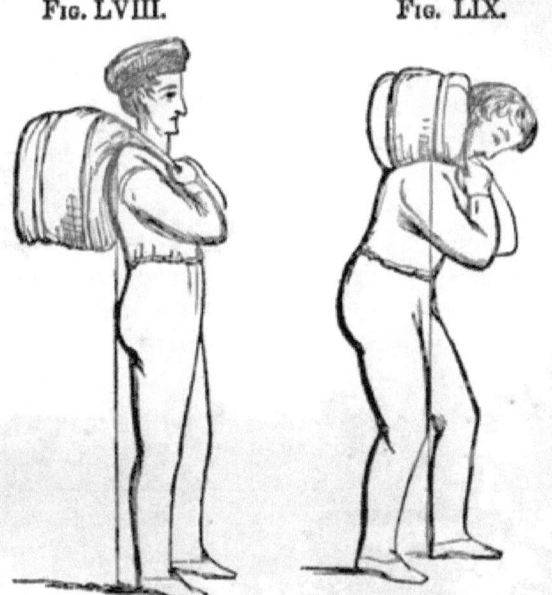

Fig. LVIII. Fig. LIX.

For the same reason, the hod-carrier leans to the left when he carries his burden on his right shoulder, and to the right when he bears it on the left shoulder. When a boy carries a heavy pail of water with one hand, he leans as far as he can to the opposite side; but, as he does not lean far enough to bring the centre of gravity over the point of support without spilling the water, he throws out the other arm, and carries it in a horizontal position, to create a greater weight on that side, and balance the weight of water. And often he finds it easier to divide his water into two pails, and, by carrying one in each hand, brings the centre of weight over the point of support, with his spine perfectly erect.

723. Upon the same principle of carrying the centre of gravity over the centre of support, every one should carry his head erect and his back-bone straight. It is not necessary for this to obliterate the natural curves of the spine, but to carry its line of support vertical from its base to the top. This will bring the head directly over the lower end of the spine. In this attitude, the weight of the head, trunk, and whatever burdens are borne, resting upon the bones, very little muscular action is required; and the bones of the lower limbs, and the general course of the spine, are in the same line. When these bones are in this upright direction, and the upper balanced upon the lower, it requires but little muscular exertion to hold them in their places. But if the lower bones are turned, or the spine is bent to either side, it requires a constant exertion of the muscles on the convex side of the joint or the spine to prevent it from bending farther. Whatever muscular strength is expended in maintaining the attitude, cannot be devoted to any other purpose.

724. *This attitude is not only the easiest, but the most graceful.* Stooping the body, or bending the head forward, when walking or standing, interferes with the elegant flexibility of the spine, and is awkward and uncomfortable to the person. Among those who carry burdens upon their heads we find the most frequent instances of graceful attitude and gait. Captain Ball, in his " Seven Years in Spain," says, " It is wonderful to see the amazing burdens that the Spanish women carry on their heads, and walk at so rapid and safe a pace without the least accident. It is remarkable that the female peasantry in Spain have a more graceful and comely style of walking than the ladies, which I have repeatedly heard accounted for by the burdens that they carry on their heads requiring a certain degree of steadiness to balance."

CHAPTER XIX.

Erect Attitude best for Walking. — For Labor. — For Mechanics and Farmers. — For great Exertions. — For Speakers. — Spine curves from Front to Back. — Becomes bent by much stooping. — Position of Students and Writers raises the Shoulders and curves the Spine from Side to Side. — Curved Spine frequent among Girls, but not among Boys. — Injures Spinal Cord.

725. "*In walking, it is all-important that the body be held as upright as possible*, the shoulders being kept back, and the breast projected somewhat forward, so as to give the chest its full dimensions. The lungs being, by this means, allowed sufficient room to expand fully, breathing is rendered free and easy, and every vital action is performed with vigor. The attitude thus assumed in walking, places all the organs of the body in their most natural position, and frees them from all constraint." *

726. The wielding of the heavy sledge of the blacksmith, and planing of the hard wood of the wheelwright, are done with comparative ease if the body is kept erect. Such operations of agriculture as hoeing, mowing, ploughing, are generally easiest in the same position. Two men mowed side by side during a summer. One of these men was rather tall, large, and very muscular, and was reputed the strongest man in his town. The other was rather under the common height, of slender form, but very active. When mowing, the strongest bent his body down, and struck his scythe with all his might. The other stood erect, and, without much apparent effort, swung his scythe as one swings a cane. These two men, making such different efforts, performed equal work in the course of the day. But when they went home at night, the strong man was wearied and almost exhausted; the other was somewhat fatigued, but lively and elastic.

* Journal of Health, Vol. I. p. 120

727. This erect position is particularly attended to by those who wish to exert the most effective force, and to strike the heaviest blows. Soldiers, who endure fatiguing marches and fight with the greatest energy, are especially directed to maintain the erect attitude. So the prize-fighters are taught to stand in the struggle; their back is straight, their shoulders thrown back, and breast forward; then their arms are completely under their control. The best public speakers and readers stand in the same position. Their chests are free to expand, and they inhale large portions of air; the head is erect, and the windpipe is not compressed; then the air can be thrown forcibly from the lungs, the voice is full, and the articulation easy and effective.

728. The healthy spine, in its natural position, curves from front to back and from back to front, but not from side to side. When left to itself, it assumes this shape; yet its structure of alternate bones and cartilages allows the column a great variety of motions and positions; but the elasticity of the cartilages tends to restore it to its natural direction, after having been bent to either side. So that, after leaning and bending the spine to the left or to the right, by force of the muscles, when these cease to act, the side of the cartilages which was flattened springs upward, and throws the spine again into its natural position. But if the column be turned to one side frequently, and continued for a long period, and at the same time be pressed down by a weight, the compressed side of the cartilage wanting opportunity to regain its usual form, will become permanently thinner, and the opposite side thicker; and then the whole pile would, without any external aid, remain curved.

729. *This great flexibility of the spinal column allows it to bend in any direction, and, for a short period, without danger of permanent curvature, if afterwards the upright posture be assumed and maintained.* But, if we continue these positions for a long time, the spine does not easily recover its proper shape. If the student bends his back and leans his head down, to bring the eyes nearer his book, if the seam-

stress bends her chest forward over her sewing, or if the engraver or watchmaker has his bench so low that the spine must be curved forward to bring himself near to his work, and if they sit in this manner for months and years, the cartilages are compressed beyond the power of reaction. The front part is flattened and the back part is thickened; it becomes wedge-shaped, and consequently the back-bone is permanently crooked, and the person stoops or is round-shouldered.

730. The same law applies to the lateral line of the back-bone, and similar habits of compression of the cartilages bring on curvatures from side to side. The position assumed at school while writing, (Fig. LX.) and often while studying, throws the spine out of its straight, lateral line, and bends it to one side or the other. The table or desk for writing or

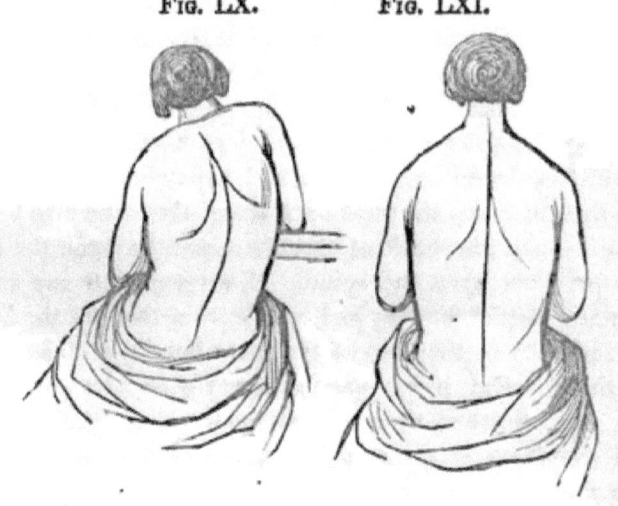

Fig. LX.　　　Fig. LXI.

drawing is usually higher than the elbow, as it hangs from the shoulders of the pupil sitting on the seat. In order, then, to write or draw, the right arm and elbow are raised and rested upon the elevated table. This raises the right shoulder, and, in raising it, bends the upper part of the spine over from the right to the left, and depresses the left shoulder. Then,

27*

in order that the head should still be over the base or point of support, the upper curvature is balanced by an opposite lower curvature. While the upper part of the spine is curved to the left, the lower part, at the loins, is curved to the right, and the whole column assumes somewhat the shape of the letter S in its lateral direction. At the same time, the lowest and the highest portions are nearly straight, and the head is vertically above the base of the column. (Fig. LX.)

731. The same effect follows from a position sometimes assumed in reading. The table is higher than the suspended elbow, and the reader does not sit directly facing it, but rather obliquely, and, lolling sidewise, raises the elbow upon the table, and rests the head upon the hand. This raises the shoulder, bends the spine, and produces the same result that comes from the unnatural posture in writing and drawing. If these positions are frequently changed, if one shoulder is raised as often as the other and neither elevation is continued for a long time, no curvature of the spine will follow. But if either bent position be assumed frequently, and maintained for a long time, the cartilages will lose their elasticity, and become compressed on one side and expanded on the other. In the natural form, the shoulders are of the same size; they both rest upon the back of the chest, and lie upon the ribs, which are fixed upon the spine. If we examine any active boy, or any playful, healthy girl, we shall see that the shoulders are exactly alike; they are of the same height, and have the same shape. But, if we examine many girls who are pursuing or have finished their education, we shall find that one of the shoulders is grown out, and is higher, and projects farther forward than the other.

732. The habits of school children, and especially of girls, of students, clerks, draughtsmen, and of some others, create a fearful frequency of this spinal distortion. Dr. Warren says, "In the course of my observation, I have been able to satisfy myself that about half the young females, brought up as they are at present, (1845,) undergo some veritable and obvious change of structure; and, of the remainder, a large

number are the subjects of great and permanent deviations; while not a few entirely lose their health from the manner in which they are reared." And again, "I feel warranted in the assertion, already intimated, that of the well-educated females within my sphere of experience, about *one half are affected with some degree of distortion of the spine.*"* Dr. Warren substantiates his opinion by that of Lachaise, a French author upon the spine, who, in speaking of the lateral curvature, says, "It is so common, that, out of twenty young girls who have attained the age of fifteen, there are not two who do not present very manifest traces of it." A fashionable mantua-maker, of extensive experience and observation, says that she has been obliged to stuff with cotton a large proportion of the ladies' dresses, on one side or the other, to make them exactly symmetrical.

733. *Nature has given to all — to both female and male — sufficiency of bone and muscle to sustain them in the most graceful and healthy position ;* and when these are faithfully used, and their strength developed, they fulfil their purposes, and keep the form straight. The lateral curvature of the spine is very rarely found among boys. Their various and free exercise strengthens all their muscles, and prevents it. But it is very common among females, who exercise less, and wear external supports, which are intended to take the place of their natural framework and muscular power, and sustain the body. But these substitutes not only fail of their purpose, but sometimes bring on the very deformity they were intended to prevent. Although nature has provided all the proper supports for the spine, yet, when they are not used, they become weak, and then the spine bends to one side. This lateral deformity is rarely found among laborious farmers or mechanics employed in the heavy trades, or among porters, or even hod-carriers, who carry heavy burdens on their shoulders or heads, but in sedentary persons, who lift the least, and whose work is the lightest, the muscles of

* Preservation of Health, p. 13.

whose backs have no other employment than to hold the spine erect.

734. *The curvature of the spine not only injures the symmetry of the frame and lessens its height, but it distorts the chest and diminishes its capacity, and interferes with the free motions of the ribs.* Accordingly, the lungs have less space for rest and less room for expansion, and therefore can receive less air at each inspiration; then, imperfect purification of the blood, and, lastly, a deficient nutrition of the body, must necessarily follow. Sometimes serious diseases of the lungs are brought on by this curvature of the spine, and Dr. Hope says, "The majority of hump-backed persons are ultimately attacked by disease of the heart."

735. The rings in the vertebræ (Fig. XXXIV.) being placed one upon another, form a channel or tube through the whole spinal column. This channel is of the utmost importance in the animal structure, for it encloses the great nerve called the *spinal marrow*, that extends from the brain to the trunk and the lower part of the system, and supplies all these parts with the principle of life. This great nerve begins in the brain, and reaches to the end of the spinal column. In the course of its descent, it sends out on each side nerves to the heart and lungs, to the organs of digestion and of motion. All the parts, therefore, of the trunk and all the extremities depend, more or less, upon the healthy condition of this great nerve for their fulness of life and freedom and energy of action.

736. Any change in the shape of the spinal column, or in the relations of these bones to each other, must diminish the capacity of this canal, and, of course, press somewhat upon this great nerve. This pressure upon this nerve interrupts its freedom of action, and interferes with the communication between the brain and the parts of the body which receive nerves through this channel. This must be the natural consequence of all distortions of the spine, all permanent curvatures from side to side, and of all unnatural curvatures from front to back, or from back to front.

CHAPTER XX.

Day is Time for Labor. — Experiments. — Soldiers. — Miller. — Sleep. — Quantity of Sleep. — Night proper Time for Sleep. — Deficient Sleep causes Weakness. — Circulation feeble, and Heat less in Sleep. — Difficult Digestion disturbs Sleep.

737. *The day is the time for labor, and the night is the time for rest.* This seems to be the almost universal law of nature. During the light of day, the air is more pure, and respiration is better sustained, the changes of particles are more easy, and consequently the muscles are better strengthened. The light of the sun has, in some way or other, a great influence upon the energies of the body and the mind. The effect of a long series of cloudy days upon the spirits is familiar to all; we then become dull and querulous about the weather, and the return of the sunshine is received with a burst of joy, as if it brought back new life. Miners, who spend most of their daytime within the earth, become bleached and dull. Mechanics and shopmen, who work or transact business in imperfectly-lighted shops, have a lower degree of energy and health.

738. Night labor is attended with the double disadvantage of bad air and darkness. The evil consequences of this were shown in the experiment of two French regiments. "One of them, although it was in the heat of summer, marched in the day and rested at night, and arrived at the end of a march of 600 miles without the loss of either men or horses; but the other, who thought it would be less fatiguing to march in the cool of the evening, and part of the night, than in the heat of the day, at the end of the same march had lost most of the horses and some of the men."*

739. A similar experiment is partially tried by individuals, almost every where, with the same success. Milkmen and

* Art of Living Long and Comfortable, p. 172.

market-men in the neighborhood of cities, and workmen in tide-mills, spend a part of the night in their business, and make up their loss of sleep in the day; but in a few years they are very glad to discontinue this course, and confine themselves to daylight labor. Mr. G. owned and worked a tide-mill, which could run only for a few hours succeeding a full tide, which came as often in the night as in the day. In course of a few years, this frequent night work wore so much upon him as to compel him to exchange his tide-water power for steam power, which he could use to suit his own convenience. Having given up night work, and limited his labor to the day, from a feeble he has become a robust man, and is able to accomplish more in the new than in the old system.

740. *It is a law of nature that all animals shall suspend their actions, and sleep.* The alternations of day and night harmonize with this want of the living animal body, and afford seasons of activity and of rest. Man needs to follow this natural indication, and alternate his sleep and wakefulness daily. Sleep is nature's restorer of exhausted power, and, though we retire wearied, we awake refreshed and strong; the expended energies are recovered, the strength brought back, and we are again ready for action. In the state of sleep, all motion of the voluntary muscles is stayed, and the brain suspends its active functions; but the involuntary functions go on as when awake; the chest moves, the lungs breathe, and the blood is purified, the heart beats, the blood circulates, and the system is nourished.

741. The quantity of sleep that is necessary is varied by so many circumstances, that no rule can be established for all. The time of life and the peculiarities of constitution make a difference. The sluggish and the lymphatic need more sleep than the active and the nervous. Some sleep very much, and are not refreshed, nor ready for action, if they are deprived of their usual quantity of rest. Others take very little sleep, and cannot obtain more. Without giving any precise rule, it is sufficient to say that men and women who have arrived at adult years, and developed their full

strength, need from seven to nine hours' sleep. The habits of sleeping create a difference of necessity, for a time at least, and cannot easily and suddenly be broken.

742. *The sleep in the day does not compensate for the loss of the night sleep.* The soldiers who rested in the day, and marched at night, had as much sleep as the others, who slept at night; and yet they suffered much more from sickness and exhaustion. (§ 738, p. 321.) The most perfect sleep and refreshing rest is obtained in the stillness of darkness, when all nature reposes; and it is all in vain that any one struggles against this law of his being. He may sleep in the day, and labor or watch in the night, but his waking hours are then not so bright, nor is his energy of life so great, as otherwise it would be.

743. There are few who do not, now and then, devote a night, or a part of a night, to some labor, to travel, to parties of pleasure, or to watching with the sick. None of these escape the penalties that always follow the violation of the law of rest. Whatever may be the time required by habit, or the constitution, for sleeping, for the recovering of exhausted power, that time cannot be shortened without impairing, in some measure, the strength and activity of the next day. With less than the required quantity of sleep, the body is not completely refreshed, nor has it the full energy for action. If each day is expected to accomplish its entire work, each night must have its complete rest; and whatever is taken from the sleep must be taken from the power of labor. If any one cannot retire at his accustomed hour, and still wishes to have his usual power of action on the next day, he must protract his rest in the morning as much as it was shortened at night.

744. During sleep, the circulation is more feeble, the respiration is slower, and the heat is generated less rapidly than in waking hours, (§ 436, p. 187,) and, consequently, we are less able to resist the effects of cold; and if they exposed to a current of air, we are more liable to suffer than when we are awake. We therefore sleep under more clothing in the

night than we wear about the business of the day. If possible, one should not go to bed cold, for it is difficult, when sleeping, to recover the heat that has been lost.

745. Sleep is the most refreshing when taken in large and airy chambers. These rooms should, therefore, be ventilated daily, and at all seasons, by opening the windows, or otherwise. The lodging-rooms should never be used for any other purpose — for sitting, working, cooking, or eating. The bed and bedding should be opened and thoroughly aired every day. A very hard bed affords but few points of support for the body, which is, therefore, not so well rested while lying upon it. A very soft, downy bed allows the body to sink within it, and keeps up too great a heat, and debilitates rather than strengthens the sleeper.

746. Nutrition goes on during sleep. But the food should be digested before retiring to rest. (§ 121, p. 59.) Sleep is disturbed with unpleasant dreams after a late or indigestible supper. The stomach works with difficulty, and the man dreams of being in difficult situations, or of attempting purposes which he cannot accomplish; still greater oppression at the stomach produces distressing dreams and nightmare; and in neither case is the natural and complete refreshment obtained from the sleep.

PART VII.
BRAIN AND NERVOUS SYSTEM.

CHAPTER I

Nervous System.—Coverings of Brain.—Spinal Cord.—Nerves of Sensation and Motion.—Distribution.

747. *The nervous system consists of the brain, the spinal cord, and the nerves.* The brain is in the head: its size and shape correspond to the size and shape of the skull. Its external surface is not smooth and level, but it is broken into

Fig. LXII. *Brain. External Surface.*

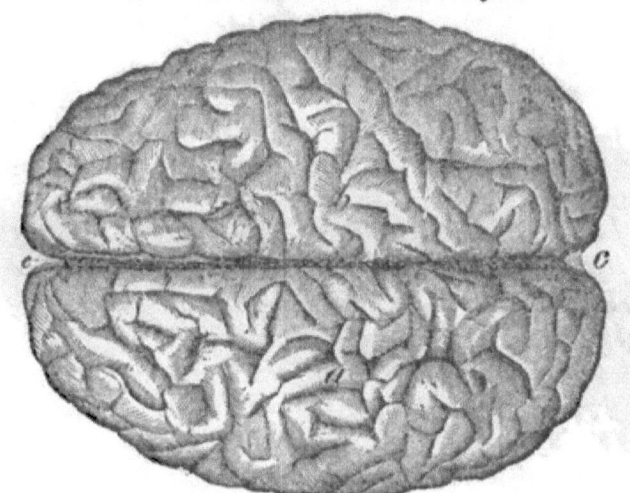

a, Right lobe.   *b*, Left lobe.   *c, c,* Division of the lobes.

various parts called *convolutions*, giving the organ the appearance of a collection of small lobes, with depressions between them, (Fig. LXII.) The substance of the brain is soft and somewhat pulpy; it is of very delicate texture, and can be easily separated by the fingers. The outer portion is of a gray color, and is called the *cortical* part. The inner portion is of a white color, and called the *medullary* part.

It is supposed, by some, that these two parts of the brain have different offices; that one is the organ of the mental operations and the affections; that, through it, the mind acts and the feelings operate, while the other is supposed to be the organ of sensibility, and, upon that, impressions from the sensory nerves are received, and in it sensations are created and perception takes place, and that it holds the communication with all the rest of the body. But how far this supposition is true is not shown, nor is it necessary for us to know.

748. *The brain is covered and held together by three membranes.* The inner and the middle of these membranes are very delicate, and give the brain a soft cushion to lie between it and its bony enclosure. The outer membrane is thick and very strong, and would hold the brain in its position and retain its shape even when removed from the skull. These membranes surround the brain on all its sides, above and below. The inner and soft membranes dip into the brain between the convolutions or little lobes, and separate them superficially.

The brain is divided into two parts, called the right and left lobes, (Fig. LXII. *a, b,*) which are exactly alike on the two sides. These are separated by a partition wall, or a wing (Fig. LXII. *c, c*) of the same membranes that cover the organ. This partition runs from the front to the back of the skull, and almost to the bottom of the brain. It supports the two lobes in their position, and prevents them from pressing upon each other when we lie down.

749. The brain is also divided into two other parts — the greater or *cerebrum*, and the lesser or *cerebellum*. The greater occupies almost the whole of the cavity of the head above and in front. The lesser is behind and below, just above the neck. A wing of the membranes extends across the skull from side to side behind, and separates these two parts of the

brain. This wing is attached to the bone, and gives support to the brain, and protects it from the injury that might come from jars; and it also prevents the upper and larger organ from pressing upon the smaller organ below.

750. *The spinal cord extends from the brain through the whole length of the back-bone*, (Fig. LXIII. *b*.) In the bottom of the skull there is a large hole, which is placed directly over, and opens into, the channel in the spine. This channel is formed by the rings of the successive vertebræ. It is closed on all its sides, and gives a sufficient and secure place for this great nerve or extension of the brain. This spinal cord is composed of pulpy, nervous matter, like that of the brain, and is protected by the same delicate and strong membranes that cover the organ above.

751. *The brain sends nerves to the whole body.* There are holes in the base of the skull through which twelve pairs pass outward. These are alike on the two sides. The optic nerves pass forward to the eyes. The auditory nerves pass sidewise to the ears. The others pass through other holes to the face, and to some other parts of the body. Twelve pairs go directly from the brain, (Fig. LXIII. *a*,) and thirty pairs go from the spinal cord, (Fig. LXIII. *c, c, c*.) These nerves divide and multiply until their branches reach every part of the body, and every organ, muscle, and blood-vessel is connected with the brain by its appropriate nerve. The nerve of the face passes out from the skull below the ear, (Fig. LXIV. *a*,) and sends its branches and filaments over the

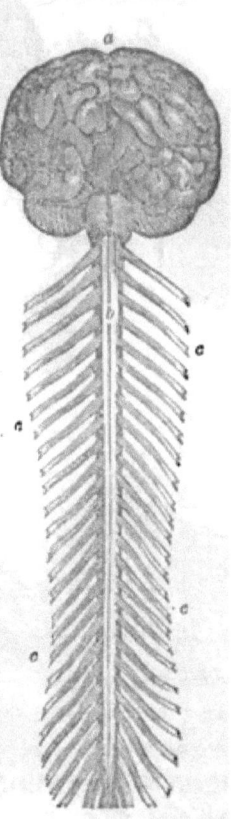

Fig. LXIII.
Brain and Cord.

a, Brain.
b, Spinal cord.
c, c, c, c, Roots of spinal nerves.

whole face. Other nerves are in like manner spread over the neck, (Fig. LXIV. *c*,) the arm, and every part of the body.

Fig. LXIV. *Nerves of the Face and Neck.*

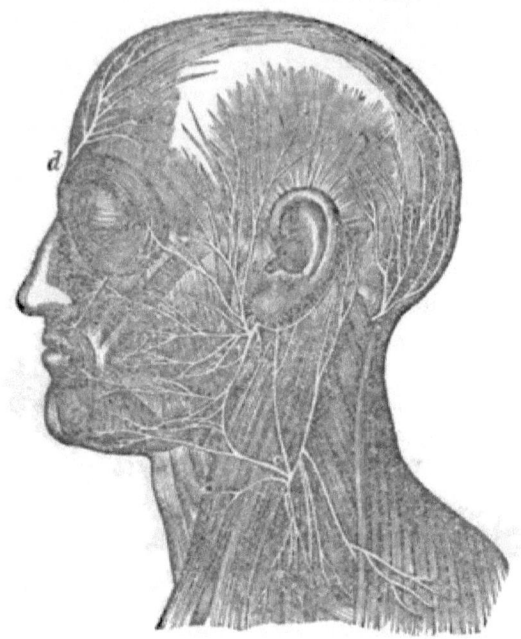

a, b, Nerve of the face.   *d,* Nerve of the forehead.

752. *Two kinds of nerves extend from the brain and spinal cord to the body —* those of sense and sensibility, called *sensory nerves;* and those of motion, called *motory nerves.* The sensory nerves receive the external impressions, and convey them to the brain. This class includes both the nerves of special sense, as the optic, auditory, gustatory, and olfactory nerves, which go to the eye, ear, tongue, and nose, and also the nerves of general sensibility, by which we feel pleasure and pain, heat and cold. The motory nerves convey from the brain to the muscular texture the stimulus of motion. The muscles are supplied with both classes of nerves, and, therefore, have both the feeling and motory power.

753. *These two kinds of nerves — the motory and sensory —*

perform separate offices; each effects its own purpose, and no more; and each one neither interferes with, nor can take the place of, the other. If, from any cause, the nerve of motion alone is disordered, separated, or pressed, the muscles cannot move, but the power of feeling remains. But if the other nerve—that of sensibility—is injured, there is no feeling, but a power of motion. If the motory nerve that connects the brain with the muscles of the jaw and lips is divided in any animal, he may still smell his food with his nostrils, and feel it with his lips, but he cannot open his mouth to take it in and masticate it. But if the sensory nerve is divided, the animal cannot feel the food with his lips, although he can move them to take it in.

754. These nerves have separate roots in the spinal cord; the sensory nerve arises in the posterior, (Fig. LXV. *d,d,d,d,*) and the motory nerve in the anterior part. (*e,e,e,e.*) These nerves are united (*c, c, c, c,*) soon after they leave the spine, and are for some distance included in the same sheath, yet their branches are not equally distributed to all the parts of the system. The skin has no power of motion, but it has great sensibility. It is, therefore, very largely supplied with sensory nerves, but not with motory nerves. On the other hand, the muscles, being exclusively organs of motion, are very largely endowed with motory nerves, but with a limited supply of the sensory fibres. In consequence of this, when the surgeon amputates a limb, great pain is suffered while the knife is cutting through the skin, and comparatively little when the muscles are divided.

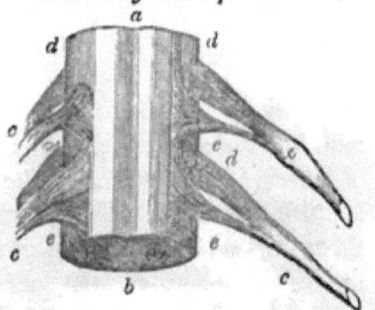

Fig. LXV.
Section of the Spinal Cord.

a, b. Section of the cord.
c, c, c, c. Spinal nerves.
d, d, d, d. Posterior or sensory roots of the spinal nerves.
e, e, e, e. Anterior or motory roots of the spinal nerves.

CHAPTER II.

Sensation is in Brain. — Produced by Impressions on outer Ends of Nerves, and carried on Nerve to Brain. — No Sensibility nor Power of Motion in Part which does not communicate with Brain. — Cutting Nerve, or Pressure on Nerve, paralyzes Parts to which it is distributed. — Foot asleep. — Injury of Spinal Cord paralyzes Parts below.

755. SENSORY nerves carry impressions from their outer extremities in the organs of sense, and in the flesh, to their ends in the brain, where the sensation is excited. The trunk of the healthy nerve has no feeling, and receives no impressions; it is merely a messenger to carry the impressions from the points where they are made, to the brain, where they are recognized. Sensation is not in the outer end of the nerve, nor in its trunk, but in the brain, at the inner end of the nerve. There are, then, three things in this work of sensation: 1st, the extremity of the nerve, which first receives the impression; 2d, the brain, which perceives the impression; and, 3d, the connecting line of nerve between them; and if either of these be wanting, or injured, there can be no healthy sensation. The power of motion requires the same three things — the brain, through which the mind determines or wills the motion; the nerve, to carry this volition or direction to the muscle; and the nervous termination, which imparts the stimulus to the moving texture.

756. *The power of motion, and the sensibility of any part, require this constant and uninterrupted nervous communication with the brain;* and if, from any cause, this connection be suspended, — if any nerve be cut, or divided, or pressed, — the power both of motion and of feeling is destroyed, or impaired, in the part where the nerve terminates. The familiar circumstance of the *foot being asleep* is caused by the pressure upon the nerves that lead down the leg to this extremity. The communication between the terminations of the nerves below and the brain above is thus interrupted, and then the

foot can neither feel nor move. In this state we try in vain to walk, for the muscles cannot act; and if we strike the foot, we feel no pain; but when the pressure is removed, and the communication restored, sensibility returns to the foot, and the power of contraction to the muscles.

757. Most of the nerves of sensation and of motion do not pass directly from the brain to the trunk and the extremities, but from the spinal cord. (Fig LXV. *e, e, e, e*.) The upper part of the cord sends nerves to the arms, and to the chest and its organs, the heart and the lungs. The middle part supplies the abdomen, and the lower part supplies the lower limbs. These several organs and parts of the body hold their communication with, and receive their nervous life from, the brain, through this nerve, or rather bundle of nerves, in the backbone, and the branches which pass from it.

758. *There are thirty pairs of nerves, or branches, which go from this cord to the body and the limbs.* These parts must not only have free nervous connection with the spinal cord, but, through the cord, they must have uninterrupted communication with the brain. If this communication be interrupted or broken off in the cord, all the parts of the body which are supplied with nerves from it, below the point of obstruction, will be deprived of their power of sensation and motion. This palsy of the muscles or parts of the body, and interruption of the regular operations of the organs, happen occasionally from such accidents to the spine as produce pressure upon its great nerve, and sometimes from distortion, or curvature, which, in a lesser degree, produce the same effect upon the cord.

759. Mr. J. fell, in the year 1830, and struck the hollow of his back on some stones, and injured the spine about the middle of the back. The cord was injured or pressed at that point, and free communication between the lower parts of the cord and the brain interrupted. All the parts of the body below the injury were palsied. But the power of motion and of sensation was restored as the cord recovered from the effects of the accident, and the pressure was removed, or the

injury healed. In another case, the paralysis was more extensive, having been produced by an injury at the lower part of the neck. There was, at first, a total loss of voluntary power over the lower extremities, trunk, and hands, slight remaining voluntary power in the wrists, rather more in the elbows, and still more in the shoulders. The muscles of the ribs were also paralyzed, and the breathing was carried on entirely by means of the diaphragm.* If the injury, in the last case, had been higher in the neck, above the origin of the nerve which leads to the diaphragm, this muscle also would have been paralyzed, and death would instantly have taken place, for want of power of respiration.

760. The higher any injury occurs to the spinal cord, the more extensive must be the bad consequences; that is, the nearer the root the interruption happens, the greater number of its branches must be affected. An injury, or a curvature, may cause pressure upon the whole or a part of this nerve, or upon a part of its branches only, and thus interrupt or interfere with the communication between the brain and the organs to which these branches lead. In this way the lungs and the stomach are sometimes disturbed or enfeebled, and difficulty of breathing or dyspepsia produced.

761. At first, Mr. J. (§ 759, p. 329) suffered great pain and palsy of the lower limbs. But the injury was not permanent; the pressure on the cord was gradually and slowly removed, the pain was relieved in all the parts, and the power of motion returned to the muscles successively; and, finally, he regained the use of all his muscles, except those which lift the feet. These were palsied and useless to him; and during the twenty-five remaining years of his life, though he could move his thighs and legs, and press his feet downward, he could not bend them upward on the ankle; and when he walked, the foot hung down, and the toes struck the ground first, instead of the heel. Probably a fibre or branch of the motory nerve, that leads to the muscles which bend the ankle, was injured beyond recovery.

---

* Carpenter's Physiology, § 178.

762. It is not to be presumed that the pressure upon the spinal marrow, from a curvature or distortion of the spine, will produce so sudden or perceptible injury as Mr. J. suffered; but his case illustrates the connection between the condition of the spinal marrow and the health and power of the organs and systems which derive their nerves from it. These cases (§ 759, p. 329) show also how the organs may suffer from an injury to the spine, or interruption of the action of the great nerve which connects them with the brain. Whatever may be the cause of this injury, the consequence of impaired life and diminished power must follow, in those parts or organs which receive their nerves from the spinal cord below the point of pressure.

CHAPTER III.

If Nerve be injured or diseased in its Trunk, Pain is felt at its outer Terminations. — Injury of Optic Nerve excites Sensation of Light. — Arrangement of the Brain and Nerves like that of Bells and Servant in Hotel.

763. The impression being made on the outer extremity of the sensory nerve, and the sensation being in the brain at the other end, the nerve is a mere channel, or highway, through which the impression is carried inward. The healthy nerve receives no impressions and originates no sensations in any part of its course; it only carries those which it receives at its end; and the brain recognizes and understands no other power or function in the nerve than that of receiving impressions at its outer extremity. It therefore refers all feelings and impressions, which come to it through the nerve, to those extremities.

764. If, by accident or disease, any impression is made upon the trunk of the nerve, — if we touch or irritate it, in any part of its course between the outer and inner ends, — this impression is conveyed to the brain, but that organ refers it. not to the point where the impression is actually received,

but to the end of the nerve, where impressions should be received; and there, at the extremity, the irritation or sensation seems to be. If there is injury of the nerve, the pain is not felt at the wounded place, but at the minute extremities. Thus, when we **strike the elbow against a table, and hurt the trunk of the nerve that leads to the fingers, we do not feel pain at the elbow, which was struck, but a tingling** pain at the fingers, where the terminations of this nerve are distributed.

765. Likewise, the pain of any local disease of the nerve is felt at its extremities. The excessively painful nervous affection of *tic douloureux* is felt on the surface of the cheek; but the cause is not, as is supposed by those unacquainted with the cause, in the skin or flesh, but in the trunk of the nerve leading from the face to the brain. This cause may be situated any where in the course of the nerve, between its outer ends in the flesh and the inner end in the brain. Wherever it may be, the effect is the same, and the painful sensations seem to be at the terminations of its branches which go off from the nerve below the diseased point.

766. The disease may be in a nerve very near the brain, and yet the pain is felt at its remote extremity. Miss W. complained of very severe and sharp pains in the arm and hand. It seemed to her, she said, as if thousands of needles were incessantly running through the flesh. For this, all sorts of applications had been made to the seat of the pain, and all without effect. Suspecting the pain had a remote origin, her physician examined the back-bone at the place where the nerve went from the spinal cord to the arm, and there discovered great tenderness — pressure on this spot increased the pain in the arm and hand. Blood was then taken from the back at this point, and the remote distress in the arm and hand was immediately relieved. She afterward had similar pains, at different times, in the side, the stomach, the lower extremities, and the feet, and these were relieved by cupping, or by the application of leeches, or a blister, over that part of the back-bone where the nerve of these several suffering parts originated.

767. The nerves of special sense are subject to the same law as the sensory nerves; they receive natural impressions at their extremities, but not in their course. They can convey to the brain, not common feeling of pleasure or pain, but such impressions as are made on their outer terminations, which are expanded in the several organs of sense. The optic nerve conveys the impression of light, the auditory conveys sound, the gustatory conveys taste, and the olfactory nerve conveys the impression of odors. When these nerves are irritated, or touched, or diseased, they still excite similar sensations. If we close the eye and press the ball upon the optic nerve, an impression is made upon the brain similar to that caused by light. If we strike suddenly upon the eye, or even the temple, so as to jar this nerve, the brain sees flashes of light. Dr. Howe has often tried these experiments with the blind, — both with those who were born in this condition, and had never seen light, and with those who became blind after birth, — and they all saw flashes and stars.

768. When we receive a blow on the side of the head, so as to jar the auditory nerve, or when the ear is diseased, we hear a ringing in the ear. In some states of disease, men complain of bad taste on the tongue. In vain they wash and purify the mouth — still the offensive taste remains; for it is not an impression made upon the tongue and carried thence to the brain, but the impression of some disturbance or derangement of the trunk of the nerve excites a disagreeable sensation in the brain, which refers it back, not to the spot which is disturbed, but to the termination in the tongue.

769. Whatever excites any nerve in its course will produce an effect upon the brain similar to that which is produced by impressions made upon its terminations, and the sensations will be referred back to the ends of the nerve as their seats. An electric shock, if passed through the nerve of the ear, will give the sensation of sound; and if through the nerve of the eye, the sensation of light. If we apply a piece of zinc and copper to the upper and lower surface of the tongue, and let their edges touch each other, they excite a sense of unpleasant taste.

770. This arrangement of the brain and nerves and their terminations, or points of impression, with their relation to each other, is similar to that of the bells in a hotel and the servant who watches them. The wires extend from the several rooms to the corresponding bells in the central room, or the servants' hall. Whenever the occupant of any room, as No. 66, wants any thing, he pulls his wire, and the bell No. 66 rings. The servant, seeing this, immediately recognizes a want in No. 66, and refers this to no other room. His only conception is that of the connection of bell No. 66 with room No. 66. Now, if any one should hit the wire between these two points, and ring the bell, the servant would have the same conception of a want in No. 66. Possibly this room might be cut off, and the wire and bell remain; if then the wire is drawn and the bell rings, the conception is still the same of a want in No. 66. So the brain, when it receives any impression at the inner end of the nerve of the finger or the eye, has no other sensation than of something pleasant or painful in the finger, or of light in the eye; and even though the finger or the eye be lost, if the nerve remains and is irritated, the brain still has the same sensation.

CHAPTER IV.

Pains in amputated Limb. — Motion excited by touching Motory Nerves. — Rapidity of Nervous Action. — Voluntary and Involuntary Organs. — Involuntary Motions fatigue less than Voluntary.

771. THE last section will explain some singular facts in regard to amputated limbs. Even after the nerves are divided, or cut off, the remaining parts, if irritated, may excite in the brain the same sensations as if they were entire. Sometimes men, after a leg has been amputated, complain of suffering great pain in the feet and the toes of the separated and buried limb; and some, believing there was a mysterious connection between the body and the lost limb, have caused it to be taken up and examined, to see if there

were any cause of suffering pressing upon it, while others have laughed at the fallacious imagination. A gentleman, whose arm was recently amputated by a surgeon on account of a cancer in the hand, said that when the end of the nerve was touched in the stump, he felt his old pain in the hand. In these and similar cases, there was no mistake in the sensation; the remaining trunk of the nerve, being pressed or irritated, carried this impression to the brain, which referred the pain to the separated limb, and to no other. This has sometimes happened many years after amputation. The reasoning faculty corrects the error, yet the sensation remains.

772. The same law holds in regard to the nerves of motion. The muscles yield obedience to the mandates which they receive from the brain. When the mind wills to move the finger, the brain sends the volition and the stimulus of contraction along the motory nerves to the muscles on the fore-arm, and they contract and bend the finger. But the muscles yield to every stimulus they receive through the motory nerve. If, therefore, we prick or irritate one of the nerves of motion, the muscles to which it leads will contract; if we apply a galvanic shock to it, it will produce the same effect; and even the limbs of the dead body, for a short time after death, can be made to move by the powerful application of galvanism to the nerve of motion.

773. *False sensations are produced directly in the brain*, independent of the nerve, by some disease which disturbs, or excites, or impresses the brain, at the points where the nerves terminate. When this impression is received in the brain, from the outer end of the nerve, a true sensation is excited; but when it is made directly upon the brain, by disease or disturbance, without the intervention of the nerve, a false sensation is the consequence.

The communicating-bells in the hotel (§ 770, p. 336) may be rung by any jar, or any thing moving them, independent of the wires that should pull them; then the servant would have the usual idea of some one ringing at the farther end of the wire, and something wanting in the chamber where the wire ends. Whether the bell be rung by some one in the chamber, or by some one touching the wire in its course, or by any jar acting on the bell itself, the same thought is produced in the mind of the servant whose business it is to watch them. In like manner, whether the impression be first received at the outer termination of the nerve in the organs of sense, or in the flesh, or on its trunk, or at its inner termination in the brain, the same sensation is excited of sound in the ear, sight in the eye, or feeling in the flesh.

774. These false sensations are created in some diseases. The insane sometimes have false hearing and vision. They seem to hear the sounds of voices, and even articulated language. When the imagination is highly stimulated, one sees visions. The timid see frightful apparitions, which seem to assume as distinct a form and color as the real objects of day. Sometimes false vision and false hearing exist together in the same person. He not only sees the form of the object, but he hears the sound of the voice. Men, when suffering from *delirium tremens*, are troubled with these false sensations. They are often frightened by the voices of spirits or the sight of enemies. The error in these cases is not in the eye, nor in the ear, but in the brain itself. The disease which disturbs this organ produces the same sensations in it that would be produced if the impressions were first made on the eye or ear, and then conveyed by the nerve to the brain.

775. Communication is very rapid through the nerves — so rapid as often to seem to be instantaneous. If we tread upon a thorn, or a heated iron, the sensory nerve conveys the impression to the brain; this organ recognizes the impression, and then directs that the muscles of the limb lift the foot out of danger, and sends this mandate through the nerve of motion. These two processes — the passage of the painful impression from the foot to the brain, and of the volition from the brain to the muscle of the foot — seem to be simultaneous, yet they are successive.

776. Some of the organs of the animal body — the hands, the legs, neck, &c. — are subject to the exclusive control of

the will; they only move when directed by the brain. These are *voluntary organs*. There are others, such as the heart, stomach, &c., which are not under the control of the will. They do not depend on our attention or volition to set them in motion, and no wish of ours can stay their actions. These are *involuntary organs*. The organs of respiration are both voluntary and involuntary. They act without our cognizance, and yet we can accelerate their motions, or entirely suspend them, by efforts of the will. The involuntary are supplied with nerves, both of sensation and of motion, as well as the voluntary organs; but these nerves are not subject to our command.

777. There are some motions which, though they are usually under the exclusive control of the will, yet at times are involuntary. We snatch the hand from burning by an effort of the will; yet if the fire be applied to it when we are asleep, and the action of the will is suspended, we snatch it away with a movement as involuntary as that by which the heart beats or the chest expands. The motions of the hand, the mouth, and the lower limbs, are ordinarily voluntary, and require a distinct volition for their execution; yet, in certain states of nervous disease, they become involuntary. They are then beyond the control of the will, and sometimes take place even when the will is opposed to them. In cases of epilepsy, in the St. Vitus's dance, and in convulsions, the muscles are contracted, sometimes with great force, without volition, and the limbs are thrown about, although the sufferer struggles to resist it.

778. There are other motions which are only effected by the direction of the will, which yet, by practice and discipline, become apparently, if not really, involuntary. Walking and the playing upon a violin, an organ, or piano, require the constant attention of the mind of the beginner to excite and direct his movements. But after practice he can walk upon familiar paths, and play familiar tunes, without the exertion of the will, and even while his attention is partially given to other matters.

779. When motions become so familiar as to be executed without the attention of the mind, they fatigue the body less

than when they require the aid of mental action. Thus, when we walk, if we lift the foot and place it down at each step by a special effort of the mind, we shall be more wearied than when we walk, as we usually do, without a special volition for each step, and have the mind free to attend to other matters. It is easy to see this difference of effect of exertion in the beginner and the practised performer of any work or art. The self-possessed dancer, who moves with careless ease through his figures, is less exhausted than the timid, cautious dancer, who anxiously watches every step. Both of these may move with the same energy, yet the one who added mental action and care to his muscular exercise drew more upon the nervous energies than the other, who exerted the muscles alone.

780. *Every function of every organ is dependent upon the nerves for its power.* The tongue has a nerve of taste, one of common sensibility, and another of motion. The eye has one nerve of sight, another of motion. The secretions of the saliva, of the tears, and of the gastric juice, &c., are dependent upon their peculiar nerves to stimulate the processes. Each one of these operations must have its own nerve, and that must be in good health, and connected with the brain directly or through the spinal cord. If this nervous connection is disturbed or suspended, the function is impaired or fails.

CHAPTER V.

Brain presides over all Organs and Functions. — When it is impaired, all Organs and Functions impaired. — Not sensitive. — Subject to Laws of Body. — When pressed, general Sensibility and Power suspended. — Subject to Growth and Decay. — Has large Supply of Blood. — Change of its Particles increased by Exercise. — Fatigued with Action. — Sleep.

781. The brain is the presiding genius over all the powers and actions of life. It stands above all and over all, giving energy to, and directing the motions and operations

of, all the organs of the animal body. It is of the utmost importance in our structure, and, on this account, very great pains have been taken to provide for its well-being, and to defend it from injury. The thick bones of the skull are arranged so as to give the greatest strength to the arch, and make it capable of bearing very heavy weights without suffering.

780. The human brain fills all the cavity of the skull, and corresponds with its shape. It is the seat of sensation, of thought, and volition, and the organ of the mind. Through this we recognize the impressions of external things. These impressions, which are made upon the organs of sense, and carried inward through the nerve, are not perceived if the brain is wanting, or if its power of action is suspended. When it is unnaturally oppressed with blood, as in apoplexy, or with water, as in dropsy, the light that shines upon the eye is not recognized by the brain, and consequently the mind receives no idea of external objects which the light reflects.

783. *Although the brain is the organ of sensation, it is not of itself sensitive.* It will bear pricking, or even cutting, with less pain than the fingers. Many experiments tried upon lower animals show that these creatures do not manifest signs of pain when the brain is cut, and even a part taken out. The same has been observed in man in cases of accidents. A child fell from a tree and fractured the skull; a part of the brain protruded, and the surgeon cut it off without occasioning apparent pain. This has frequently been done with the same result, and when the mind was perfectly clear, and capable of attending to the impressions communicated to it through the brain.

784. Whatever provisions are made by the Creator for the action and support of the brain, are as necessary for its well-being, and for the health of the mind, as the provisions made for the heart and lungs are for their well-being, and for circulation and respiration. It must have room for action. The cavity of the skull — its resting-place — is just large enough for it. It needs so much space, and no more, and does not safely bear any diminution. If this room be diminished by

any thing which crowds the brain, the organ suffers. A sudden effusion of water, or pressure of bone from fracture of the skull, or blood from apoplexy, immediately suspends its action, and then its functions, and those of the voluntary organs and the mind, are suspended; torpor and heaviness overwhelm the whole system. In other cases, where a tumor grows, or water is effused very slowly, the skull may expand, and still leave room for the brain; but in most of these instances, however slow their progress, the brain, sooner or later, suffers from the pressure, and then the physical and mental powers are impaired or suspended.

785. Sir Astley Cooper relates a remarkable instance of a man whose skull was broken on a British man-of-war, in June, 1799, in the Mediterranean Sea. He was found in a state of insensibility, and incapable of voluntary motion, and he remained in this condition till May, 1800, when he was carried to St. Thomas's Hospital, in London. There the surgeon found a piece of bone forced in, and pressing upon the brain. When he removed this bone, the man recovered his sensibility, and was soon restored to health and activity, after having lived nearly a year in a state of unconsciousness.*

786. *The brain is subject to the law of growth and decay of its atoms.* It requires nourishment of new particles and removal of the old, as well as all the other organs. It is therefore provided with an apparatus for nutrition and absorption, and arteries that bring the new blood, and veins that carry off the old blood and the wasted particles of matter. It seems to require more blood for its nourishment, and for the supply of its waste, than other organs of the same size, for it receives a much greater proportion than any other part of the body. The human brain receives from one fifteenth to one tenth of all the blood that flows in the body, and yet it weighs only about one fortieth of the whole frame. Its arteries, being a part of the general circulatory system, beat in the same manner and with the same frequency as the arteries of the wrist.

* Lectures on Surgery, Vol. I. p. 233.

787. *Liebig says that action of the brain implies change of particles and waste, and, therefore, a greater supply of blood.* To meet this increased want, the arteries beat with greater force, and send more blood, whenever the mind and the brain are excited or active. Sir Astley Cooper saw this in a young man who had lost a portion of the skull. "His brain could be distinctly seen beating, through the opening of the skull;" and, whenever he was irritated by any opposition, the pulsation was much more violent, and it became more quiet when he was calm, and his mind was easy.* There was a girl in Montpellier, France, who had lost a large portion of the scalp and skull. Her brain could be seen for a considerable extent of surface. "When she was in a dreamless sleep, her brain was motionless, and low within the cranium; but when her sleep was imperfect, and she was agitated with dreams, her brain moved" and beat, more blood was sent to it, the arteries expanded, and the brain protruded through the hole in the bone. This protrusion was greater in active than in calm dreams; and when she was awake, the same difference was observed, consequent upon the activity and the quiescence of her mind. If she was in vigorous thought, the brain swelled, and protrusion was very observable.†

788. The eye becomes weary with long exposure to light, and seeks rest and relief in shade; and, if not thus relieved, it finds it difficult to discriminate objects. The muscles, also, are fatigued with long and continuous exertion, and are incapable of contraction. Rest restores this power. So the brain is fatigued with long and uninterrupted attention to subjects of deep thought, and incapable of fixing its attention upon matters of a grave nature; then it wants, and must have, opportunity of rest to recover its energies; and, if this is not granted, the brain and the mind will be weakened or disordered.

789. *Sleep is the natural rest of the brain.* It gives rest to the mind and to the voluntary organs of the body. The

---

* Surgical Lectures.
† Combe, 255. Annals of Phrenology, No. I. p. 39.

brain, and the nerves of sensation, and those that convey volition, sleep; but the motory nerves of the heart and arteries, and the lungs, never sleep. When a man dreams, his brain is not in complete rest; and, in the cases of sleep-walking, the sleep of the brain and of the nervous system is still less perfect, and of course the system is less refreshed by it. That sleep is most refreshing, and the best recruits the frame, in which there is neither dream nor motion, but absolute inaction of all the voluntary powers.

CHAPTER VI.

Night is the proper Season for Rest of the Brain. — Brain gains Power by Exercise. — Weakened by Over-Action. — Connected with other Organs; with Lungs, Stomach, Muscles. — Effect of Alcohol on Brain and Muscles; on Mechanical Skill; on Use of Tools.

790. *There is a natural connection of the action and inaction of the brain with the alternation of day and night.* The brain, as well as the muscles, has more vigor and a greater power of action in the light of day, and is more prone to rest and to recruit itself in the darkness of night. The day is, then, the appropriate time for mental, as well as bodily labor, and the night the proper season for sleep of the mind and the brain.

791. The stillness of the night, when the busy world is quieted, and we are secure from interruption, seems to invite the student to his books. Then the mind is not disturbed with other claims upon its attention, and there is then better opportunity for concentration of thought upon any subject. Therefore some attend to external matters during the day, or even sleep during some portion of it, and reserve their mental labor — the toil of the brain — for the night. It is an unprofitable habit of some clergymen to write their sermons in the evening or night before they must be delivered, and then, by concentrating the whole energy of

the brain upon their proposed labor, accomplish their work. This is frequently done, and with temporary success; but it is at the cost of the permanent power of the brain, as well as of the general health of the body. Naturally, the brain has the greatest power of labor, and the mind the greatest energy of thought, in the early part of the day, and less in the latter part, and least in the night; and consequently all mental labor exhausts the brain more in the evening and night than in the morning and the bright hours of day.

792. *The brain gains strength by moderate and appropriate exercise*, when this is interchanged with rest; and, if frequently called upon to exert itself, it acts with greater energy, just as the muscles become strong with use, or the skin hardy with exposure. It is easy to see the difference of mental power in the laborer, — who uses his muscles, and not his brain, and works only under the direction of another, — and in the employer, whose brain is ever active with his plans of business. *But excessive mental toil exhausts the power of the brain.* Long-protracted labor of the brain, with insufficient or no intervals of rest, waste and weaken it; and any over-exertion, for even short periods, is injurious. The brain can bear excess of action no better than the muscles. The blacksmith or the stone-cutter can as safely do two days' work every Saturday afternoon and night, as the clergyman can write his sermons in the same time.

793. Although the brain is placed above and over all the other organs of the body, to give them life, and energy, and direction, yet it is dependent upon them for its own health and power. It has intimate connections and sympathies with each one of the others; it is strong with their strength, and weak with their weakness; it enjoys their pleasures, and suffers with their pains.

794. *The brain is constantly connected with, and immediately dependent on, the heart.* It must have a large and unfailing supply of blood at any time. If the heart is diseased, and cannot admit the return of the venous blood, the brain is crowded with it; then pain and confusion, and even

insensibility, may follow. It needs and receives more blood during its action and excitement, and less in its quiescence and calmness. The heart must therefore be in good health, and able to supply the greater want of the active and excited brain. But, when the heart is diseased, it cannot send this increase of blood; and, if it attempts it, it struggles in vain, and great distress, and even death, may follow. Persons who suffer from disease of the heart, cannot safely bear any mental excitement. For this reason, a celebrated surgeon, who had this disease, continually guarded himself against any irritation of temper or agitation of mind, yet was suddenly excited on an occasion, and immediately died.

795. *The connection of the brain with the lungs is not less apparent than with the heart.* The brain needs not merely a large quantity of blood, but that of the purest and the best quality. Whenever the blood is imperfectly purified of its carbon, either from defect of the respiratory apparatus or from want of pure air, the brain feels it immediately. If the waste and dead particles are not carried off from the blood in the lungs, the impure blood is sent again through the heart and arteries to supply the body; the brain suffers more than the other organs, and becomes inactive, and often painful, and the mind dull. The audience of a close and crowded lecture-room, and children in an unventilated school-room, lose their mental energy and their power of application. (§§ 383, 384, p. 165.)

796. *The sympathies between the brain and stomach are familiar to us.* Most men have been compelled to know how these two organs suffer together in sick headache. The frequent pains in the head are generally to be referred to the derangement of the stomach, and, when this organ is relieved, the brain is usually well. Dyspeptics complain of much headache, and, on the contrary, those who suffer from disease of the brain are often troubled with digestive disturbance. A blow on the head will often occasion vomiting, and excess of action of the brain will sometimes suspend the action of the organs of digestion.

797. The brain controls the actions of all the muscles, and supplies them with the stimulus of motion through the nerves which connect them together. The muscles are thus completely dependent on the brain for their vitality and their energy. They can have no more life and energy than that controlling organ has to give them. If it is impaired, the power of muscular contraction is diminished. When the brain is pressed and incapable of action, the muscles are palsied and the limbs motionless. An injury of the head, which destroys sensibility, paralyzes the frame; and then the sufferer can neither stand nor move, for all voluntary action is suspended. It is the brain that directs every contraction of the muscles, and, by its complete discipline, harmonizes their actions so that the desired motions are produced in the limbs; and thus the movements of the musical performer, the skilful mechanic, and of the writer, are executed with beautiful precision. (§§ 636—640, p. 276.) But if any thing disturbs the brain, and suspends or impairs its balance of action, it loses this control over the muscular actions, and they become irregular.

798. *Alcohol, in any of its forms, excites the brain and nervous system, and disturbs their actions,* and impairs or suspends their control over the muscles. For this reason, the drunken man reels or staggers; he is unable to direct his feet and put them in the appointed places. Some of the muscles may contract too much, and others too little, and carry his feet too far, or not far enough, to one or the other side; or they may not contract even sufficiently to hold him up, and then he falls to the ground. When he attempts to work with his hands, he finds the same difficulty, the same want of control over the muscles that move the arms, hands, and fingers. He cannot direct them and the tools which he holds with the desired precision, and therefore often fails of striking the proper point. He may strike where he least intended, and do injury to his work or to his own person. The muscles of his tongue, also, suffer in the same way, and he articulates indistinctly.

799. After the brain and nervous system have been frequently excited, and their control of the muscular actions interrupted with stimulating spirits, they do not recover the complete command of the muscles when the fits of intoxication pass away. Therefore old drunkards, even when sober, walk with a faltering step and work with an unsteady hand They lose their power of skilful workmanship. If they are nice mechanics, they impair their skill by their intemperance, and are then compelled to apply their hands to coarser work; and some are obliged to give up their handicraft altogether, and betake themselves to the rudest of common labor.

800. There was a very skilful worker in iron. He was remarkable for the dexterous use of his hands, and the beauty and fitness of his manufactured articles. But he became intemperate, and, after some years, lost the exact command of his hands, and the power of exact adaptation of his tools to the material on which he worked. He gave up his nice work, and manufactured coarser articles. In process of time, his muscles became less and less under his control, and he gave up his shop and trade altogether; and, for the rest of his life, he sawed wood, dug in the ground, carried the hod for masons, doing nothing but the roughest work, which required the least discipline of the brain and command of the muscles.

801 In this undisciplined condition of the brain, and absence of command of the muscles, the intemperate man loses the power of self-protection, and consequently meets with more accidents than other men. When he walks, he makes missteps, he loses his balance, and stumbles over small obstacles. If he drives a nail, he is not sure to direct the hammer so as to strike the head; he may often hit his fingers. If he uses sharp tools, he may strike in a wrong direction, or his instrument may slip and cut his own flesh. A mechanic, when he wounded his knee with an axe, complained of his very frequent ill luck. He said he was always meeting with accidents. But a short time before, he bruised his finger with a hammer while driving a nail; then he cut his foot; now he had cut his knee. He was intem-

perate; and, though not always intoxicated, — probably not so even at the time of his injuries, — yet he had lost the perfect control of his muscular actions, and could not direct his blows safely

---

CHAPTER VII.

Brain Seat of Mind, Affections, Passions. — Power and Action of Mind limited by Power of Brain. — Stimulation of Brain stimulates Mind, and, on the contrary, Mind subject to Liabilities of Brain. — Impaired by Indigestion; by Excess of Eating; by Hunger. — Moral Feelings affected by Stomach. — Effect of Cheerfulness.

802. *The brain is the only avenue which the mind has to the outward world.* It is the organ through which the intellect acts in regard to other minds and to external things. It is the seat of the passions, of the affections, and of the moral feelings. The immortal mind — the spirit itself — is, indeed, something more than, and different from, the physical brain; yet the Creator has so connected these together in this life, that we know of the operations of one only through the medium of the other. As the eye is the organ of sight, and the ear the organ of hearing, so the brain is the organ of perception, of thought, and affection. The eye is not sight, though there is no sight without it. The ear is not hearing, though there is none independent of it. So the brain is not mind, though there is no mental operation without it.

803. The brain being the only instrument with which the mind plays here, they are indissolubly connected together in this life. Their powers of action have equal limits. Whatever we may say about the illimitable power of the expansive mind, it can move no farther nor faster than the brain can go. Whenever the brain is weary, the mind is weary. Whenever the brain wants rest and sleep, the mind needs the same. The brain can make only a definite amount of exertion, and work only a definite number of hours, and then it must suspend all labor, and lie down to rest in complete inaction. The mind can do no more. Precisely at

the point where the brain is fatigued, the intellect is fatigued, and when the brain sleeps, the mind falls into a state of unconsciousness.

804. *Whatever excites the brain excites the mind.* The wine that first stimulates and then oppresses the nervous system, also at first quickens mental action, and then overwhelms the mind with stupor. Whatever stimulating substance sends more blood to the brain, excites the mind to quicker action, and gives it a greater grasp of thought. On the contrary, whatever excites the mind to unusual action, occasions unusual flow of blood to the brain. The arteries of the brain were seen to beat (§ 787, p. 343) when the passions were irritated or the mind excited, and even when the sleeper was troubled with dreams. The blood flows in unnatural abundance through the brain of a man in passion, and of the scholar while he is studying with all his mental activity. The reverse happens when any emotion oppresses the mental energies. When one is appalled with fear, or depressed with sorrow, his countenance is pale with the absence of blood, and the brain is supplied in the same imperfect degree.

805. *The mind is subject to all the liabilities of the brain.* It shares in all its sympathies with other organs of the body. When the brain is pressed with blood, as in apoplexy, or with a piece of broken skull, the mind is stupid. When the head aches from breathing foul air, the mind is dull; and it is torpid when one breathes charcoal gas. When alcohol gently increases the flow of blood in the head, the mind is more active, and the spirits more lively; and when this flow is a little more increased, and the brain stimulated too highly, the mind loses its balance, the thoughts run, and the tongue talks wildly; and a still further increase of blood stupefies the brain, and then the mind is torpid, and insensibility follows.

806. I lately saw a child whose brain seemed to be torpid, but whose whole body was writhing in convulsions. The brain was oppressed, and the child was senseless; she could

neither hear, nor see, nor understand; but the motory nerves were excited, and the muscles thrown into violent action. The child had eaten great quantities of unripe fruit, which the stomach could not digest. But as soon as the stomach was relieved of this unnatural load, the convulsions ceased, and the consciousness returned. But she had no recollection of what had passed. Here was decided proof that the brain and nerves were connected with the digestive organs.

807. *Indigestion and nausea, which create pain in the head, impede mental action.* When suffering from dyspepsia, the student cannot apply his thoughts to weighty subjects, and the accountant is unwilling to attend to his figures, and make his calculations. Recently, a merchant, whose dinner oppressed him, found his mind so confused that he left his counting-room and went home to recruit his powers. Having a taste for grave matters, he attempted to read history; but he could not confine his attention to the subject. He then tried to read an exceedingly interesting biography — the life of a friend; but even this required too much mental exertion; and at last he betook himself to one of the lightest of tales, — Valentine Vox, — which he said required all the energy of mind that he could then exert.

808. The same indisposition to mental exertion follows after eating an excessive meal. Great eaters, who keep their digestive organs constantly at work, have usually but little intellectual activity. They prefer sleeping to thinking, and study is a burden to them. On the contrary, hunger is none the less an enemy to mental labor, (§ 154, p. 73,) and in cases of extreme hunger, the mind cannot apply its powers to any matter of thought or business; and in starvation it loses its self-control entirely, and insanity sometimes follows.

809. The moral powers and affections are influenced by the state of the digestive organs. A sour stomach produces a sour temper, and men are usually thought to be cheerful and good-natured while at dinner. Dyspeptics are frequently irritable and suspicious. They are sometimes gloomy, and look upon all about them with fear and distrust. A friend,

who is naturally of a kind and generous temper, but subject to occasional fits of painful indigestion, said, a few weeks since, that the world, at different times, wore two entirely different aspects, varying according to the state of his stomach. "Now all is bright and promising; I have an abundance of friends, and every body is kind; I see nothing to mar the present, and feel no doubt of the future. But last week, when my stomach was in trouble, and my food oppressed me, every thing was as different as darkness from light. Then it seemed that I had no friends, and no one cared for me; the world was selfish, and gave me no sympathy nor encouragement; their actions and their speech were hostile to my character and peace. I put an unfavorable construction upon what was said to and concerning me. The very language that now seems to be that of kindness, seemed then to be injurious. The present was then full of doubt and fear, and the future promised nothing better. The cloud has now passed away, the sun shines brightly again in my heart and my prospects. It is all owing to the state of my stomach."

This is not an uncommon case. It may be frequently found, though perhaps in a less degree, and sometimes in a greater degree in the world. The connection between the indigestion and depression of spirit or suspicious temper is not so clear; but it is none the less certain.

810. *On the contrary, the states of the mind and feelings affect the stomach and the other physical organs and their functions.* Cheerfulness excites the respiration, and favors the purification of the blood. (§ 322, p. 143.) It aids the action of the heart (§ 235, p. 108) and the nutritive process throughout the body. But in sorrow and care, the respiration is languid, the purification of the blood imperfect, the heart moves heavily, and nutrition is sparing. Muscular power is increased and diminished by the same causes. The languid limb goes with the heavy heart, and the laborer works feebly whose spirit is weighed down with sorrow or discontent. (§ 713, p. 308.) Those who gain add more and more energy to their exertions, and those who labor unsuccessfully make

fainter and fainter efforts. These principles have been known from the times of old, when Solomon said, " A merry heart doeth good like medicine, but a broken spirit drieth the bones."

809. The work of digestion goes on best in company with the warm and gentle affections, where love is predominant, and tenderness animates the soul. (§ 166, p. 78.) There the food is best converted into blood, and the blood into flesh, and this has most permanent power of action. But bitter and harsh feelings impair this work in all its processes. Fat people are usually supposed to be cheerful and contented with themselves and the world. They are not easily disturbed by the ordinary affairs about them. But the lean are usually more anxious and careful; they worry and fear more; they are not so easily satisfied, and are more affected by the mischances of every-day life.

812. *Thus there is a remarkable and a beautiful harmony between the flesh and the spirit.* Cheerfulness and love add to the physical powers; and, on the contrary, robust health and bodily vigor aid in the buoyant flow of spirits. Melancholy people are, therefore, usually less healthy and strong, as well as less happy, than the cheerful; and, moreover, they are more unprofitable workers.

CHAPTER VIII.

Brain superintends physical and mental Operations; sustains these well when it is vigorous, and several at the same Time, if they are easy, but not if they become difficult. — Mind works best when Body is easy. — Uncomfortable Seats interrupt Study. — Bad Light and Temperature, and Fatigue, have the same Effect.

813. THE brain superintends, or is connected with, the operations of all the organs of the body; both those which are involuntary and beyond the control of the will, and those

which we direct by our volition. The movements of the limbs by the contraction of the muscles, the circulation of the blood, digestion, nutrition, respiration, the development of heat, the secretion of all the fluids, depend upon the brain. The mental and moral actions, the reception of knowledge through the organs of sense, thought, volition, the feelings, and the passions, are also connected with the brain. Every living action depends on the life of this organ, and every correct action depends upon its soundness. This organ and its powers are placed partially under our control; we think, we feel, we indulge in passion through it, and we direct its energies to muscular action. But it gives its energies to the involuntary actions without our will. We are responsible for the use of its power in all voluntary actions, and we may so use it as to interfere with its control of the involuntary actions.

814. The brain performs all these offices well when it is fresh and vigorous; but, whenever the nervous energies are exhausted or reduced, the brain works languidly, and all the functions of the other organs also languish. When we are fatigued with muscular effort, digestion is feeble. (§ 162, p. 77.) When the powers of the nervous system are diminished by excessive mental labor, the animal heat is sparingly generated, (§ 434, p. 186;) or in whatever way the sustaining power of the brain is reduced, it is less able to support any of the bodily or mental operations, until it shall renew its power by sufficient rest.

815. The brain sustains and superintends some of the voluntary, and all the involuntary, operations at the same time. We can walk, and breathe, and think, at the same moment, when neither of these requires any great effort or attention of the mind. But the brain cannot concentrate its power, so as to make any unusual exertion, upon more than one thing at a time. If, therefore, any one of these operations becomes difficult, or if we perform it with unusual energy, and consequently demand of the brain extraordinary power of direction or exertion, it can give but sparing energy and power to the

performance of the other operations, (§ 708, p. 307,) or perhaps it must suspend them altogether.

816. If any of the involuntary operations, which require the direction of the brain, but not the consciousness of the mind, becomes difficult, and requires mental effort, the brain can do little more than attend to it. Thus we maintain respiration and carry on the other operations in conjunction with it; but in paroxysms of asthma, or in croup, when the whole nervous power is concentrated in the effort to breathe, all labor of body, and all thought must be suspended, and the whole nervous energy devoted to respiration.

817. An accomplished musician can play several parts of a tune with his fingers upon the organ, and read and sing the words of the song, or call the figures of the cotillon; but if the tune is not familiar, — if it requires a special effort of the attention to read the notes, — he can neither sing the words of the song nor call the figures of the dance.

818. If the brain is occupied by any other efforts, or by any disagreeable sensations, it cannot give its full attention to any mental operation. When a man wishes to give the undivided energy of his mind to any subject, he places himself in such a position that his whole frame is most easy. If he sits, he selects a chair suited to the form of his frame, so that his body and limbs are supported without effort; and regulates the temperature of the room, so that he is neither hot nor cold; and the light, so that it is neither painfully glaring nor insufficient for the easiest perception. Some prefer a standing posture, and others will walk their rooms; but most prefer the sitting position while they are in intense thought; but, in either arrangement, nothing external calls the brain from the subjects of study.

819. It is the fault of many school-rooms, that they are so constructed that a very considerable portion of the attention of the pupils is taken up with their uncomfortable physical sensations, and with their endeavors to obtain relief. Instead of the comparatively easy chairs, somewhat adapted to their forms, which children are accustomed to enjoy at home, these

school-rooms are usually furnished with seats of one kind, and nearly of one size, and made without regard to the human shape. These seats are sometimes without backs, and sometimes with backs so square and perpendicular as to give no comfortable support. They are often so high that the smaller boys find no rest for the foot, or so low that the larger boys have not the usual support for the thigh. They are often built on an inclined plane, upon which the foot tends to slide forward and downward, and the child is then continually reminded of his position by his uncomfortable feelings, and is compelled to make constant exertion of his muscles to prevent his feet sliding forward.

820. Some school-houses are so situated as to be protected neither from the severest storms of winter nor from the burning sun of summer. They are often imperfectly warmed, and the temperature is unequal in the various parts. There are many whose chilly feet, in the cold season, make irresistible drafts upon their attention; and, in the warm season, in the absence of both shade-trees and blinds, or curtains, the heat and the glare of the sun make equal claims upon their feelings, and withdraw their attention from their books. It is all in vain that their teachers urge upon them to study vigorously, and forget their discomforts, and that the good scholar, who is anxious for his lessons, does not regard these external matters. The physical sensations will come first; they will have the first care of the brain and the mind; and it is only by great mental discipline — such as few children possess — that they can be resisted and forgotten. And whatever attention these suffering children give to the physical sensations is manifestly not given to their lessons.

821. *The same incompatibility exists between great fatigue and mental labor.* While the nervous energies are devoted to restoring power to the muscles, they cannot be given to thought or reflection. The boy who is wearied with hard work before the hours of school, has not then the free command of his brain for his mental action. Very laborious men are apt to fall asleep when they take up a book,

or when they attend to a lecture or to a sermon in church, especially in the summer, when their toil is the most exhausting. This is a natural and necessary consequence. The mind cannot have the use of the brain when it is occupied with the restoration of exhausted physical power; for there is only a definite quantity of nervous energy, and, if this is expended in muscular action, it is gone, and cannot be given to mental labor.

CHAPTER IX.

Moral Feelings interfere with mental Action. — Anxiety and Fear prevent Attention to Business and Study. — Fear, and misdirected Hope, improper Motives for mental Action. — Best Motive.

822. THE mind cannot give its full and undivided attention to observation or reflection while it is distracted or disturbed by any moral feeling. One who is anxious or in fear cannot easily study. Hence some become confused, and lose their self-possession, when they are in danger; they do not then concentrate their thoughts, and see clearly the actual circumstances of their case, or the means of relief. For the same reason, when one is riding with an ungovernable horse, he may not command his muscles in the best way for his safety. Failing to perceive his true condition and means of escape, so as to direct his movements to this purpose, he may do the very things that increase his peril.

823. Anxiety absorbs much of the energy of the brain, and prevents mental concentration; consequently, one cannot easily study, or give his mind to ordinary business, while he is anxious for the life of a parent or child, who is dangerously ill. For the same reason, when men devote themselves to hazardous speculations, politics, or gambling, they often neglect their usual engagements, and lose, not only the confidence of their employers, but even the power of successfully managing their customary affairs.

824. The action of the same principle is seen in the school-room. When the boy is anxious about his play, or when any great and desirable purpose is before him, and especially if it be a matter of doubt whether he shall be permitted to enjoy it, his lessons may suffer. If a pleasant excursion is proposed for the afternoon, provided the weather permit, and a cloudy forenoon render it uncertain whether a rain may not keep him at home, he cannot study well in this fear. The attention which the mind gives to the anxious doubt must all be at the cost of that effort which otherwise might have been devoted to his books.

825. Whatever may be the kind of moral feeling, if it is strong, and absorbs the attention and power of the brain, it interferes with that concentration of the mind that is necessary to the study of books, or learning any other matters. The homesick boy, away from home, cannot give the full energy of his mind to his books, nor even to any labor. An active boy was sent from a very pleasant home to learn a desirable trade in an unpleasant situation and unkind family. Instead of giving his whole thoughts to his new business, he brooded over the joys and comforts that he had left behind him; his heart was oppressed with sadness, and his yearnings for home occupied his mind. His employer thought him dull to learn, and lazy at work; and, after several months of ineffectual trial, by common consent of his father and his master, he was taken away. At home, he again manifested his former activity and desire to learn the same trade, and was then sent to another and more satisfactory place and family, where every thing was kind and encouraging. He there showed great interest in his work, learned the art rapidly, and became an unusually skilful and active workman.

826. There are some seeming contradictions to this principle; for many men have studied and become accomplished scholars when oppressed with pecuniary trials or bodily pains. Some of the best works in the language were written under the stimulus of poverty; and the late Robert Hall arose from his bed of acute distress to preach his most eloquent and

powerful sermons. Some are so absorbed by their business or anxiety that they give no heed to impressions that would excite physical, and even painful sensations; while thus engaged, they may feel neither hunger nor cold, and forget the hour of their meals, (§ 87, p. 46,) or their chilled flesh.

827. These persons, however, are exceptions to the general law. They had power of concentration sufficient to withhold their attention from the causes of physical and moral suffering. For the time, they forgot their painful sensations, or resisted the absorbing influence of their distress, and concentrated their nervous energies upon mental action. They have extraordinary power or discipline of mind, or are governed by an extraordinary motive to study or think amidst such counteracting influences. Nevertheless, these men bear a double burden — one in the disturbing cause, and the other in the intended labor of the brain; and, though they think and study much, they could do more if their minds were entirely free.

828. It requires more mental discipline to study amidst these counteracting or disturbing influences. It needs a greater power of the will over the feelings to abstract the attention from all that would excite agreeable or painful sensations or emotions, and there are but few who possess this power in full degree. Yet it is to some extent necessary; for, though one can study better when the body is perfectly easy, and the mind free from care, and the heart from pain, yet this condition is not always attainable.

829. Even the motive offered as an inducement for action may become a disturbing cause, and absorb so much of the energy of the brain as to prevent, in some degree, the very effort it was designed to encourage. In this respect, both the motive of fear and of misdirected hope are often injurious. When the iron rule prevails in school, the boy's constant fear that he shall be caught idle, or that he shall fail in his lessons, or some unexpected accident or unpremeditated misdemeanor may happen and subject him to punishment,

makes some demand upon his brain, and withdraws so much attention from the subjects which he is required to learn.

830. The motive of misdirected hope and of undue reward is often held out exclusively as a stimulus to greater and greater mental exertion. Whenever the reward does not grow naturally and necessarily out of the subject of study, it may interfere with its own purpose, and divide, rather than concentrate, the power of the brain and the attention of the mind. If, as an inducement to commit a lesson, or write a legible manuscript, the reward of a silver medal, or of a book, or an opportunity of declaiming before a public audience, is proposed, the effect of division of thought and weakening mental effort follows.

831. But the motives for mental exertion which belong to, or grow out of, the subject to be studied, not only withdraw none of the energies of the brain and the mind from the proposed object, and therefore neither divide nor weaken their exertions, but aid in concentrating all their force upon the single point of study. The value or the usefulness of the knowledge, or the advantage that must result directly from it, and, above all, the mere pleasure of learning, are, therefore, the most effective motives for, and auxiliaries to, mental labor. For this reason, boys who study with the idea that they are thereby to fit themselves for usefulness, respectability, and happiness, men who acquire professional knowledge, or learn science, as a means by which they shall obtain their support, or fortune, or station, or do good to others and, above all, naturalists and others, who study for the love of the sciences to which they give their attention, — are the most successful scholars.

CHAPTER X.

Various Powers of Mind. — Strengthened by Exercise. — Education adapted to Powers. — Education of Children. — Mind cannot be prematurely strengthened. — Action of Brain needs Attention to other Organs. — Ill Health of Students.

832. *It is plain that there are various faculties of the mind,* or the mind has power of application to various purposes; and the commonest observation will show that these faculties are not equally strong in all persons. As the muscular strength is unequally distributed to the several limbs, so the mental and moral power is unequally distributed to the several faculties of different persons. Thus one man is strong for one purpose and weak for another. He may have a genius for mathematics, but little power to comprehend languages. He may excel in music, painting, or mechanics; he may be a skilful machinist or financier, and make great proficiency in any one of these subjects, while, in all others, he may not be above the average of men. In regard to these, as well as all other subjects to which the human mind is applied, there is a great difference in the mental power of men. This difference is partly native, and partly the result of education.

833. As the physical powers grow and become strong by proper use and exercise, and as any one of these becomes stronger than the others if it is more used than they are, so the mental and moral powers may be strengthened by similar means. All proper education is progressive, and is adapted to the state of the brain and mind which are to be educated. It begins with the strength and knowledge already acquired, and uses these as the means of acquiring more. Perfect education brings forth and strengthens all the mental and moral faculties, and gives them equal power. If these are originally unequal, they will require unequal care and exercise for their development. This plan of education of the various powers is also adapted to the natural order of their appear-

ance. In this order, the appetites appear first, next the muscular power, then the senses, and lastly the moral and mental faculties; and a person successively eats, and moves, and observes, and reflects.

834. The child enjoys the use of his senses. He wants things visible and tangible. His perceptive faculties are developed before his reflective. He observes before he reasons. He learns better from things that he can see and touch, than from descriptions of things which are not present to his senses. He can better give his attention to insects, flowers, and other natural objects, than to any abstract principles of which he may not see the application or the use. When these simple matters are taught, and the child learns the uses and the relations of such things as he can see and feel, and when these studies are sufficiently varied and interchanged with muscular exercise and recreation, the brain is not fatigued, but, on the contrary, grows stronger, and able to undertake higher and more abstruse matters.

835. In early life, the brain and the mind are feeble, like the other organs and powers, and are subject to the same laws of exercise and rest. Children love action, but they have little power of endurance. They dislike to confine their attention long to one subject. They are fond of change, for a variety of subjects exercises different powers. They are soon weary of one kind of play or work, and want another; they like to change their studies frequently; they prefer small books and short stories. But the most agreeable change for them is that of the powers and systems which are put in action. They love to use the brain awhile, and then the muscles, and then these both together. They like to study, then play, then work.

836. The human brain, being subject to the same laws that govern the whole physical system, cannot be prematurely strengthened and applied to labor, with more safety than the arm or the stomach. Its growth, from the earliest infancy to the maturity of manhood, is naturally slow and gradual; and it would be as injurious to attempt to force the

development of the infant mind, or induce it to make extraordinary exertions, as to impose upon the child's hands a degree or weight of labor beyond its years.

837. The brain and nervous system of most precocious children are unusually active; but often they are stimulated by injudicious education. In these cases, the nervous energies, that should sustain the nutritive and the other systems, are absorbed in the mental labor. The mind may grow a while at the expense of the physical organs, and the child make great advancement in learning; but, in proportion as the brain is unnaturally excited, the other organs are deteriorated, and the health falters, day by day, and the child may sink in death from over-stimulation of the brain, in accordance with that universal law which would have put an end to life if the stomach or the muscles had been unduly excited and exerted to the same degree.

838. *In youth and manhood, the brain cannot work long and vigorously without the health of the rest of the body.* The organs are all linked together, and dependent one upon another. Each must have its due supply of nervous energy; if any one has more than this, — if it be stimulated to extraordinary exertion, — the others must suffer. Owing to inattention to the physical organs, in connection with mental action, many students, and men in the sedentary professions, suffer from ill health. Ministers, lawyers, and teachers, are frequently obliged, from this cause, to suspend or give up their callings, and devote their whole attention to the recovery of lost health.

839. *Health fails more frequently among students than among men in the more active employments.* More men leave college or quit their professions than leave any other callings on this account. This ill health among literary men is not the necessary result of their employments; it comes from the irregular distribution of the nervous power, and want of due coöperation among the various organs and systems that go to sustain life.

840. There are many examples of the happiest and most

vigorous longevity in the pursuits of literature and science. Some ministers, lawyers, and physicians, have attended to their professional responsibilities until they were even more than fourscore years old. Most of the men whose vigor and usefulness were thus prolonged, manifested great activity of body, as well as energy of mind. Their mental powers were never idle. They were laborious in their vocations, and stood among the foremost as scholars. But with their great labor of the brain they judiciously combined due attention to the other organs and functions, and thus sustained their physical health.

CHAPTER XI.

Mental and physical Powers unequal in various Persons. — May be equalized by Education. — Inequality of mental Powers often increased by Education and Pursuits of Life. — Some excel in one Thing and are deficient in others; in mechanical Arts; in Morals. — Any mental or moral Power may be developed and strengthened.

841. *In some persons, the several systems have originally various degrees of power.* The nervous, nutritive, or muscular system may be strong and active, while the others are weak and inactive. This inequality may be removed, partially or entirely, by judicious training, by exercising and strengthening those which are weak, and allowing the stronger to rest. For this purpose, the young man who has naturally strong and active brain, and weak muscles and digestive organs, needs the exercise of physical labor for his equal development, but is injured by much mental excitement; while, on the contrary, the robust and vigorous, whose brain is sluggish, needs the stimulus of study, and can bear the physical inaction of a student's life.

842. The inequalities of the mental and moral powers may be removed by a similar principle in education, which exercises and develops the weak, and leaves the stronger faculties more at rest. But, by a mistake in the purposes

of education, a contrary principle is often adopted, and the strong faculties are made stronger and the weak weaker. Progress and acquirement seem to be the great object with some, and therefore they apply the main force of their minds to the subjects in which they make the easiest and most rapid advancement. If they have a taste for, or an extraordinary power of understanding, music, mathematics, mechanics, or general affairs, they give their attention to them, and neglect the others, for which they have less taste, and which they learn with difficulty.

843. Whatever may be the cause that only a part of the faculties are exercised, and the others dormant, it is certain that those which are in active employment will be quick and comparatively strong, and give pleasure when in action, while the others are slow and weak, and act unwillingly, and even with pain. If the whole force of the brain has been directed to the cultivation or the action of one or a few faculties, the dulness and weakness of the others are about in proportion to the energy and activity of these. Hence, in the division of labor, a man becomes a more perfect workman within his narrow sphere; but his range of knowledge is limited.

844. Some men learn and perform, during their whole lives, only one operation in the mechanic arts. In this limited sphere of action, they show exquisite workmanship; but beyond this they have neither knowledge nor power, and in the subjects of general interest they manifest great weakness and inactivity of intellect. In the manufacture of pins there were formerly twenty processes, and a man performed only one of these. From the beginning to the end of his working life, he exercised only the faculty of making one twentieth part of a pin, and if required to do any other work, or attend to any other subject of thought, he did it unskilfully, and with reluctance.

845. The same is shown in the intellectual processes. Some men cultivate their memory to a remarkable degree, without a corresponding cultivation of the reasoning powers.

Their minds are grand storehouses of facts, which they do not know how to apply to useful purposes. Some have great reasoning powers, but neither carefully observe, nor remember the facts that are presented to them. Some are very sagacious in some kinds of business, to which they have given particular attention, and seem lost when affairs of other kinds are presented to them.

846. *The moral powers, individually, grow or suffer by the same treatment.* If cultivated, they are strong and active; and, if neglected, they are weak and sluggish. Some men are rigidly honest and sincere, but they are harsh and unkind. Others are the very reverse of these — gentle, affectionate, and full of benevolence, while they fail in justice and truth. The mind may be so exclusively devoted to one interest as to lose sight of the worth of others. A philanthropist may be so intensely absorbed in one kind of human distress or one means of relief, as to think all other sufferings light, and other plans of relief unworthy of notice.

847. In any time of life, the weak faculties may be strengthened, and the strong ones made stronger. However difficult any mental action may be at the beginning, it becomes familiar and easy by frequent repetition. The faculty of memory, for instance, may labor hard at first, but, after a time, a man will commit pages with the same effort that he had exerted in acquiring as many lines. A young minister, taking charge of a church in the western country, — where clergymen usually preach without notes, — and yet not being accustomed to preach extempore, determined to write his sermons, and then commit them to memory, and thus avoid the use of his papers. At first, it cost him nearly as much labor to commit as to write his sermons; but, after three years' practice, the same work required only about an hour's attention.

848. The power of observation is very greatly quickened and strengthened by being constantly called into use. One's eyesight becomes sharp, and he learns to recognize the matters which he looks for. The practised seaman discovers a

sign of a storm, the hunter a track of game, and the botanist a flower, which escape the notice of the unpractised observer. A teacher of botany carried, in June, a box full of new flowers to a new but zealous pupil, who was much gratified with the sight of such a variety, and asked where he had found so many. "On the road to the mill." "I walked over the whole of that road," said the scholar, "and looked, as I thought, carefully for flowers, this morning, but could find only two."

849. Any one or more of the moral powers may be educated in the same manner. Self-command in times and scenes of peril comes by education, and is confirmed by habit. The new sailor climbs fearfully to the mast, and has hardly command of his muscles to assist in the management of his sails; but, after a few voyages, the same man will run over the rigging, and work there, even when the ship is violently rocked in the storm, with as much self-possession as if he were working in a shop on land. A painter's young apprentice crawls up the ladder with fearful agitation, and uses his brush with such timid caution that he touches over only a narrow surface on either side. But constant practice dispels all fear, and then he runs up the longest ladder without hesitation, and paints on either side to the farthest reach of his arms.

850. In this development of the moral and intellectual faculties, teaching is the guide, and shows the way, and no more; but exercise gives them power. Learning the principles alone will not make a man a musician, a mechanic, or a philanthropist. It will not fill his heart with love, cheerfulness, or self-denial. The brain and the muscles, the mind and the hand, must be accustomed to the practice of these arts and principles, in order to make one skilful or virtuous. He who would be truly benevolent, must accustom himself to do kind acts; he must not only know how, but he must be actually employed in relieving distress. Cheerfulness must be established by the same practical law. Men must not only believe in this rule, but they must habitually take

cheerful views of life, and always look hopefully upon the future.

851. True politeness, which regards others' feelings, and attends to their wants, becomes, by use, so ready a habit as to be almost a part of our nature. No principle nor motive can supply the want of this practice. If it is not a familiar habit, the politeness is artificial and awkward. It is very easy, in company, to see the difference between the cold and ungraceful manners of one who is unkind, and selfish, and clownish at home, but assumes gentle airs in society, and the easy and unassuming manners of another, who is ever the same, whether at home or abroad. The suavity which is assumed only for the public eye cannot conceal the harshness and coarseness of domestic habits.

CHAPTER XII.

Habitual Actions easy and agreeable. — Retired People averse to general Society. — Strength, gained by Exercise, preserved by same Means. — In perfect Men, all Powers developed. — Concentration of Mind. — Brain, when exhausted, needs Rest; cannot be overworked advantageously. — Vacations of Schools.

852. HABITUAL actions are not only easy, but agreeable, but those actions to which we are not accustomed exercise faculties which are not frequently employed, and are both difficult and painful. We therefore perform them unwillingly. For this reason, many men prefer to associate and talk with others of the same pursuits, interests, or views of life; for conversation with them calls for the use of powers that are habitually active, and association with men of different character would call for the use of powers that are usually dormant.

853. Men who are unused to society are averse to general visiting, because miscellaneous conversation demands the exercise of inactive powers. They are therefore timid, and fear to trust themselves in the discussion of subjects with which they are not familiar, and in which, perhaps,

they may falter. A few families, living in a retired district of the country, have associated almost exclusively with each other for two or three generations, and have had very little intercourse with the world, either at home or abroad. Their exclusive and familiar association has made their conversation upon the topics and interests of their little neighborhood easy and agreeable. But it has limited their ideas and feelings to their vicinity. Other feelings are dormant, and other ideas are strange to them. Consequently, conversation with men from other places calls upon their brain to attend to such matters, and make such exertions, as are neither familiar nor pleasant. They are, consequently, timid in presence of strangers, and suspicious of those whose habits of thought and notions do not harmonize with their own.

854. The strength of the brain and the mind, or of any of its faculties, which is gained by exercise, must be preserved by the same means. If this habitual activity of the mental or moral power is not kept up, it becomes again weak. The musician loses his skill, and the accountant his facility of reckoning, when out of practice. The bold man in danger becomes more timid after living a long time in secure places; the bold man in society loses his self-possession when he retires to obscure life; and the hospitable man entertains his friends with anxious hesitancy after he has ceased to keep open house.

855. As in the perfect body all the organs are equally attended to, and all the muscles exercised, so in the perfect mind all the mental and moral faculties are developed, exercised, and strengthened, in due proportion. Washington was a remarkable instance of this equality and completeness of physical, mental, and moral character. None of his powers were dormant, and none had excessive growth. All were subject to the control of his will. This mental and moral fulness and discipline give men command of their resources, and great power in every emergency. They are thus prepared for the various chances of life. They meet

with few difficulties, and always overcome them. They are, therefore, the most successful and the happiest men.

856. If all the mental faculties are faithfully cultivated and exercised, the mind acquires its greatest strength and power of universal application. But one thing more is wanting; that is, the power of concentrating the whole mental energies upon a single subject, to the exclusion of all others; that is, the power of the will over all the physical organs, over the instincts, and the passions, and the mental faculties.

857. This mental discipline subdues not only all the physical and moral, but the mental powers, to the control of the will, and enables us to concentrate the whole force of the nervous system upon one object, and exclude all others; otherwise, the mind is frequently wandering from the subject proposed to it. The possession or want of this control of the mind — this power of concentration — constitutes one great difference between the strong and the weak mind — the rapid and the dull scholar.

858. It is the misfortune of many students that, while they fix their eyes upon their books, their thoughts are afar off, upon their play, their home, or their pleasures. They look upon their lessons in school; their eyes run over a page of history at home; every line, every word, is presented to the organ of vision; but no sensation is excited in the brain, the mind receives no ideas. Another, with better mental discipline, withdraws his mind from all but the subject before him; and, while he is reading or studying, he thinks of nothing else, — all other thoughts are excluded, and his whole mental force is directed to the matter about which he is reading; and thus he loses no time, and wastes no mental effort. Every exertion aids him in his advancement, and he is therefore a successful scholar.

859. The brain has the same desire and enjoyment of exercise as the muscles. The child runs and plays, and observes and talks; and the man, if he has no occasion for motion, will walk for exercise; and, if he has no call for

thought or observation, he will yet read,— perhaps only the lightest books, such as require the least exertion of mind, — or he will talk, if of nothing more important, he will tell of the news of his little neighborhood, or he will sit at his window and watch the passengers in the highway. In some way or other, the brain is frequently exercised, though often in the gentlest way. There are few who sit long in entire listlessness, without a thought or an emotion. This would be as painful as to sit immovably still for any considerable period, without moving a limb.

860. The brain has the same liability to fatigue from labor, and exhaustion from excess of exercise, as the muscles. It has a definite power of exertion, beyond which it cannot pass, without leaving it enfeebled. If its action is confined within this due limit, and sufficient opportunity is given, at proper intervals, for its recovery, it will never fall below its average standard of effective labor, and it will be able on each day to do a full day's work. But, if this limit is exceeded, its strength is reduced so far below its own standard, that it is not recruited in the allotted time of rest, and consequently is unable to perform even the usual work on the following day.

861. A gentleman in 1842 had occasion to write several hundred letters, and wished to finish them as early as possible. With the average industry he could write thirty a day, without being fatigued beyond the power of the night to restore his mental energies. But, feeling over-anxious to finish the work, he began one morning at six o'clock, and sat at his desk until twelve at night. Within these eighteen hours he wrote fifty letters, and then, exhausted in mind, he retired to rest, but did not readily sleep — the brain, being much excited, was not easily quieted. The next day he was too weary to write, and wrote none. That day was entirely lost; and even on the third day he fell short of his thirty letters; consequently he lost more than he gained by this vain attempt to overwork the brain.

862. When the brain has thus been in long and active

labor, its excitement does not subside readily with the cessation of work. It is then useless to attempt to sleep. The nervous energies, thus stimulated, must have some vent, either by light reading, conversation, or some gentle muscular exercise. An eminent lawyer in Massachusetts, who was employed, during the sessions of the court, in an almost continuous succession of cases, and whose brain was excited to unremitting labor from morning till late at night during this period, found it impossible to sleep immediately after his labors ceased, although he was much fatigued. He therefore accustomed himself to walk for some time, after leaving the court-room and before going to his chamber. With this preparation, he slept comfortably, and awoke on the following morning refreshed and prepared for renewed labor.

863. It would be well so to arrange our business and studies, that the brain should be required to work and expend no more energy in each day than it can recover at night. In this way, it would be able to perform as much on the day following, and on each successive day thereafter, and need no long periods of rest throughout the whole of life. But this is not usually done. Our business, our schools, and colleges, are arranged upon the erroneous plan of doing more than a day's mental work in each day during the busy season or term time, and of having intervals of rest in vacations. This arrangement of mental action and rest is ordinarily made, not in reference to health and power of the mind to labor, but to convenience, or pleasure. Some colleges and schools have long terms and long vacations, to allow the pupils to visit their distant homes. The mind becomes weary, and works languidly, in the latter weeks of these long terms, and the habits of study are broken up in the long vacations; and neither is so advantageous for mental action and health as short and more frequent periods of labor and rest.

CHAPTER XIII.

Digestive and Mental Powers vary. — If this be disregarded, Digestive and Mental Disorder may follow. — Mind disordered by Dyspepsia; Cold; Heat; Over-Action

864. It is well known, that the digestive organs are not alike in all persons. One cannot eat some kinds of meat, another cannot digest some kinds of vegetables, a third is sickened with some kinds of fruit, while a fourth eats of all these, and obtains nutriment and comfortable health from each. So long as these persons avoid that food which injures them, they eat freely and maintain their health. But, if any one eats that which he does not digest easily, he suffers from pain; and, if he perseveres in eating it, his stomach becomes deranged, and then digests nothing easily.

865. So it is with the mind, which is not equally strong for every purpose. If its attention and actions are confined to the purposes which it can accomplish, and if it avoids all others which it cannot master, it manifests no disorder. But when it is required to attend to or comprehend such subjects as are beyond its power, or assume responsibilities which are impossible for it to bear, it struggles to do this with pain, and is wearied with the effort; and, if the attempt is persisted in, the mind is weakened, and sometimes becomes deranged.

866. Some men transact their usual business, and fulfil their responsibilities, discreetly and successfully, but when they go out of their ordinary paths, and engage in political strife or religious excitement, or when they suffer from grief or from the unprosperous turn of their worldly affairs, their minds lose their balance and become deranged. Many others pass through the same excitements, or are afflicted with similar troubles, without mental disorder. Those who fall, had some previous weakness of mind which prevented their enduring what the others endured in safety. They

became insane, therefore, in circumstances which were harmless to their associates.

867. The immediate sympathy between the other organs and the brain compels it to feel their ills, and to be oftentimes deranged with them. Pains of the head and confusion of mind are connected with the sickness of the stomach. *Insanity sometimes rises from dyspepsia.* In such cases, even during convalescence, the mental disorder is brought back by renewal of the digestive trouble. Any error in diet, any improper food, too hearty meals, or gas in the stomach, excites the brain; then the old delusions again return, and the mind suffers acute distress until the stomach is relieved.

868. The suppression of evacuations to which the system has become accustomed sometimes disturbs the brain, and causes mental derangement. Even the closing an ulcer which has been running for a long time may produce the same effect.

869. Very great cold confuses the brain, and deranges the mind. Captain Parry, in the journal of his voyage to the Northern Ocean, states, that, when his men were exposed to extreme cold, they seemed to have lost their power of mind, and upon one occasion, when some of his men returned from an expedition in which they had suffered from great severity of weather, they were confused, and stared vacantly and wildly. They could give no account of themselves, nor of their late conduct; but, after they recovered their natural temperature, they regained their clearness of intellect. Similar instances are given in Fremont's journal of his second expedition over the Rocky Mountains. A high, as well as a low temperature, usually affects the mind unfavorably. Mania is sometimes caused by exposure of the head to great heat.

870. *Very frequent causes of insanity are connected with the abuses of the mental and moral powers.* As dyspepsia arises from errors in diet, — from the wrong purposes to which the digestive organs are applied, or from the excessive bur-

dens imposed upon them, — so insanity follows the wrong application of the powers or the excessive labors of the brain. When the mind is required to attend to and comprehend subjects beyond its powers, or manage affairs beyond its control, or when it is compelled to work too long upon any one subject, which it can manage to a moderate extent, it must falter, and be liable to irregular action.

871. When we fix the eye for a long and uninterrupted period upon any single object, the organ becomes weary, and fails to receive clear impressions, and convey them to the brain. If this is done frequently and perseveringly, the eye becomes weak or diseased, and ceases to perform its functions. In the same manner, the brain, by over-exertion, is wearied, and refuses to give its attention. When this undue labor has been continued for a long time, without proper intervals of rest, the mind becomes exhausted, and it loses its self-control and its power of direction, and cannot be roused to any satisfactory exertion. It is then permanently weak and uncertain in its operations.

CHAPTER XIV.

Insanity, from misdirected Education, and false Hopes; from unfounded Expectations; religious Anxiety; perverse mental Habits.

872. MISDIRECTED education, the preparation for purposes which cannot be attained, or stations which cannot be filled, impose upon the mind an excessive burden, and involve it in a profitless struggle, and often entail upon it weakness, and sometimes disease. All wrong notions of life necessarily end in disappointment. They are based upon false views of the world and of the relations of society, and lead men to look for such events as will not happen in the circumstances which must surround them. The flattered child of popular favor, who expects to find amid the

responsibilities and cares of riper age the same adulation and caresses that come to the careless joyousness of earlier years, and who confidently expects that the future will bring him a measure of success and prosperity for which he is not now making an adequate preparation, must meet with disappointment, and suffer the consequences of sorrow and perhaps of mental disturbance.

873. All those expectations which are founded in hope rather than on calculation, which depend upon uncertain and inappreciable chances for their gratification, and excite the imagination strongly, must often fall short of their fulfilment. Speculation in property of variable value, in confidence of great profits; mining in those regions where no human sagacity or foresight can tell whether the ore can be found, or obtained at profitable cost; the doubtful struggle for situations of honor or profit, — are all attended with great anxiety; and when they fail, as they frequently must, the spirit sinks and the mind may wander.

874. Religious anxiety, or the struggle of the mind in the transition from old to new conditions or opinions, is occasionally productive of mental disorder. When the mind has given up the old foundations on which it rested its hopes and its confidence, and before it has adopted the new, upon which it can securely stand, it seems to be unloosed from its hold, and thrown upon uncertainty. Persons in this condition lose their self-control; the world and its cares and enjoyments, and the future, with its promises and its threatenings, change their aspect, to them, and they may be overwhelmed with distress. If then they can see their way clearly to a better life, the excitement passes away, and the mind is easy; but if this light does not appear to them, they may fall back to their former condition, or become a prey to more abiding and painful derangement.

875. The natural appetites may be so pampered, or artificial appetites may be created and may grow so strong, that we cannot control them. The appetite for intoxicating

drinks is among the most powerful in its influence over those who indulge it. This is not a natural want. It is artificially created and encouraged, until it is strong enough to take care of itself and compel its gratification. It is then in vain that the man thinks and says, he will not desire to drink; the stomach and the nervous system are so changed as to kindle and keep up this burning thirst for alcoholic stimulants. The man may have moral force sufficient to prevent the gratification, but one may as well say that the parched tongue of fever shall not be dry and crave cold water, as that the stomach, excited and disordered by intemperance, shall not thirst and crave its accustomed indulgence.

876. The irregular habits of the mind sometimes overpower the will. Some individuals exhibit a propensity to do strange things, and to utter startling opinions. They have a fondness for attracting attention by their oddity of manners, or thoughts, or language. At first, these singularities are assumed and put off at pleasure; but, if they are allowed to come often, they establish a habit which cannot be resisted. The man is then compelled to exhibit his oddities at times when he would be glad to appear like other people. In other matters, his brain and his mental operations may be manageable, and he may be sound in mind; but so far as he cannot or does not regulate his thoughts, his mind is not sound.

CHAPTER XV.

Day-Dreaming. — Fits of Passion. — Intoxication. — Fright may cause Insanity. — Various Grades of mental Health between Sanity and Insanity. — No sound Mind without sound Body. — Most Causes of mental Disorder within our Control.

877. Some take great pleasure in day-dreaming. They love to abstract their minds from the facts and things about them, — from subjects of real existence, — and, creating an

imaginary world, surround themselves with a train of circumstances from which unpleasant things shall be excluded, and nothing but the agreeable be near them. In such a world, they determine what they would do and say, what principles should govern them, and what impressions they would make upon their associates. Thus they revel in this delightful revery, where all is beautiful and satisfactory.

878. This habit of revery grows more easy and inviting, so that the mind insensibly falls into it when not otherwise occupied. Then the will loses its power to exclude it, and direct the thoughts to the mixed realities of life, until, at last, the dreaming becomes irresistible, and the dreamer can no longer control his wandering mind, nor see things as they are. For a period, — perhaps for years, — he governs his imagination in presence of others, and conceals his dreams from the world. But the habit grows stronger, and finally, regarding neither men nor circumstances, it will speak out; and when the dreamer talks as he thinks, and uses strange language, and perhaps exhibits strange conduct, he is acknowledged insane.

879. When the mind has been strongly excited, the law of continuance prevents its immediate return to rest, and we think of those matters that intensely interested us, after the time of their action has passed away. We cannot dismiss immediately strong sensations, and feelings, and thoughts; and, if they were violent, they may remain for a long period. The agitation of fright continues after the cause is removed. As blindness, partial or total, may arise from the glare of intense light, so the mind, when confused with terror, or disturbed with powerful irritation, may not recover its clearness and self-control, but remain disordered.

880. Men under the excitement of passion lose their self-control, and perform acts, and give utterance to language, which, in their calm moments, they would not willingly allow in themselves. If these passionate excitements are not checked, they gain more and more power to subdue the will, while the resolution to resist grows weaker. Gradually, the

subjection of the will to the excited feelings increases, the mind is made insane from slighter causes, and the derangement is longer continued, until this disorder is fixed, and the passionate man is a maniac.

881. Whether the brain be unduly excited by mental action, or by any physical stimulant, the result is the same — mental derangement. Intoxication with alcohol creates a powerful operation on this organ, which is plainly an irregular and uncontrollable one. The drunken man has no more power over his thoughts than the maniac. Usually, after a fit of intoxication passes away, the brain recovers its self-possession, and the mind is restored to health; yet sometimes this complete restoration does not take place, and the mind of the drunkard continues weak and irregular in its action, and he is then a lunatic.

882. It is not to be supposed that all these causes produce insanity, or that this disease must always follow these violations of the natural laws. But there is a wide difference between the clear and well-disciplined mind, that can be directed at will, and understand and reason correctly, and is buoyant with cheerfulness, and the mind that is totally deranged with lunacy, or overpowered with melancholy. And in this wide interval there are all grades of mental health and power. The mind that is excited with alcohol or passion, or depressed with fear, is incapable of the clearest perceptions of the true and the reasonable. The brain that is torpid after an excessive dinner, or that is in pain from dyspepsia, is, for the time being, deprived of its full power of action. Anxiety, grief, disappointment, and day-dreaming, absorb some of the nervous power, and prevent the free and untrammelled range of thought. These, and all other habits and conditions that diminish or absorb any of the nervous energy, so far as they lay any tax upon the strength or the labor of the brain, or interfere with its free operations, oppress or excite the mind.

883. It will now be plainly seen that there is no soundness of mind without a sound brain, and that disorder of any

or all of the other organs may derange the nervous system, and produce mental derangement. It is also manifest that the abuse or neglect of any of the passions, propensities, or mental faculties, may produce the same result, in greater or less degree, of disordered mind. Most, if not all the causes of partial or total insanity, come within our cognizance, and are originally within our control. The brain and the mind are as subject to fixed laws as the other organs, and it is left for us to see that the conditions of their life are fulfilled, and that we enjoy, not only general and open mental health, but, under every circumstance, and in every moment of our lives, in secret as well as in public, we possess full mental strength, and the clearest power of thought, and the most perfect control over our feelings and passions.

EYE.

CHAPTER XVI.

Eye. — Situation. — Composition. — Humors. — Lens. — Coverings. — Iris. — Pupil. — Effect of Light. — Lids. — Tears. — Lachrymal Apparatus. — Muscles. — Cross-eye. — Optic Nerve.

884. *The eye is placed in a deep, bony socket in the skull.* This socket extends far backward at the base of the brain, and defends this tender organ from blows and accidents on every side except the front.

885. *The eyeball is composed of three substances.*

The *aqueous*, or *watery humor*, (Fig. LXVI. *b*,) is a clear, transparent fluid, and stands in the front of the eye.

The *vitreous humor* forms almost the whole of the globe, (Fig. LXVI. *d*.) This is a transparent substance, and soft like a jelly. It is enclosed in a very delicate membrane, which covers its outside, and, extending through it, forms many cells, which contain this humor.

The *crystalline lens* stands between the vitreous and the aqueous humors, (Fig. LXVI. *c*.) This is a double convex lens, much more dense than the vitreous humor, and holds its shape without any covering. It is composed of concentric

FIG. LXVI. *Section of the Eye.*

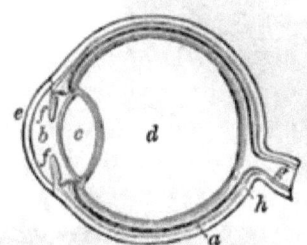

a, Coats of the eye.
b, Aqueous humor.
c, Crystalline lens.
d, Vitreous humor.
e, Cornea.
f, f, Iris.
g, Optic nerve.
h, Retina.

layers like those of an onion, which can be removed one from another. In the fish, this lens is globular. In man, it approaches flatness.

886. Three membranes or coats enclose these humors, and retain the eye in its globular shape.

The outer, or *sclerotic* coat, like the dura mater of the brain, is very firm and strong, and able to resist considerable force without being broken. It covers almost the whole eye. It has a large aperture in front, in which the cornea is placed, (Fig. LXVI. *e*.) The *cornea* covers the front of the eye. It is transparent and strong. It projects in the shape of a watch glass, and covers the aqueous humor.

887. The middle, or *choroid* coat of the eye is very delicate and soft. It contains a black pigment, which absorbs such rays of light as are not needed for vision.

888. The third or inner coat is the *retina*, which is principally the expansion of the optic nerve. This receives the rays of light from the objects which are presented to the eye.

889. In the front part of the eye are, 1st. The cornea. 2d. The aqueous humor. 3d. The iris and pupil. 4th. The crystalline lens, and then the vitreous humor.

The *iris* is a very delicate circle, or continuation of the middle or choroid coat, (Fig. LXVI. *f, f*.) The *pupil* is an aperture in the centre of the iris. The iris is expansible and contractile: when it expands, it extends toward the centre, and lessens the diameter of the pupil; and when it contracts, it draws back from the centre, and enlarges the pupil. By this means, the amount of light received into the eye is regulated. When we are in a dark place, the iris contracts, the pupil is enlarged, and more rays are admitted. When the light is increased, the iris expands, the pupil is contracted, and fewer rays are admitted. When we first go from a bright light, as from a well-lighted room to the darker air abroad in the evening, we see with difficulty, because the pupil is so small that few rays can enter the eye. But soon the pupil enlarges, more rays enter, and we see with ease. On the contrary, when we go suddenly from a dark to a very light place, the pupil being large, much light enters, and the eyes are dazzled; but soon the iris expands, the pupil diminishes and fewer rays enter, and we bear the light without inconvenience.

890. *The lids protect the eyes in front.* They are composed of cartilages adapted to the shape of the eye, the skin without, and the lining membrane within. The lining of the lids is continued over the front of the eye. It prepares and throws out upon itself a thin mucous or glairy fluid, that oils the surface and allows the lids to glide smoothly over the ball. One circular muscle surrounds the open part of the eye, (Fig. LII. *b*,) and closes the lids when it contracts. Another muscle, attached to the upper eyelid by one end, and to the bone of the socket by the other, opens the eye.

891. *The tears wash the eye and keep its surface clean.* The apparatus for this purpose consists of the lachrymal glands, tubes, ducts, and canal. The lachrymal glands are placed in the upper and outer corner of the socket, (Fig. LXVII. *a.*) They prepare the tears, which then flow through the ducts (Fig. LXVII. *b*) under the upper lid into the eye. By the motions of the lids the tears are spread over all the surface of the eye, and wash away any particles of dust. Then they fall into a little groove or trough in the upper edge of the

lower lid, and flow along to the inner corner of the eye. There they are received through very small apertures into the lachrymal canals, and then they pass through the nasal duct (Fig. LXVII. *d, d*) into the nose.

Fig. LXVII. *Lachrymal Apparatus.*

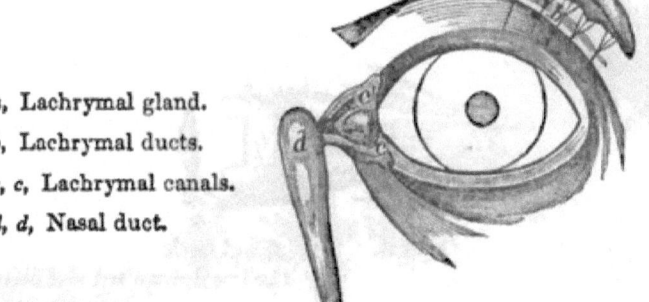

a, Lachrymal gland.
b, Lachrymal ducts.
c, c, Lachrymal canals.
d, d, Nasal duct.

892. The lachrymal canals are sometimes inflamed and closed, and the passage for the tears into the nose is thus stopped. The tears then find no outlet, and flow over upon the cheek, causing some irritation.

893. *This apparatus sympathizes with the moral affections.* The tears are prepared in the gland, and flow more abundantly than they can be received in the canals, in grief, and sometimes in joy, and then they flow over the cheeks.

894. *The eye is rolled by a set of muscles peculiar to itself.* These are attached by one end to the bony socket, and by the other to the eyeball. By their contractions they roll the eye in every direction; and, by their coöperation, both eyes are directed to a single object.

895. *In cross-eyed persons*, these muscles do not work in harmony; some one acts more powerfully than the corresponding muscles, and draws one eye to one side more than the other: this is most commonly inward.

896. *The optic nerve* (Fig. LXVI. *g*) passes from the base of the brain forward through the socket and into the eyeball. After passing the outer and middle coats, it is spread out on their inner surface, and forms the retina, which receives the rays of light.

Fig. LXVIII. *Muscles of the Eye.*

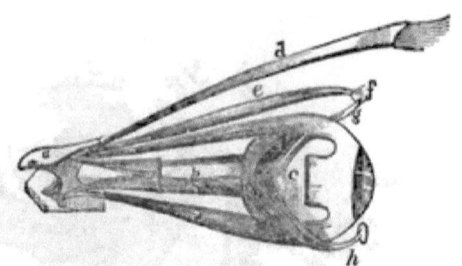

a, Part of the bony socket.
b, Optic nerve.
c, Eyeball.
d, Muscle that raises the upper lid.
e, g, Muscle that passes from the socket at *a*, through the loop *f*, and back to the ball. It rolls the eye downward and outward.
h, Muscle attached to outer edge of the bony socket, and to the side of the ball. It rolls the eye upward and inward.

897. The eye is thus complicated, with many and various parts, all of which are arranged and harmonized together, and all adapted to the action of light without, and to the perceptive power of the brain within.

The light is reflected from objects, and passes through the transparent cornea and the pupil into the ball. The humors and the lens refract these rays, and give them such a direction that they fall upon the retina, where they form the image of the object. This impression is carried along the optic nerve to the brain, and there perception takes place and the object is seen.

CHAPTER XVII.

Near-sightedness. — Spectacles to be worn cautiously. — Eye-Glasses injurious. — Far-sightedness. — Eye suffers with other Organs. — Needs Cleanliness. — Bathing. — Pure Air. — Sufficient Light. — Rest.

898. THE eye is subject to very many and various derangements, which impair vision in various ways and degrees.

Near-sightedness is one of the most common defects of vision. This arises from various causes. It is most frequently produced by the habit of looking at very near objects; as in reading, writing, engraving, sewing, &c., when the books, papers, or work are held close to the eye. In persons so employed, the eye so habitually adapts its focus to these near objects, that it is difficult, or even impossible, to adapt it to objects at a greater distance.

This defect may be avoided or lessened by being much abroad and accustoming the eye to look at distant objects, landscapes, scenery, &c., and also by holding the books or the work as far from the eye as possible.

899. In this disorder, the lenses are supposed to be too round. They refract the rays too much, and concentrate them, and form the image, before they reach the retina.

Concave spectacles obviate near-sightedness. They give a different refraction to the rays, and throw the image upon the retina. When they are used, and especially when they are worn constantly, the eye makes no effort to accommodate its focus to distant objects, and remains permanently near-sighted, and frequently the difficulty is increased. But if spectacles are omitted as long as possible, and then used only occasionally and for seeing distant objects, leaving the eye to its own resources for all near and household objects, the evil would not tend to increase, and the eye would enjoy a wider range of vision.

900. Spectacles covering both eyes affect them equally, and give them the same focus. But eye-glasses being used for only one eye, makes that more near-sighted than the other and these organs, therefore, have unequal power of vision.

901. *Far-sightedness* is a defect of age, when the eye loses the power of adapting its focus to near objects. The lens loses its convexity in some degree, and the rays are not concentrated upon the retina. This evil is obviated by the use of convex glasses, which give the rays the proper refraction, throw the image upon the retina, and enable the eye to see near objects distinctly,

902. *The eye suffers with the rest of the body.* The sight is best in vigorous health, and is impaired by many diseases. Some disorders of the eye have their origin solely in disorders in distant organs. A troublesome affection, called *muscæ volitantes*, or flying flies, is sometimes caused by indigestion merely. The dyspeptic then sees flies or motes, or little clouds, that seem to be flying before his eyes. These are owing to the state of the retina, which is frequently caused by the state of the digestive organs; and when the stomach is restored to health, the flying flies are gone.

903. *The eye wants the utmost cleanliness for its health.* It should, therefore, be bathed and kept free from dust and other matters. It is benefited by the bath as well as the skin. It is well to dip the face every morning, with the eyes open, in cold and clear water, and then to move the lids and thus wash the surface. This should be done daily, and oftener when exposed to dust or other offensive matters.

904. *The eye needs fresh and pure air.* Those who live in the foul air of crowded dwellings and shops, or in the smoke of some rooms, often have disordered vision.

905. *The eye is made for, and should be accustomed to, the light.* Those who work in dark shops, or live in dark streets or houses, or in parlors closely darkened with curtains and blinds, and women who wear veils to shut out the free light of day, have comparatively weakened vision.

906. *The eyes need light for vision*, and suffer or lose their power in some degree when required to labor in insufficient light. Thus they are injured when used for reading, sewing, or examining any minute objects by twilight or moonlight, or in any insufficient light, by day or night. All imperfectly

ighted apartments, counting-rooms, houses and shops in dark alleys, or with insufficient windows, weaken the vision of those who study, write, or work in them.

907. *The eyes suffer from protracted exertion* in the same way as the brain and the muscles. They become wearied, and even sometimes disordered, from looking long at objects that require minute attention, as reading fine print, engraving, miniature portrait painting, sewing, &c. Those who are employed in such things would do well to give their eyes change of occupation and rest.

EAR.

CHAPTER XVIII.

Composition. — External Ear. — Position. — Not to be covered. — Ear-Wax. — Membrane of the Tympanum. — Eustachian Tube. — Bones of the Ear. — Labyrinth. — Nerve. — Requisites of hearing. — Air. — Healthy Ear. — Attention. — Deafness. — Causes. — Hearing may be cultivated. — Ear for Music.

908. The organ of hearing includes the external ear, which is on the outside of the head, the passage to the tympanum or drum, and the internal ear, which is within the drum.

909. The outer ear is composed principally of a somewhat stiff cartilage, that retains it in its shape.

The shape of the outer ear is that which is best adapted to catch sounds and transmit them to the internal ear. This form has been adopted by skilful mechanicians, to gather sounds in rooms and transmit them to other and distant places.

The ears of the lower animals are differently shaped, according to their different purposes. The human ear is scarcely movable; but, in some other animals it is moved to catch sounds in different directions.

388   PHYSIOLOGY AND HEALTH.

910. *The human ear, in its natural condition, stands out from the head* at a considerable angle. This position gives it the greatest advantage for catching sounds. But the custom of wearing caps and other head dresses, and the manner of dressing the hair, press the ear near, and in some persons close to, the head, and thus diminish their acuteness of hearing.

911. The entire external ear stands open for the reception of sounds; but when any of its parts, or the whole, is covered with the hair or any dresses, the access of sound is obstructed, and the hearing somewhat impaired.

FIG. LXIX. *Ear.*

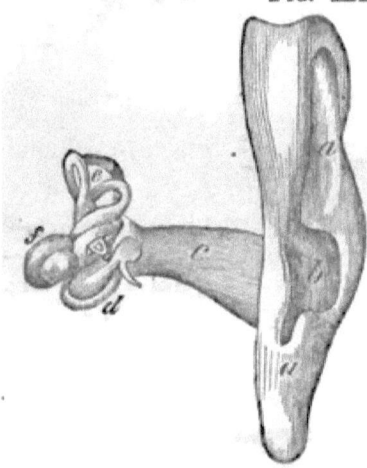

a, a, External Ear.
b, Opening to the internal ear.
c, Canal leading to the drum.
d, Membrane of the tympanum.
e, Semicircular canals.
f, Snail-shell, or cochlea.

912. *The external canal or meatus,* (Fig. LXIX. *c,*) opens from the external to the internal ear. It is about an inch long. Its course is not straight nor direct, but somewhat forward and curved. There are many little cells in its lining, in which the ear-wax is prepared. There are, also, hairs about this canal. The wax and the hairs protect this canal from the entrance of insects.

The ear-wax is sometimes secreted in so great quantity as to fill the canal, and prevent the access of sound.

913. *The membrane of the tympanum, or covering of the drum of the ear,* is spread across the bottom of the canal and

closes it. This very delicate membrane separates the canal, or middle ear, from the internal ear.

914. *The cavity of the internal ear* is behind the membrane of the tympanum. This cavity is filled with air. It has no outlet to the external ear. But there is a passage or tube, called the *eustachian tube,* which leads from the back part of the mouth to the cavity of the internal ear. The air has free access from the mouth to the inner ear through this tube.

The air may be forced from the mouth through this tube into the internal ear, by closing the lips and the nostrils, and pressing the air from the lungs through the windpipe. The air is then felt pressing into the ear with a sound, and sometimes with a loud sound. The acts of gaping and swallowing have a somewhat similar effect; the latter creates a distinct murmuring, and the former a sort of explosive sound in the inner ear.

This tube holds the same relation to the drum of the ear as the hole in the side of a martial drum does to that musical instrument. When the vibration of the air strikes upon the membrane of the tympanum, the air within receives the impression, and partly escapes through the eustachian tube, and thus the impression is modified.

915. *There are within the drum of the ear three small bones*, which are so arranged as to connect the membranous covering with the labyrinth, &c., where the auditory nerve is spread, and to convey the impressions, which are made by the undulations of the atmosphere on the outside of the membrane, to the nervous filaments within.

916. There are also *three semicircular canals*, (Fig. LXIX. *e*,) and the *cochlea*, (Fig. LXIX. *f*,) which have their use in the function of hearing, but precisely what use, it is not easy to explain. These are placed within the parts of the bone at the side and the base of the head.

917. *The auditory nerve* passes from the brain through a hole at the bottom and side of the skull, and is spread about in the labyrinth of the ear.

918. All the several parts of the ear are adapted to receive

the impressions made by sonorous bodies, and to convey them to the brain. The sonorous body causes vibrations in the air. These vibrations strike upon the membrane of the tympanum. The membrane acts upon the series of bones, and through them upon the internal parts and the branches of the auditory nerve, and then along this nerve the sonorous impression is conveyed to the brain, where sensation is caused, the sound is perceived, and the noise is heard.

919. It is necessary, for hearing sound, that there should be a sonorous body to create it, air to convey it, the healthy ear to receive, and the brain to perceive it.

920. *There can be no sound where there is no air.* If a bell be rung in an exhausted receiver of an air pump, no noise is made. The sound is more or less loud according to the state of the air. It is conveyed more distinctly and farther in the direction of the wind than in the opposite course, or in any direction when the air is still. Thus we hear the sounds of bells, &c., when they are at the windward better than when they are at the leeward from us.

921. *It is necessary that the parts of the ear should be sound for perfect hearing.* When the outer canal of the ear is filled, or the membrane of the tympanum is covered, with wax, hearing is impaired.

922. The eustachian tube is sometimes closed. Inflammation of the throat, from colds, may extend to the lining of this tube, and prevent the free passage of air. When this happens from this or other causes, we feel an uneasy fulness and pressure within the ear, and noises have an unnatural and unpleasant sound.

When this tube is closed from slight and temporary causes, it can be opened by gaping, or by pressing the air into it from the lungs.

924. *Hearing is impaired, and deafness, in various degrees, arises, from very many causes*, and from diseases in other organs as well as those within the ear. Worms in the digestive organs, scarlet fever, measles, small-pox, and influenza sometimes produce this effect.

925. Hearing requires the active attention of the brain and the mind; and deafness or imperfect hearing may be caused by the mere habit of neglecting impressions received by the ears.

926. The faculty of hearing may be cultivated to a very high degree. The practised hunters and the American Indians, who are trained to attend to and catch very slight and distant sounds, can hear the natural voice of animals and men, or their footsteps, or even their breathing, or other noises indicating their existence, when others hear nothing.

927. *The ear for music* or power of distinguishing harmonies of sound, is partly a natural gift, and partly a matter of cultivation. Almost all have it in some degree, and some in a very high degree. And there are very few in whom it may not be increased by education.

CONCLUSION.

Man responsible for Care of his Health. — Strength and Weakness, and Length of Life, given according to Man's Faithfulness. — Intention of Nature that we live happy and long. — Power lost by Sickness. — Life shortened. — Errors in the Management of Health. — Constitution impaired in various Ways. — Effect of Education and Circumstances on Constitution.

928. THE human body, with its complicated structure and organs, is left in the charge of man. He is appointed to take it as it comes from the hands of the Creator, and develop and exercise it, direct its actions, supply its wants, and govern all the appetites, according to the requirements of life. These conditions are exact and unyielding, and the good or evil consequences are certain to follow their fulfilment or neglect. In ratio of our obedience will be the fulness of life, its strength, its comfort, and its duration. We can have no health except so far as we obey the law. We can relax in

no required exertion, omit no necessary supply, and indulge in no wrong appetite or propensity. However small the error, the ever-watchful sentinel of life visits it with a proportionate punishment, either of positive pain or lessened enjoyment.

929. Various powers are given to us, and all are necessary to our being and happiness. The animal powers and wants, the appetites and propensities, give pleasure when used and gratified in suitable degree. The moral and intellectual powers give a higher enjoyment. As the mind needs the body for its earthly home, so the body needs the mind as a director. The bodily health is preserved by acting in obedience to the intellectual and moral faculties, and the mental exercise required for this management of the body is necessary for the health of the brain.

930. These, then, are the intentions of Nature — that we lead long, full, and happy lives; that, from the beginning to the end, we have neither sickness, nor weakness, nor discontentment; and that our bodies attain their fulness of strength, and preserve it to a good old age; that all our faculties be developed and strengthened in the performance of the duties of life, and every day be filled with uninterrupted faithfulness or unalloyed pleasure. It is plain that we fall short of all these blessings of life.

931. Between complete life and death there is a wide interval, in which there are many degrees of health and strength; and so accustomed are men to the lower degrees, that they seldom look for the higher, but seem generally content with less. But there is a point in which there is a fulness of physical, intellectual, and moral power. This, and his alone, is perfect health.

932. It is rare that any one passes any considerable period without some sickness so severe as to compel him to suspend his usual employment, and give himself up entirely to the work of recovery. Sickness and weakness, in one form or other, seem to be expected as the occasional lot of all; and much of our time, power, and comfort, is thereby lost

But the whole amount of these which we lose by sickness that prostrates us, is much less than the amount of those lost by the many lesser ailments or debilities which impair our energies and diminish our ease in small degrees, and for short periods, and thus lay light, but very frequent taxes upon our vitality. There are hours or days when we have colds, headache, pain or stiffness in the limbs; when we are heavy and inert from indigestible or over-abundant food, or other causes; when we are timid or irresolute, irritable, peevish, or melancholy; when we have not the full control of all our faculties, because the body or the mind does not willingly, or cannot, direct all its energies to our intended purpose These, individually, make but slight deductions from the force, the productiveness, and the enjoyments of life; yet, when added together, their sum is very great.

933. Not only are the power and the value of life very materially diminished in its course by the greater, and lesser sicknesses and indispositions, weakness, and languor, but life itself is shortened by these and other causes connected with our existence. The natural period of human life, in favorable circumstances, is supposed to be seventy years; yet comparatively few reach that term. The average duration of life differs in different countries. According to the bills of mortality, the average age of those who died in England was 30 years and 10 months; in Sweden, 30 years and 8 months; and in Russia, of the males, 22 years. In Massachusetts 12·2, in England 13·8, in Ireland 10·6, in Sweden 14·6, and in France 18·9 per cent. of all who died were over 70 years old. There is still a wider difference in the duration of life in the various classes of society even in the same place. In England, the average duration of life of the families, including the parents and children, among the most favored classes, was, in Liverpool, 35; in Rutlandshire, 52; and in Wiltshire, 60 years; and among the poorer classes it was, in Liverpool, 15; in Rutlandshire, 38; and in Wiltshire, 33 years. Wherever the same examination has been made in this country, a similar difference in the duration of life has been shown.

934. Thus we see that the most favored people fall short of the full period of their earthly existence, and the poor in some places do not average one fourth of it. If we add to this abbreviation of life the deductions made by the lighter and temporary indispositions, and the severer and protracted sicknesses, and deduct the whole from the allotted period of threescore and ten years, it is manifest that a large part of mankind receive but a small share of the amount of active and productive life that seems intended for them.

935. This great abridgement of life is not caused by imperfection of the Creator's work. There is nothing in the healthy organization that indicates the necessity of disease, debility, or early death. Nature has not made the mistake of giving man a set of organs, all of which may continue in successful operation seventy years, with the exception of the lungs, or stomach, or brain, which will wear out, or become disordered and fail, in half that time. These are not the mistakes of nature; for, with the exception of hereditary diseases and imperfections which some parent has engrafted on his own constitution and transmitted to his children, most men are born with perfect and equal organization, with equal power of action and endurance in all the parts of their frames.

936. Few die, at the end of their full period, from exhaustion of all their physical powers by proper and regular action through the whole period. Most persons die, before the natural term is completed, from the failure or disease of some of the organs rather than from general decay in old age. In every 1000 who died in Massachusetts during the 22 years preceding 1864, 314 died of diseases of the lungs, 137 of diseases of the digestive organs, 78 of diseases of the brain, and only 54 of old age.

937. Here is a very small portion — but little more than one twentieth — that died because the machinery of life was worn out. The great majority died from the disease or failure of some one of the organs to sustain itself and perform its part in the work of life. If these organs had

originally equal power, and were prepared to perform equal work, there must have been some variation from, or failure in, the conditions of being. This failure of any one or all of the organs may arise from one of two causes — from some deficiency of the building up the body, in the development and strengthening of its organs, or from some mistake in the expenditure of its powers.

938. Strength should be constantly added, by means of food, air, exercise, &c., and a portion of this strength may be expended through the muscles, or the brain and nervous system. We may err in the building up the body, by supplying it with insufficient, innutritious, indigestible, or excessive food, or with impure air. Or we may repair the vital machine with too much cost of nervous power, or with the wear and the waste of the organs of supply. To this daily repair of the body some strength must be given. This repair must be attended to before any other matters; and, if it be faithfully made, it will generate more strength than it consumes, and leave a surplus portion to be devoted to other purposes.

939. The energy, or power of the body for its self-sustenance, and for its action beyond itself, is what is called *the constitution*. This vital constitution, in regard to health and action, may be likened to capital in trade, and the surplus power of action may be considered as its income. That amount of income, or surplus strength, which is gathered daily, and no more, may be daily expended. But if the expenditure exceeds the income, and more strength is expended than is gained, it draws so much upon the capital, or the constitution, and then the body must lie still and rest, to regain its loss.

940. There are manifold ways in which the gathered power may be expended, — in mental labor, or muscular exercise, in grief, irregularities and intemperance of every sort, excessive action of the digestive organs, stimulation of alcohol, excitement, passion, exposure to cold, &c. In most of these ways, a limited amount of power may be

expended, and leave the capital unimpaired; but any excess, however small, like an excess of expenditure of money over the income, must be taken out of the constitution. All failure in the building up, all privations of nutriment, of sleep, or of due exercise, or bathing, however small, inasmuch as the body is thereby strengthened less, produce so much less income, and create a deficiency of the vital power.

941. The excess of expenditure of strength, in every way, over the daily income, and all deficiency in strengthening, then, wear upon the constitution. In these many ways, the deterioration may be very slight and imperceptible at the time; the evil consequences may not be great enough to call our attention to them; yet the power of life is diminished, there is less energy in the action of the organs, and less power to resist causes of disturbance. Each one of these errors diminishes the capital of life in proportion to its extent. One takes a little, and another a little, and yet the loss is unnoticed until the whole, added together, weakens the constitution, impairs the health, and wastes the strength so much, that some other cause creates a perceptible disorder or pain, and this we call *disease*. This may be fatal, not because of its own force or violence, but because the vital force had been previously so much reduced, that it could not resist this cause of disturbance. Thus the system is not only laid open to attacks of disease, but its power of overcoming it is lost; as men's affairs are sometimes embarrassed apparently by a new debt or loss, but really because their capital had been diminished so much by previous misfortune or mismanagement, that the new obligation is an insupportable burden.

942. The natural and artificial varieties of human constitution are variously affected by education, habits, circumstances, employments, and localities. These influences may be so used as to diminish, and often remove, these inequalities, or, on the contrary, to increase and establish them. If they are carefully regarded in the training of children

and youth, in the selection of occupations or places of residence, the weak may become strong, or a part or organ that cannot be strengthened will not be compelled to bear a burden beyond its powers. But, owing to neglect of this principle, the circumstances of life are often so used that the weak organs become weaker; the inequality is thereby increased, and the health is impaired. The robust and the feeble, the nervously-excitable and the lymphatic, obviously need different employments. Those who have weak lungs, and inherit predisposition to consumption or asthma, cannot safely engage in the same pursuits, or inhabit the same localities, which would be beneficial to one of more perfect organization. Thackrah, in his valuable work on the Influence of Employments on Health and Longevity, says that not fifty of the fifty thousand who annually die of consumption in Great Britain, would fall by this disease, if proper occupation and habits were adopted. The dyspeptic needs an active, and not a sedentary avocation; and the nervous suffer if the brain is called into excessive exercise by study, or the anxious cares of business; and those who are subject to catarrh and asthma are made worse by working in the dusty trades or places.

943. Out of our own organization, and with the external means offered to us by a generous Providence, we are to sustain our health and prolong our life. For this purpose, as a judicious engineer first learns the structure, and uses, and power of his machine, and then supplies all its materials, adapts the surrounding circumstances to its wants, and governs its movements, and applies its powers precisely to its intended purposes, so, in the management of our vital machine, we must first learn its structure, powers, and wants, and then supply the one and direct the other precisely according to the law of life. This responsibility for the care of the body and the mind comes upon every one, in every condition; and whosoever discharges it with intelligence and faithfulness, will increase his powers and his enjoyments and have length of days on earth.

QUESTIONS

ON

JARVIS'S PRACTICAL PHYSIOLOGY.

PREPARED BY

EV. SOLOMON ADAMS,

OF BOSTON, MASSACHUSETTS.

PART I.

DIGESTION AND FOOD.

CHAPTER I.

1. What changes take place in the animal body from birth to manhood?
2. What law is impressed on all animal beings?
3. What makes food necessary?
4. What is the difference between food and living flesh?
5. What is the process of this change?
6. What organs constitute the digestive apparatus?
7. What offices do the several parts of the mouth perform in the digestive process?
8. What teeth have carnivorous animals?
9. What teeth have herbivorous animals?
10. To what kinds of food are the teeth of man adapted?
11. How many teeth, and what kinds, has man?
12. How are the teeth set in the jaw? Of what are they composed? What causes their decay?
13. How may the cause of decay be prevented?
14. Why are decayed teeth painful?

CHAPTER II.

15. When do the glands of the mouth secrete saliva?
16. When do the glands refuse to perform this office?
17. What must be done to food before it is swallowed?
18. Describe the second chamber of the mouth.

QUESTIONS ON

19. How many passages open from this chamber? What are they?
20. Where is the mouth of the windpipe? What is the epiglottis?
21. What are the offices of the epiglottis?
22. Why can we not breathe when swallowing?
23. Describe the œsophagus or gullet.
24. How does the œsophagus move the food towards the stomach?

CHAPTER III.

25. Describe the stomach? Why is it always full?
26. On what does the average size of the stomach depend?
27. What is the texture of the stomach? Of how many coats is it composed? Describe the outer or peritoneal coat.
28. Describe the middle or muscular coat.
29. The inner or mucous coat.
30. What familiar illustration of these coats?
31. What office do these coats severally perform in the digestive process? By what is the food dissolved in the stomach?
32, 33. What effect has the gastric juice on all proper kinds of food?
34. What opportunity had Dr. Beaumont?
35, 36. What observations of Dr. Beaumont and others have made known the steps of the process?

CHAPTER IV.

37. Do Dr. Beaumont's observations explain the process? or only reveal the several stages of digestion? What advantage results from a complete mastication?
38. In what quantity is the gastric juice secreted?
39. What is the limit of this secretion? What is the consequence of this limit?
40. Have we any measure of the amount of food which we ought to take at a meal? How can this measure be ascertained?
41. What condition is necessary to make this measure a guide?
42. What is hunger? When felt?
43, 44. Give the illustration.
45. When does the work of digestion begin? How long is the gastric juice secreted?

CHAPTER V.

46. What is the relative position of the stomach, lungs, and diaphragm?
47. How does respiration keep the stomach in motion? What effect has this motion on digestion?
48. To what substance is all the food reduced in the stomach?
49. What temperature does digestion require?
50. Relate Dr. Beaumont's experiments.

51. What inferences may be drawn from these observations on the temperature of the stomach?
52. What part of the stomach first receives the food from the mouth? What is the *pyloric valve?* What is its office?
53. What power of discrimination does it seem to possess?
54. How does it treat improper food? What is our sensation at such times?
55. How does this struggle between the stomach and the pylorus end? What are the effects?

CHAPTER VI.

56. Are all articles of food digested with equal ease?
57. What is the average time which a healthy stomach requires for digestion?
58. Give the results of observations on St. Martin, in relation to the time of the digestive process?

CHAPTER VII.

59. What is the first work of the stomach in digestion?
60. What is the effect of drink taken with food? What suggestion is offered to persons who have weak stomachs?
61. What is the proper moisture for food?
62. Will the stomach act more easily on a large or on a small quantity of food? Is the quantity of nutriment always in proportion to the bulk of food?
63. With what should concentrated food be mixed? What is the practice of some rude northern tribes?
64. Is the nature of the food a matter of much importance?
65, 66, 67. Give the illustrations.

CHAPTER VIII.

68. What is chyme? When does the food become chyme? Into what organ does the chyme pass from the stomach?
69. What coats compose the alimentary canal?
70. How does the mucous membrane of the alimentary canal differ from that of the stomach? What are its structure and offices?
71. What is the office of the pancreas?
72. What duty is performed by the digestive juices?
73. What are the lacteal absorbents? What the lacteal duct?
74. What are the absorbent veins?
75. How is the digested food disposed of?
76. On what does the proportion of chyle depend? What are the remote effects of imperfect mastication?
77. Into what stages is the process of digestion divided? How should each stage be performed?

CHAPTER IX.

78. How much do we know of the digestive process?
79. How far does the agency of man go in this process?
80. When are we unconscious of the process in the stomach? When do we become conscious of it?
81. What sensations accompany healthy digestion?
82. At what stage of digestion does man's agency cease, and when does nature take care of it? What guide is needed?
83. What is hunger?
84. What incorrect suppositions are mentioned? Does hunger return as soon as the stomach is empty?
85. What creates a desire for food? What forms of disease prove this?
86. Where is the sensation of appetite? What may result from a diseased state of the nerves of the stomach?
87. Can we need food without being conscious of it?
88, 89. Give the illustrations.

CHAPTER X.

90. How is the appetite affected by various bodily and mental states?
91. What does hunger indicate? How do the wants of the body vary in a healthy state of the system?
92. Of what is appetite the usual sign? Should we eat when we are not hungry?
93. When does the desire for food fail to give evidence of digestive power?
94. What distinction exists between *appetite* and *taste*?
95, 96, 97. Relate some remarkable instances of absence of appetite under disease, or excitement.

CHAPTER XI.

98, 99. Relate some instances of extraordinary appetite for food?
100, 101. What causes may produce this extraordinary appetite?
102. When is a new supply of nutriment needed?
103. On what does the interval between the hours of eating depend?
104. How may the appetite be trained to return at regular intervals?
105, 106. Give some illustrations of the accommodating power of the stomach?
107. Will the stomach bear sudden changes in the time of eating?
108. Give an illustration.
109. What are the effects of irregular hours of eating?
110. What, in general, are proper intervals of eating?

CHAPTER XII.

111. Why may the interval between the evening and morning meals be longer than others? Why breakfast soon after rising?
112. What is recommended when the morning meal is late?
113. When does the body sustain labor and exposure best?
114. What advice is given in section 114?

115. Who especially should take early morning refreshment?
116. What usages have prevailed in regard to the time of eating dinner?
117. What faulty custom is mentioned? The common remedy?
118. Illustrate.
119. Who may properly take a forenoon lunch?
120. How many daily meals are needed? When may supper be omitted?
121. How long before sleeping should supper be eaten? Why?
122. What custom meets the wants of the body? What are the effects of more frequent meals?

CHAPTER XIII.

123. Can the quantity of food be fixed by a uniform rule?
124. What is the rule in the British navy? In the army of the United States? What the rule for emigrant passengers?
125. On what does the proper quantity depend? Do corpulency and leanness depend on the quantity of food?
126. What illustration is given?
127. How does occupation affect the quantity of food?
128. What variation in diet should the same individual make?
129. What change of diet does change of occupation require?
130. What are the consequences of neglecting this change in the quantity of food?
131. State facts which have occurred at Cambridge.
132. Why do growing persons require more food than adults? Why convalescents more than the healthy?

CHAPTER XIV.

133. What has been shown in sections 93, 94.
134. How long may we safely eat?
135. Why should we eat slowly, and masticate thoroughly?
136. What would be a monitor of health, and preventive of disease connected with eating?
137. Who will not err in his diet?
138. What sensations come from eating enough? What from eating too much?
139. Do evil consequences always follow excess of eating?
140. Why is it not economy of time to eat hastily?
141. Illustrate.
142. What habits often prevail in hotels and steamboats?

CHAPTER XV.

143. What is one proof of Divine benevolence and wisdom? What two principles are to be observed?
144. What abuse may come from the pleasure of eating?
144, 145. Illustrate, and specify a difference.
146. What is a common error in regard to eating?
147. What properties of food do most persons know? Of what are they ignorant?

148. What usages of the table are injurious?
149. What rites of hospitality violate the laws of digestion?
150. How are children often improperly indulged?
151. Is such indulgence confined to children?
152. What are the consequences of these indulgences of appetite?
153. What are the effects of deficient or bad food?
154. Give illustration.

CHAPTER XVI.

155. What happens when some one organ, or portion of the body is in action?
156. Why should other organs rest while the digestive organs are active?
157. What comes from attempting the vigorous action of two parts of the body at the same time?
158. What kind of mental action interferes with digestion? What kind does not?
159. Relate experiments on hounds.
160. What kind of exercise is compatible with the digestive process?
161. What is the effect of exercising any of the organs violently?
162. What is the effect on mental action?
163. What is fatigue? What is the process of rest?
164. Why is a short interval of repose needed between hard labor and eating?
165. State illustrations.

CHAPTER XVII.

166. What conditions are recommended during the time of eating?
167. What should be avoided during the eating hour?
168. What is lost by disregarding the laws of digestion?
169. How is dyspepsia produced?

CHAPTER XVIII.

170. In determining what kind of food should be eaten, what preliminary questions are important?
171. What question is still discussed?
172. What examples do both parties find?
173. Are there any advocates for an exclusively flesh diet?
174. What is the general belief in regard to diet? Why?
175. How do different kinds of food differ in their effect on the body?
176. What influence has climate on digestion? What is the food in the polar regions?
177. What in tropical? What usually in temperate regions?

CHAPTER XIX.

178. What differences in the temperaments of men?
179. What are marks of a lymphatic temperament? What food suits it?

180. How is the nervous temperament distinguished? What food suits it? What does not?
181. What accompanies the sanguine temperament? What food is injurious?
182. How is the bilious temperament distinguished?
183. What examples? What food is proper for it?
184. What may we observe among our associates?
185. What is the consequence of disregarding temperament in the regulation of diet?

CHAPTER XX.

186. How does childhood differ from old age?
187. What modification of diet does this difference require?
188. How should the habits of an individual modify diet?
189. What relation has diet to employment?
190. What is the consequence of neglecting this law?
191. What does a change from light to severe labor require?

CHAPTER XXI.

192. In all kinds of food what two things are to be considered? Are they identical?
193. Who especially should make this distinction? Why?
194. What are the natural effects of stimulation? Illustrate.
195. Why are condiments and stimulants injurious to a healthy stomach?
196. What is the effect of alcohol?
197. How do all stimulants affect the natural sensibility of the tongue and mouth?
198. Can the original sensibility be recovered? What has Dr. Kitchener remarked?
199. What do we learn from this examination?
200. What must every individual do?
201. Why cannot a dietetic code be framed suited to all men?

PART II.

CIRCULATION OF THE BLOOD AND NUTRITION.

CHAPTER I.

202. What becomes of the chyle?
203. What is the apparatus of the circulation of the blood?
204. Describe the heart.

205. How is the heart situated? What is its beating?
206. How is the heart divided?
207. What are the other divisions of the heart?
208. What are the valves? How do they act?
209. What are the sets of blood vessels? What relations have they to the heart? The body?
210. Describe the arteries.
211. What is the aorta? What its divisions?
212. Describe the subclavian arteries.
213. Describe the arteries of the neck, head, and face.
214. What are the inguinal arteries? the femoral? How are the arteries finally distributed?

CHAPTER II.

215. What are the veins?
216. What is the vena cava? What are its offices?
217. What are the large branches of the veins?
218. How are the veins finally distributed?
219. How are the arteries or veins arranged in respect to each other?
220. What is the capillary system?
221. What is the general circulation?
222. Where are the arteries situated? Why? When is pulsation felt?
223. Where are the veins situated? Why?
224. How does the blood pass from the left to the right side of the heart? How from the right to the left?
225. Describe the pulmonary arteries. The pulmonary veins.
226. Describe the double circulation.

CHAPTER III.

227. By what force is the blood conveyed through the arteries?
228. How and upon what does the heart act?
229. What prevents the blood from flowing backwards?
230. How are the arteries distributed? What produces their beating?
231. How is blood moved through the arteries?
232. How much blood in a man of average size? In what time does it all circulate through the system?
233. What circumstances affect the rate of circulation?
234. Whose pulsations are strong? Whose feeble? Why?
235. How do mental states affect the circulation?
236. Is the expansion of the arteries the same in all parts of the body?
237. How may we affect the circulation?

CHAPTER IV.

238. What materials of the body are obtained from the blood?
239. What elementary substances are found in the blood?
240. Are all these elements found in every texture of the body?
241. At what stage is the blood changed to flesh?
242. Which makes the largest demand on the blood, growth or change of particles?
243. How are the new atoms of flesh disposed of?
244. What is the office of absorbents?
245. Give Dr. Johnson's description.

CHAPTER V.

246. Do the particles that compose our bodies remain the same?
247. How can the atoms change, without a change of the body? Illustrate.
248. What experiments have been tried on pigs? What result?
249. How is this fact explained?
250. During what part of life is the work of the arteries and absorbents equal?
251. When does nutrition predominate? When absorption?
252. What is the effect of all exercise on nutrition and absorption?
253. What law is a physical one, as well as moral?
254. How may the relative activity of destruction and creation be disturbed?
255. How are wens and other fleshy tumours produced? How scattered?

CHAPTER VI.

256. Is the work of nutrition and absorption equally rapid at all periods?
257. Whose flesh is ever young? Whose ever old?
258. Where does our knowledge of nutrition end?
259. What elements do the nutritive organs select to form fat? hair? muscle?
260. Do the nutritive organs ever misplace a particle?
261. How does arterial blood differ from venous?
262. By what means is the venous blood renovated?

PART III.

RESPIRATION.

CHAPTER I.

263. How are the wasted particles of the body disposed of?
264. Of what parts does the venous blood consist? Why would not these nourish the body? To what process must they be submitted?

265, 266. How are the lungs situated and protected?
267. Describe the spine. 268. Breast-bone. 269. Ribs. 270. Their position.

CHAPTER II.

271. What provision for moving the ribs?
272. What motion of the ribs expands the chest?
273, 274. What is the diaphragm, and its office in respiration?
275. Describe the process of *inspiration*. 276. Of *expiration*.
277. Explain fig. VI.

CHAPTER III.

278. What is the relative position of the heart and lungs? What the substance of the lungs?
279. Describe the air-tubes; the blood-vessels of the lungs?
280. What is the windpipe?
281. What is the organ of voice? Its diseases?
282. How is the windpipe divided? What are the air-cells?
283. How are the air-vessels lined? What is coughing?
284. How may sensibility be impaired?
285. How are the minute arteries separated from the air-cells?
286, 287. What two operations constitute respiration?
288. Name the organs employed in respiration.

CHAPTER IV

289. What elements of waste matter are separated from the blood? How?
290. Into what does carbon enter and compose a part?
291. What are the constituents of the atmosphere?
292. What is oxygen, and what are its combinations?
293. What is nitrogen and some of its combinations?
294. What new compounds are formed in the lungs?
295. What is carbonic acid? Where found? Its properties?
296, 297. What interchange takes place between the air and blood?

CHAPTER V.

298. In what state does the blood enter the lungs? In what state does it leave them?
299. Why do the veins, and the flushed cheek differ in color?
300. What is the effect of respiring the same air several times? How much does our respiration change it?
301. Give Davy's experiment.
302. What is the point of saturation?
303. How may it be shown that water comes from the lungs?
304. What other matters are carried off by the lungs?

CHAPTER VI.

305. What part of the oxygen of the air does one respiration consume?
306. Will a second respiration, and a third, consume each another fourth?
307. Give an illustration.
308. Will any other proportion of oxygen, than that which is in pure air answer?
309. What besides the loss of oxygen unfits the air for a second respiration?
310. What is the limit of the capacity of air to remove offending matter from the lungs?
311, 312. Illustrate.
313. What is removed from the lungs besides carbon?
314. In what three ways is air vitiated?

CHAPTER VII.

315. How does temperature of the air affect the removal of waste?
316. What sensations are experienced in warm weather? Why?
317. How does mountain air affect breathing?
318. What impurities in some mines? What their effect?
319. Does the state of the system affect the removal of waste?
320. How do diseases of the lungs impair respiration?
321. What occurs in lung fever and some other diseases?
322. What effect have mental states?

CHAPTER VIII.

323. What does respiration imply in regard to the chest?
324. What organs are passive in respiration?
325, 326. What apparatus is active, and how does it act?
327, 328. How may the motions of the ribs be impeded?
329, 330, 331. How may the motions of the diaphragm be impeded?

CHAPTER IX.

332. What is the natural principle of beauty? What the ideal?
333. What is the relation of the chest to the body?
334. What principle of utility is to be considered in the size of the chest?
335. What is the real standard of beauty of the chest and waist?
336. When do we find the waist of natural shape? What is it?
337. What is the effect of close and small dresses on the chest?
338. What upon respiration?
339, 340. How does pressure affect the bony frame? The ribs?

CHAPTER X.

341. What disorders may impede the action of the diaphragm?
342. What is the average number of respirations in a minute?

343. What is the average capacity of a man's lungs, when not expanded? How much air is received at each inspiration?

344, 345. Does the quantity of air inspired correspond to the amount of waste to be removed?

346. What are nearly the proportions, of blood, of air, and of waste?

CHAPTER XI.

347, 348. How much air is unfitted for respiration in a minute by the loss of oxygen? What per cent. of carbonic acid gas unfits air for respiration? How much does one person unfit in this proportion in one minute?

349. How much watery vapor will air at 32° contain? How much at 65°? at 70°?

350. When is this vapor condensed? When frozen?

351. How much air will the vapor from the lungs saturate in a minute?

352. What other source of moisture?

353. What is the average amount of insensible perspiration in a minute?

354. In what three ways is air unfitted for respiration?

CHAPTER XII.

355. How much fresh air ought to be supplied to each person per minute?

356, 357. What is neglected in dwellings and public rooms? What partially remedies the neglect?

358, 359. Mention a deficiency in sitting rooms; in sleeping rooms.

360. Deficiencies in public boarding houses, &c.

361, 362, 363. What other places are still more crowded?

CHAPTER XIII.

364. How large a workshop is thought sufficient for six or eight men?

365, 366. What other places are badly ventilated?

367. What facts are mentioned respecting churches?

368. And halls?

369. What facts, respecting school-houses?

370. To what does habit reconcile us?

371. Do persons entering a crowded room, and those living in it, have the same sensations?

372. What fails to be accomplished in such cases?

373. For what do well-arranged means of ventilation provide?

CHAPTER XIV.

374. What correspondence between respiration and vital energy?

375. What illustration do hybernating animals afford?

376. What animals are most active? What most sluggish?

377. In the same class, who have a lower life than others?

378. How does consumption waste the flesh and strength?

379, 380. What is a necessary result of imperfect respiration?

381. Why does sleep sometimes fail to refresh? 382. Illustrate

CHAPTER XV.

383. What effect has corrupted air on a crowded audience?
384. What effect has an ill-ventilated school-room on the children?
385. What occurred in the Black-hole of Calcutta? What is the difference between this result and the faintness of a crowded room?
386. What effect has pure carbonic acid gas?
387. What is a common source of danger from carbonic acid gas?
388. How does drowning produce death?
389. What are some more remote effects of bad air?
390. Why are females more susceptible of consumption than males?

CHAPTER XVI.

391. How does the privation of air affect different animals?
392. What power can man acquire by long practice?
393. What necessity is imposed on all animals?
394, 395. Is there any natural deficiency of air?
396. What reciprocal offices do animals and vegetables perform?
397. When do vegetables consume carbonic acid? When give it out?
398. When are house-plants salutary? When injurious?

PART IV.

ANIMAL HEAT.

CHAPTER I.

399. What is the temperature of most animals compared with that of the surrounding medium? 400, 401. What illustrations?
402. What were the experiments of Sir Charles Blagden?
403. What tendency is almost universal? 404. Illustrate.
405. What exception to this tendency?
406. What effects would follow, if living bodies could not retain a uniform temperature?

CHAPTER II.

407. How is the heat of the living body affected by cold bodies?
408. How does the law of heat among dead substances differ from this?
409. What is the origin of the heat in living bodies?
410. Into what classes are animals divided in relation to heat?
411. From what difference of structure does this difference of temperature arise?
412. Does the same distinction occur among the inhabitants of the sea?
413 What conditions are necessary to maintain this internal heat?

CHAPTER III.

414. With what process is internal heat connected?
415. What do we understand by the term latent heat?
416. What is given us as a general law of matter?
417. What is the difference between sensible and specific heat?
418. How does combustion illustrate this distinction?
419. Apply these principles to explain animal heat.
420. What was once generally believed?
421. What is the process now generally maintained?

CHAPTER IV.

422. How does exercise increase animal heat?
423. How does impeded circulation affect the temperature?
424. How is this internal combustion maintained? How impeded?
425. What hindrances are enumerated?
426. Why do persons in a crowded room grow cold?
427. Who need most external protection from cold, and why?
428. What is needed to maintain this internal combustion besides air? How is it supplied?
429. Who can best resist external cold?
430. Which protects from cold best, alcohol or food?
431. Which warms most, flesh or bread? What necessity has nature met and supplied?

CHAPTER V.

432. What other influences affect the supply of heat?
433. How do different states of the system modify the quantity of heat?
434. What lessens the production of heat?
435. What difference at different periods of life?
436. How does sleep affect the power of producing heat?
437. On what does the amount of heat in combustion depend?
438. How much heat is generated in the body in a day?
439. What prevents an increase of temperature in the body?
440. How does heat escape from the body?
441. What active power does the skin exert? In what way does perspiration cool the body?
442. How do Blagden's experiments illustrate this principle?
443. What beautiful adaptation is mentioned?
444. What is the winter constitution? The summer constitution?
445. What were the experiments of Dr. Edwards?
446. Why is the transition from the cold of winter to the heat of summer unattended with suffering?
447. Who especially need the protection of thick clothing in winter?

PART V.

SKIN.

CHAPTER I.

448. What protects the organs of life from external agencies?
449. Describe the cuticle. 450. Is it subject to change?
451. When does it become thick and tough?
452. Will every kind of friction produce this effect?
453. What is the effect of friction gradually applied?
454. To what extent may the outer skin be made thick and tough?
455. How does new and coarse work affect the hands?
456. How are corns produced?

CHAPTER II.

457. What protection does the cuticle afford?
458. What other parts grow out of the cuticle?
459. Describe the nail.
460. What is the structure of the hair?
461. How is the scarf-skin kept fresh and new?
462. Where is the seat of color? What is the rete mucosum?
463. What produces various hues in some animals?
464. Describe the true skin?
465. When is the surface florid? What may make it more so? Less so?
466. What sense and what degree of sensibility are in the skin?
467. What is directly under the skin? Where thick? Where thin?

CHAPTER III.

468. In what form does the waste of the body escape through the skin? In what quantity? What was Sanetorius's experiment?
469. At what results did Seguin arrive?
470. How can this insensible perspiration be made manifest?
471. What is sensible perspiration? Which is constant? Which greatest in the whole amount?
472, 473, 474. Relate the experiments at the Phœnix gas works.
475. How is the weight of the body kept uniform?
476. What is the average amount of cutaneous exhalations? What produce variations?

CHAPTER IV.

477. What external circumstances modify the amount of perspiration?
478. What is the effect of the atmosphere saturated with moisture?
479. Describe the minute structure of the perspiratory organs.

480. Do all animals possess the perspiratory apparatus?
481. What provision for keeping the skin soft and smooth?
482. What properties has this oily secretion?
483. What will render the skin stiff and hard?
484. What kind of clothing is injurious? and why?
485. What kinds of hats are too close?

CHAPTER V.

486. What connection exists between the skin and the internal organs of the body?
487. What facts illustrate this connection?
488. How are the lungs and skin related? What are their sympathies?
489. What relations exist between the skin and the digestive organs?
490, 491. What internal organ is most liable to suffer when the cutaneous circulation is disturbed?
492. What may always be inferred, when the skin is dry?

CHAPTER VI.

493. When does the skin act as an absorbent? 494. Give illustration.
495. State the case of absorption in a hot bath, and that of Ann Moore
496. What other substances may be absorbed?
497, 498. What facts prove cutaneous absorption?
499. At what times is the absorbing power most active?
500. What stimulates the cutaneous absorbents?

CHAPTER VII.

501. Is the sense of touch uniform over the whole body?
502. When is cutaneous sensibility most acute? In what persons?
503. On what does the facility of cutaneous sensation depend?
504. What can education do for the sense of touch?
505. How is the loss of one sense compensated?
506. What illustrations do the blind and some others furnish?
507. What blunt the sensibility of the skin?

CHAPTER VIII.

508. How does the skin prevent the effects of heat and cold?
509. What tendency is constant? What its effect?
510. Why is perspiration a cooling process?
511. How do different states of the air affect its cooling powers?
512. In what state does the skin perform its functions well?
513. How do some states of disease affect the skin?
514. Can we measure temperature correctly by touch?
515. What illustration may two travellers furnish?
516. What bodily and mental states favor, and what impede the healthy action of the skin?

CHAPTER IX.

517. What three offices does the skin perform?
518. Do the functions of the skin require the aid of clothing?
519. What is the practical view of the case?
520. Do all parts of the skin protect equally well?
521. What illustration is drawn from the dress of the two sexes?
522. What from the Indian? From the Highlander?
523, 524, 525. Give the facts and the conclusions to which they lead.
526. What practical suggestions are given?
527. What effects do habits of dress in different periods of life produce?

CHAPTER X.

528, 529, 530. What, besides habit, should influence the amount of clothing?
531. Can a general rule be given? 532. What general directions?
533. What are the effects of too little clothing?
534. What power comes from habitual exposure?
535. Can all persons acquire this power suddenly?
536. Why do infancy and old age require more clothing than the middle period of life?
537. How can even the feeble gain their power of endurance?

CHAPTER XI.

538. What qualities should clothing possess?
539. Why are loose garments warmer than tight?
540. What are the objections to inner garments of linen?
541. What are the good qualities of cotton?
542. What qualities has silk for inner garments?
543. What qualities recommend woollen garments? What bad effects on some?
544. Against what should the skin be guarded? What is the best protection?

CHAPTER XII.

545 What fact proves the good effects of flannel?
546. What other measures were used for the health of seamen?
547, 548, 549. What do inner garments retain?
550, 551 What changes should be practised?
552. How should beds and bed-clothes be treated?
553. What objectionable practice prevails on board canal-boats?
554. What injurious necessity exists in some dwellings?

CHAPTER XIII.

555. What change of the cuticle goes on constantly?
556. What effect of the warm bath is mentioned?
557. What neglect are many persons guilty of?

558. What difference between their skins and that of those who bathe?
559. What daily practice does perfect health require? Is water alone sufficient?
560. What custom has prevailed in some nations?
561. In what are the English and Americans surpassed by some other nations?
562. Describe different kinds of baths.

CHAPTER XIV.

563. What bath is suitable in good health?
564. What may be easily used instead of a plunge bath?
565. What are a proper time and manner of taking the cold bath?
566. Is the time required for the bath lost time?
567. Who may not safely use the cold bath?
568. What is occasionally necessary?

CHAPTER XV.

569. What are the good effects of cold bathing?
570. What confirmation does Dr. A. Combe give?
571. What is recommended to the weakly?
572. What condition of the stomach should be regarded?
573. What is the best time for the bath?
574. What common notion is erroneous? 575. Give an illustration.
576. To what else may the same principles be applied?

CHAPTER XVI.

577. What effect has bathing on the nervous sensibility? Why is a burn on the skin often more dangerous than a deeper wound?
578. What neglect blunts the sensibility of the skin? How does cleanliness increase it?
579. State the case of Laura Bridgman.
580. What is the practice of Esquirol?
581. What has been shown in regard to the structure and offices of the skin?
582. What duties belong to every human being?

PART VI.

BONES, MUSCLES, EXERCISE, AND REST.

CHAPTER I.

583. What is the structure of the bones? When are they strong?
584, 585. What is the condition of the bones at different periods of life?

586. With what organs are the bones supplied?
587. How are broken bones re-united?
588. What shows the process of absorption and deposition?
589. What effect has exercise on the bones?
590. What stunts the growth? How may the bones of childhood be distorted? What injury is sometimes done in schools?
591. What causes the rickets? Who are exposed to this disease?
592. What is the internal structure of the bones?

CHAPTER II.

593. How many bones in the human frame? What their shapes?
594. Describe the principal parts of the bony frame. The skull.
595. Describe the chest; the spine; a vertebra.
596. How are the vertebrae distributed?
597. What substance between the vertebrae? How does it aid in bending the spine? Why is a man shorter at night than in the morning?
598. What is the shape of the spine?
599. Describe the pelvis.

CHAPTER III.

600. Describe the upper extremities: shoulder blade: collar bone: arm: wrist: hand.
601. Describe the lower extremity: the foot.
602. Describe composition and form of the foot.
603. How may the benefit of the arch be shown?
604. How do we avoid a jar in jumping from an elevation?
605. What is the natural shape of the foot? What the artificial?

CHAPTER IV.

606. What office do the joints perform? What are the different kinds of joints?
607. Describe the hip joint.
608. Describe the upper joints of the neck.
609, 610. What contrivance to prevent jars and friction?
611. What keeps the joints from wearing out? How are they kept moist?
612. How are the bones held together? What is the synovial membrane?
613. What is a sprain?
614. How are the bones dislocated?

CHAPTER V.

615. Describe the muscles. What is their office?
616. What is their power? What is their action?
617. How are they arranged?

618. Describe the muscles that lift the shoulder; that bend the elbow; that draw the arm forward; that bend the wrist. What is the tailor's muscle? What muscles bend the hip joint? What muscles straighten the leg?

619. Describe the Trapezium muscle. Describe the muscle that straightens the elbow; that straightens the hip; that bends the knee; that straightens the ancle.

620. Describe the muscles that draw the shoulder forward; that roll the thigh outward.

CHAPTER VI.

621. What are the shapes of the muscles?
622. How does the diaphragm act? The heart? The oesophagus?
623. What is done by the muscles?
624. How are the muscles attached? Describe some of the muscles on the face?
625. How are the muscles placed in regard to the bones?
626. What occurs in the muscles when they act? How can you feel their action?
627. When can the action of certain flat muscles be perceived?
628. How many sets of muscles has each hinge joint? What is the office of each set?
629. How can you perceive the alternate working of the muscles that move the fingers?
630. How is the variety of motions produced in the ball and socket joints, and in some others?

CHAPTER VII.

631. What muscles act at disadvantage? 632. Illustrate in fore-arm.
633. Why is this loss of power made? How is it provided for?
634. Give another illustration.
635. Specify instances of rapid muscular motion.
636. Specify instances of action by a concert of muscles.
637. What power do performers on the piano-forte acquire?
638. What precision does a skilful violinist acquire?
639. What illustration may be drawn from the act of writing?
640. What does this control over muscular action enable men to do. Give an illustration.

CHAPTER VIII.

641. To what is the strength of muscles generally proportioned? What exceptions?
642. How does the strength of man compare with that of a flea?
643. To what does muscular power correspond?
644. For what purpose is muscular power given?
645. How do the muscles of the active and inactive differ?
646, 647. Give the first illustration. Second. Third.
648. What employments develop the muscles best? Why?
649. What employments develop some muscles disproportionately? Why? How is this shown in other animals? in birds?

CHAPTER IX.

650. Why are some persons strong in one part and weak in another?
651. Give illustration.
652. How are the size and strength of the muscles preserved?
653. Illustrate.
654. What collateral advantages result from muscular exercise?
655. What connection has digestion with exercise?
656. How does exercise aid the circulation? 657. How respiration?
658. How does exercise increase animal heat?

CHAPTER X.

659. How does a muscle gain size and strength by exercise?
660. What other benefit of exercise is mentioned?
661. What difficulties may be removed by exercise? 662. Give the history of the robust boy?
663. How is each one to determine his proper degree of exercise?
664. What results indicate too much action? What result indicates enough?
665. What is said of the permanence of muscular action?

CHAPTER XI.

666, 667, 668. How should the feeble begin to exercise? Give an illustration.
669. When do debilitated students, &c., derive no benefit from labor?
670. What is recommended to invalids going to sea for health?
671. Why have not gymnastic exercises produced the expected result?

CHAPTER XII.

672. What kind of exercise is best? 673. How should we walk?
674. What exercises are allowed to boys? 675. How are girls restricted?
676. What custom prevails among English women?
677. What employments may be combined with walking?
678. In what circumstances is dancing a good exercise?
679, 680, 681. Give the suggestions relating to the time of taking exercise.

CHAPTER XIII.

682. Where should exercise be taken? Why?
683. What kind of weather may prevent exercise abroad?
684. Are any exempted from the need of exercise?
685, 686. What, besides the quantity of exercise, demands attention?
687. Are the consequences of neglect sudden? or remote? Are they the less certain?
688. What laws are established?

CHAPTER XIV.

689. Is the exercise which health requires, the limit of muscular power?

690. What questions are important for the laborer?

691. How can the laborer know he has overworked? How long may he increase his exertions?

692. How can he maintain the fulness of his strength?

693. Has man an indefinite power of endurance?

694. How is premature old age sometimes induced?

695. What is a common effect of overworking? Illustrate.

696. Among what class is the length of life shortest?

697. May not other circumstances account for the shorter life of the day-laborer and sailor?

CHAPTER XV.

698. Does excessive labor for a short period produce the same kind of result?

699. What permanent injury often results? Illustrate.

700. Who suffers most from great exertion?

701. When does a man acquire full strength? What is the order of development of power?

CHAPTER XVI.

702. Why does labor require healthy organs of digestion and nutrition?

703. What kind of food does labor require? Why?

704. What kind of food should a laboring man eat? Why?

705. Can a man labor well whose skin and lungs are not in a healthy condition? Why?

706. How do diseases of the heart unfit for labor?

707. How do the state and health of the brain and mind affect the ability to labor?

708, 709. What is the proper state of the brain and mind while taking exercise?

CHAPTER XVII.

710. What motive does labor for profit require?

711. What are the effects of hope and despair on labor?

712. What of confidence, and of doubt?

713. What is the effect of cheerfulness and of melancholy?

714. What is the effect of alcoholic liquors?

CHAPTER XVIII.

715. What is the natural form of the spine? How is the head held?

716. Describe the muscles of the back?

717. How does a curvature of the spine affect the muscles of the back?

718. What is said of the strength of the back? What illustration?
719. What is the best position of the spine for burdens on the head?
720. How is the head carried most easily? Explain.
721. What is said of the centre of gravity in carrying burdens? Give illustrations.
722. What farther illustrations?
723 Explain this farther.
724. What is said of the grace of this attitude?

CHAPTER XIX.

725. What is the best attitude for walking? Why?
726. What for labor? What illustration?
727. What other persons act best in this position?
728. How is the lateral curvature of the spine induced?
729. How is the forward curvature produced? Who are liable to it?
730. Who are exposed to this curvature? How?
731. How is the spine affected by the position of persons writing? or reading?
732. What is Dr. Warrens' opinion?
733. Why are boys less subject to curved spine than girls?
734. How are the lungs affected by curved spine?
735, 736. How is the spinal cord?

CHAPTER XX.

737. What law of nature is almost universal?
738. What disadvantages has night labor? Illustrate.
739. What is the experience of milkmen and others who devote a part of the night to their business?

CHAPTER XXI.

740. What natural indication does man need to follow?
741. Can the quantity of sleep be fixed by a general rule?
742. What shows night to be the season of sleep?
743. What is the uniform effect of loss of sleep?
744. Why does sleep require an increase of clothing?
745. What are the conditions necessary for refreshing sleep?
746. Does the digestive process go on well in sleep?

PART VII.

BRAIN AND NERVOUS SYSTEM.

CHAPTER I.

747. Describe the brain.
748. Into what portions is the brain divided?
749. What are the offices of the membranes which divide the brain.
750. What is the spinal cord, and its position?

751. How is every organ connected with the brain?
752. What two kinds of nerves, and what their offices?
753. How is it shown that they perform separate offices?
754. How do these two kinds of nerves differ in their termination?

CHAPTER II.

755. What three things are necessary for sensation?
756. What happens when the foot is asleep?
757. How do the nerves pass from the brain to the trunk and extremities?
758. How do injuries to the spine affect the nerves? 759, 760, 761. Give illustrations.
762. What does the case of Mr. J. illustrate?

CHAPTER III.

763. Why does the brain refer sensations to the end of the nerves?
764. If the trunk of the nerve is irritated or injured, to what part is the impression referred?
765. Where is the seat of disease in the tic douloureux? 766. Illustrate by Miss W.'s case.
767. Where do the nerves of special sense receive impressions? How can an impression like that of light be made on the brain, when the eye is closed?
768. What illustrations are derived from the sense of hearing, and of taste?
769. What is the effect of exciting a nerve in any part of its course?
770. Illustrate.

CHAPTER IV.

771. What fact does the last section explain?
772. How is muscular motion excited naturally? How artificially? Illustrate.
773. Where is sensation? What is its immediate cause? Where is a true sensation excited? Where a false one? How do the chamber bells in a hotel illustrate true and false sensations?
774. What instances of false sensations are mentioned?
775. Is the communication through the nerves rapid?
776. What organs are voluntary? Why so called? What are involuntary? Are these supplied with nerves?
777. Do the voluntary organs ever act involuntarily? State cases.
778. What voluntary motions become apparently involuntary?
779. Why do familiar motions exhaust less than others? Illustrate.
780. On what does the power of every organ to act depend?

CHAPTER V.

781. Why is the brain carefully guarded from injury?
782. The brain is the seat of what?
783. Is the brain sensitive? Illustrate.
784. What is the effect of pressure on the brain?

785. Relate a case recorded by Sir Astley Cooper.
786. Is the brain subject to growth and decay? With what apparatus is it furnished? What part of the blood of the body does the brain receive?
787. What effect has the action of the brain on its blood-vessels?
788. When does the brain require rest?
789. What is the natural rest of the brain? What kind of sleep refreshes the brain most?

CHAPTER VI.

790. What is the proper season for mental labor? For mental repose?
791. What do some students practise? With what effect? When has the brain naturally the greatest power for labor?
792. What effect has proper exercise on the brain? What effect has excessive exercise?
793. What relation exists between the brain and other organs?
794. What between the brain and heart? 795. Between the brain and lungs? 796. Between the brain and stomach? 797. Between the brain and muscles?
798. How does alcohol affect muscles? 799. Is this permanent?
800. Illustrate.
801. Why is the drunkard, when sober, exposed to accidents?

CHAPTER VII.

802. How does the mind communicate with the external world?
803. What fixes a limit to the action of the human mind?
804. What reciprocal influences are exerted by the brain and the mind?
805. Mention some specific influences.
806. State the case of a child.
807. What is the effect of indigestion and nausea?
808. How does excessive eating affect mental action? How does hunger?
809. How are the moral powers and affections influenced by the stomach? Illustrate.
810. What effects are ascribed to cheerfulness? What to sorrow and care?
811. What feelings favor digestion? What feelings retard it?
812. What connection has health with cheerfulness?

CHAPTER VIII.

813. What operations does the brain superintend? How far is it under our control? How far are we responsible for its action?
814. When does the brain perform its offices well? When not well?
815. When can the brain superintend more than one operation at a time? When only one?

816. When can the brain superintend only the involuntary operations?
817. Explain this by the musician.
818. In what conditions can the mind be concentrated on a subject?
819. What faults in many school-rooms interrupt study?
820. What further faults are mentioned?
821. Why is fatigue incompatible with mental labor?

CHAPTER IX.

822, 823. How do moral feelings affect mental attention?
824. What illustration does the school-room furnish? 825. What other illustrations?
826. What seeming exceptions are mentioned?
827. How are they reconciled with the general principle?
828. What do these disturbing influences require?
829. What motives for study may interrupt it? 830. Illustrate.
831. What motives aid the mind in fixing attention on a subject?

CHAPTER X.

832. What differences exist in the various faculties of the same person? Of different persons? What is the cause of this difference?
833. How are the mental and moral powers strengthened? What is the purpose of education? In what order do the powers appear?
834. How is the child prepared for abstruse subjects?
835. What are some characteristics of childhood?
836. Can the mind be prematurely strengthened?
837. What is the effect of premature mental exertion?
838. What dependence among the organs? Illustrate.
839. Why does the health of students often fail?
840. Is mental effort compatible with health and long life?

CHAPTER XI.

841. Are the several systems equally strong in the same person? How can this diversity be remedied?
842. How can mental irregularities be remedied? What is often done at variance with the proper remedy?
843. What will be the state of the faculties which are exercised? What of those not exercised?
844. What illustration is drawn from the mechanic arts?
845. What is the effect of exclusive cultivation of some of the intellectual powers?
846. Does the same principle apply to the moral powers? Illustrate.
847. What facts show that any faculty can be strengthened at any time of life?
848. How does constant use affect the power of observation?
849. Show how the same principle applies to the moral powers.
850. What part of this work belongs to teaching? What part to practice?
851. How can true politeness be distinguished from the assumed?

CHAPTER XII.

852. Who are a man's chosen associates? Why?
853. Why are men unused to society often averse to it? What is the character of people long excluded from the world?
854. How is the acquired strength of the mind to be preserved?
855. What constitutes completeness of character? What are its advantages?
856. What is mental concentration?
857. What is one difference between a strong and a weak mind?
858. On what does the progress of a student mainly depend?
859. Does the exercise of the brain afford pleasure?
860. To what degree must the brain be exercised, that it may do a full day's work every day?
861. Is any thing gained by overworking the brain? Illustrate.
862. Does a high mental excitement subside, as soon as the exertion ceases? Illustrate.
863. How should our business and studies be arranged? What is the fault of the actual arrangement?

CHAPTER XIII.

864. What diversity is found in the digestive organs? What is the effect of eating indigestible food?
865. How does the same rule apply to the mind?
866. How are some men affected by intense application to subjects out of their common course?
867, 868. How do the ills of the other organs affect the brain?
869. How does great cold affect the brain? Great heat?
870. What abuses may lead to insanity? 871. Illustrate.

CHAPTER XIV.

872. What is the effect of misdirected education?
873. What expectations may produce the same effect?
874. What is the effect in some cases of religious anxiety?
875. What is the tendency of natural, or artificial appetites, if indulged?
876. What may be the effect of irregular mental habits?

CHAPTER XV.

877. What is day-dreaming? 878. To what results does it lead?
879. What is the effect of strong emotions? Of fright? 880. Of uncontrolled passion?
881. How do physical stimulants affect the brain?
882. Is insanity the uniform result of these causes? Between what extremes do they leave the mind?
883. To what extent is man responsible for mental health and strength?

CHAPTER XVI.

884. Where is the eye placed?
885. What is the aqueous humor? What the vitreous? The crystalline lens?
886. Describe the sclerotic coat; the cornea.
887, 888. Describe the choroid coat; the retina.
889. What are the parts of the eye, beginning in front? Describe the iris; the pupil, and its action.
890. Describe the lids.
891. Describe the apparatus for the tears.
892. What happens sometimes to the lachrymal canals?
893. What sympathies has the lachrymal apparatus?
894. What move the eye?
895. What is the cross-eye?
896. Describe the optic nerve.
897. How is vision produced?

CHAPTER XVII.

898. What is near-sightedness? How may it be avoided?
899. What is the state of the lens? What is the effect of concave spectacles? How should they be worn?
900. What is the effect of single eye-glasses?
901. What is far-sightedness?
902. What are some of the diseases of the eye?
903. How should the eye be treated for cleansing?
904. How is the eye affected by air?
905. Who have weakened vision?
906. How are the eyes sometimes injured?
907. What is the effect of protracted use of the eyes?

CHAPTER XVIII.

908. What does the organ of hearing include?
909. Describe the outer ear.
910. What is the natural position of the ear?
911. Should the ear be covered?
912. Describe the external canal.
913. Describe the membrane of the tympanum.
914. Describe the internal cavity. The Eustachian tube. What office does this tube fulfil?
915, 916. What are within the drum?
917. What is the auditory nerve?
918, 919. How is sound produced? What is necessary?
920. How do we know that air is necessary?
921. What is the effect of unsound ears?
922. What happens sometimes to the Eustachian tube?
923, 924, 925. What causes deafness?

926. Is this faculty always the same?

927. What is said of the ear for music?

928. What is man's responsibility in regard to his body? What is the consequence of violations of the law?

929. What relations exist among the bodily, intellectual, and moral powers?

930. What are the intentions of nature in regard to man? Does he realize them?

931. What is perfect health?

932. What seems to be expected as the common lot of all? What causes a greater loss than prostrating sickness?

933. What is supposed to be the natural period of human life? Is this often attained? What is the average duration of life in Massachusetts? In Sweden? In Russia? Among different classes in England?

934. To what conclusions do these facts lead us?

935. Is this abridgment of life owing to natural defects?

936. What are the most common causes of death?

937. What proportion in Massachusetts died because the machinery of life was worn out? From what cause may the premature failure of one, or of all the organs arise?

938. How should strength be added? How may a portion of it be expended? How may we err?

939. What is meant by the term constitution? To what may it be compared?

940. How may vital power be expended? When is a sufficiency of vital power produced?

941. How does the system become exposed to disease? What increases the danger of disease?

942. What power has man over the varieties of constitution? What is the opinion of Thrackrah?

943. By what means are we to sustain health and prolong life?